Design and Development of Metal Matrix Composites

Design and Development of Metal Matrix Composites

Guest Editor

Sónia Simões

MDPI

Basel • Beijing • Wuhan • Barcelona • Belgrade • Novi Sad • Cluj • Manchester

Guest Editor
Sónia Simões
Department of Mechanical
Engineering
University of Porto
Porto
Portugal

Editorial Office
MDPI AG
Grosspeteranlage 5
4052 Basel, Switzerland

This is a reprint of the Special Issue, published open access by the journal *Metals* (ISSN 2075-4701), freely accessible at: https://www.mdpi.com/journal/metals/special_issues/92DR0O83M4.

For citation purposes, cite each article independently as indicated on the article page online and as indicated below:

Lastname, A.A.; Lastname, B.B. Article Title. *Journal Name* **Year**, *Volume Number*, Page Range.

ISBN 978-3-7258-5061-7 (Hbk)
ISBN 978-3-7258-5062-4 (PDF)
https://doi.org/10.3390/books978-3-7258-5062-4

Contents

metals MDPI

Editorial

Design and Development of Metal Matrix Composites

Sónia Simões [1,2]

[1] Department of Mechanical Engineering, Faculty of Engineering, University of Porto, Rua Doutor Roberto Frias, 4200-465 Porto, Portugal; ssimoes@fe.up.pt; Tel.: +351-220413113

[2] LAETA/INEGI-Institute of Science and Innovation in Mechanical and Industrial Engineering, Rua Doutor Roberto Frias, 4200-465 Porto, Portugal

1. Introduction

Metal Matrix Composites (MMCs) constitute a class of advanced materials distinguished by their exceptional mechanical, thermal, and tribological properties, offering significant advantages over conventional metals and alloys. These materials consist of a metal or alloy matrix reinforced with ceramic particles, fibers, or whiskers, leading to a unique combination of high strength, stiffness, wear resistance, and, in some cases, enhanced thermal or electrical conductivity. As a result, MMCs are increasingly employed in high-performance structural and functional applications in the aerospace, automotive, defense, and electronics industries [1–4].

The development and optimization of MMCs constitute a highly interdisciplinary endeavor, bridging core principles of metallurgy, materials science, and mechanical engineering with emerging trends in advanced manufacturing technologies, including additive manufacturing (3D printing), powder metallurgy, spark plasma sintering, and friction-based processing methods, which have opened new avenues for tailoring microstructures and achieving superior composite performance [1,3,4].

This Special Issue was conceived as a comprehensive platform consolidating cutting-edge research and technological developments that address both the scientific foundations and practical challenges associated with MMCs. The contributions cover a broad spectrum of topics, such as innovative manufacturing techniques, interfacial engineering, microstructural design, mechanical behavior modeling, and the development of hybrid or functionally graded composites.

The aim of this Special Issue is to highlight recent advances that deepen our understanding of MMCs' behavior under various service conditions and foster their integration into industrial applications. A further aim is to present experimental investigations, computational approaches, and case studies that demonstrate MMCs' potential in solving complex engineering problems. The selected contributions reflect the growing maturity and strategic importance of MMCs as key enablers in the design of next-generation materials and systems.

2. An Overview of the Published Articles

This Special Issue comprises ten original contributions that collectively illustrate the breadth of current research on Metal Matrix Composites (MMCs), spanning advanced processing techniques, multiscale modeling, and performance evaluation under extreme conditions.

Johanes et al. (contribution 1) present an experimental investigation of Mg/SiO_2 nanocomposites subjected to various cryogenic temperatures, revealing how cryo-treatment alters grain structure, dislocation density, and phase distribution, ultimately impacting

 https://doi.org/10.3390/met15080848

thermal conductivity and compressive strength. Interestingly, they show that ultra-low temperatures do not always lead to superior performance, challenging conventional assumptions and offering new pathways for optimizing thermal designs.

Monteiro and Simões (contribution 2) examined the fabrication and behavior of hybrid nanocomposites based on Al6061 alloy reinforced with ceramic nanoparticles. Their work details the dispersion methods, sintering parameters, and resulting improvements in hardness, tensile strength, and corrosion resistance, demonstrating how hybridization strategies can be used to fine-tune material properties, especially for lightweight applications where the strength-to-weight ratio is critical.

Kunčická et al.'s work (contribution 3) focuses on high-shear deformation applied to powder-based copper composites. Using high-resolution microscopy and mechanical testing, they show that such processing leads to significant grain refinement, texture development, and enhanced mechanical strength. Electrical conductivity is also preserved at high levels, indicating the technique's suitability for high-performance electronic components.

Walunj et al. (contribution 4) investigated titanium substrates coated with different ceramic materials—including TiN, TiC, TiCN, TaN, and NbN—using spark plasma sintering. The coatings were evaluated in terms of microhardness, frictional behavior, and resistance to wear. The authors concluded that specific ceramic phases significantly improve surface durability and biocompatibility, providing strong support for applications in biomedical implants and aerospace components.

Wang and Zhang (contribution 5) present the synthesis of $MoSi_2$–SiC composites through an innovative two-step spark plasma sintering (SPS) process. They emphasize how a controlled reaction between Mo and Si powders, in the presence of in situ-formed carbon, enables a uniform dispersion of SiC particles. The resulting composites exhibit high-temperature stability, low thermal expansion, and increased fracture toughness, making them ideal for use in turbines and thermal protection systems.

Wan et al. (contribution 6) delved into La_2O_3-reinforced A356 aluminum alloy composites. Through systematic variation of the reinforcement content, they assessed tensile strength, elongation, and electrochemical corrosion resistance. Advanced characterization techniques, such as SEM and XRD, revealed the role of La_2O_3 in refining grain boundaries and forming protective oxide films, making these composites promising for marine and automotive environments.

Liu et al. (contribution 7) explored eutectic high-entropy alloys (EHEAs) based on AlCrFeNi with silicon doping. Their work demonstrates that Si incorporation promotes the formation of dual-phase microstructures—consisting of a BCC matrix and intermetallic phases—leading to a balanced combination of high yield strength and ductility. The research bridges the gap between traditional MMCs and emerging HEA design philosophies.

Akarou et al. (contribution 8) utilized atomistic simulations to model Cu-Mo nanocomposites and examine interfacial behavior at the atomic scale. By testing multiple interatomic potentials, they identified the most accurate models for representing Cu-Mo bonding. Their results highlight how interface characteristics—such as misfit dislocation arrays and shear response—affect mechanical performance, offering insights crucial for the design of multilayered nanostructures.

Kunčická and Kocich (contribution 9) contribute a second study focused on Cu-La_2O_3 composites processed via Equal Channel Angular Pressing (ECAP). They combined experimental deformation analysis with finite element modeling to predict strain distribution and grain refinement. The addition of La_2O_3 improved strength and thermal stability, supporting the use of ECAP as a viable route for producing ultrafine-grained functional composites.

Xi et al. (contribution 10) provide a theoretical study on the Sc_3AlC MAX phase under high-pressure conditions using first-principles calculations. Their analysis includes the evolution of lattice parameters, elastic anisotropy, electronic band structure, and dislocation core energies. Their work uncovers the pressure-dependent behavior of the material, revealing its potential for extreme environments where both structural resilience and electronic functionality are required.

3. Conclusions

This Special Issue on the Design and Development of Metal Matrix Composites offers a thorough and multifaceted perspective on the latest advancements and future directions in the rapidly evolving field of MMCs. By offering a diverse array of high-quality contributions, this issue not only enhances our understanding of the underlying physical mechanisms, interfacial phenomena, and microstructural evolution inherent to these composites but also showcases novel fabrication strategies and practical applications useful for a wide range of industrial sectors.

The collected works reinforce the strategic relevance of MMCs in addressing the growing demands for lightweight, high-performance, and durable materials in critical areas such as the aerospace, automotive, biomedical engineering, defense, and renewable energy industries. Furthermore, they highlight the transformative potential of integrating computational modeling, advanced experimental techniques, machine learning, and design optimization into the development pipeline of MMCs—paving the way for smarter, more efficient, and sustainable material systems.

Importantly, this Special Issue underscores the value of interdisciplinary collaboration, wherein materials scientists, mechanical engineers, manufacturing specialists, and computational researchers converge to tackle complex challenges in MMC research and applications. The synergies between theory, processing, characterization, and performance evaluation contribute to a more holistic approach to composite materials design.

We hope that the insights and innovations presented herein will serve not only as a state-of-the-art reference for current practitioners but also a catalyst for future investigations. By encouraging dialogue and collaboration across disciplines, we hope that this Special Issue supports the continued evolution of MMCs toward increasingly sophisticated, customized, and impactful solutions for next-generation technologies.

Conflicts of Interest: The author declares there are no conflicts of interest.

List of Contributions:

1. Johanes, M.; Mehtabuddin, S.; Venkatarangan, V.; Gupta, M. An Insight into the Varying Effects of Different Cryogenic Temperatures on the Microstructure and the Thermal and Compressive Response of a Mg/SiO_2 Nanocomposite. *Metals* **2024**, *14*, 808. https://doi.org/10.3390/met14070808.
2. Monteiro, B.; Simões, S. Production and Characterization of Hybrid Al6061 Nanocomposites. *Metals* **2024**, *14*, 1206. https://doi.org/10.3390/met14111206.
3. Kunčická, L.; Walek, J.; Kocich, R. Microstructure Development of Powder-Based Cu Composite During High Shear Strain Processing. *Metals* **2024**, *14*, 1331. https://doi.org/10.3390/met14121331.
4. Walunj, G.; Choudhari, A.; Digole, S.; Bearden, A.; Kolt, O.; Bari, P.; Borkar, T. Microstructure, Mechanical, and Tribological Behaviour of Spark Plasma Sintered TiN, TiC, TiCN, TaN, and NbN Ceramic Coatings on Titanium Substrate. *Metals* **2024**, *14*, 1437. https://doi.org/10.3390/met14121437.
5. Wang, Y.; Zhang, G. In Situ Synthesis of $MoSi_2$-SiC Composites by Two-Step Spark Plasma Sintering. *Metals* **2025**, *15*, 144. https://doi.org/10.3390/met15020144.
6. Wan, H.; Shuai, L.; Ling, L.; Hu, Z.; Yan, H. Mechanical Properties and Corrosion Resistance of La_2O_3/A356 Composites Fabricated by Ultrasonic-Assisted Casting. *Metals* **2025**, *15*, 184. https://doi.org/10.3390/met15020184.

7. Liu, S.; Li, A.; Kang, K.; Zhang, J.; Huang, D.; Che, C.; Jiang, Y.; Xu, M.; Zhang, B.; Li, Y.; Li, G. Microstructure and Performance of Body-Centered Cubic-Based Dual-Phase Composite Eutectic High-Entropy Alloys Prepared by Si Doping. *Metals* **2025**, *15*, 207. https://doi.org/10.3390/met15020207.
8. Akarou, A.; Baras, F.; Politano, O. Microscopic Modeling of Interfaces in Cu-Mo Nanocomposites: The Case Study of Nanometric Metallic Multilayers. *Metals* **2025**, *15*, 282. https://doi.org/10.3390/met15030282.
9. Kunčická, L.; Kocich, R. Characterizing the Behavior and Microstructure of Cu-La_2O_3 Composite Processed via Equal Channel Angular Pressing. *Metals* **2025**, *15*, 368. https://doi.org/10.3390/met15040368.
10. Xi, J.; Wang, Z.; Zhang, L.; Ma, L.; Tang, P. Unveiling High-Pressure Behavior of Sc3AlC MAX Phase: A Comprehensive Theoretical Study on Structural, Mechanical, Dislocation, and Electronic Properties. *Metals* **2025**, *15*, 492. https://doi.org/10.3390/met15050492.

References

1. Chen, L.-Y.; Qin, P.; Zhang, L.; Zhang, L.-C. An overview of additively manufactured metal matrix composites: Preparation, performance, and challenge. *Int. J. Extrem. Manuf.* **2024**, *6*, 052006. [CrossRef]
2. Kar, A.; Sharma, A.; Kumar, S. A Critical Review on Recent Advancements in Aluminium-Based Metal Matrix Composites. *Crystals* **2024**, *14*, 412. [CrossRef]
3. Ammisetti, D.K.; Kruthiventi, S.S.H.; Vinjavarapu, S. A review on reinforcements, fabrication methods, and mechanical and wear properties of titanium metal matrix composites. *J. Eng. Appl. Sci.* **2024**, *71*, 60. [CrossRef]
4. Tamtam, A.; Abusoua, A. Recent developments in metal matrix composites with fiber reinforcement. In *Comprehensive Materials Processing*, 2nd ed.; Elsevier: Amsterdam, The Netherlands, 2024; Volume 12, pp. 374–394. [CrossRef]

metals MDPI

Article

An Insight into the Varying Effects of Different Cryogenic Temperatures on the Microstructure and the Thermal and Compressive Response of a Mg/SiO_2 Nanocomposite

Michael Johanes [1], Sarah Mehtabuddin [2], Vishal Venkatarangan [3] and Manoj Gupta [1,*]

1. Department of Mechanical Engineering, National University of Singapore, 9 Engineering Drive 1, Singapore 117575, Singapore
2. Department of Bioengineering, University of California, 306 Stanley Hall MC #1762, Berkeley, CA 94720-1762, USA
3. Department of Chemical Engineering, Faculty of Applied Sciences, Delft University of Technology, 2629 HZ Delft, The Netherlands

* Correspondence: mpegm@nus.edu.sg

Abstract: This study for the first time reports that insights into microstructure and thermal and compressive responses can be best achieved following exposure to different cryogenic temperatures and that the lowest cryogenic temperature may not always produce the best results. In the present study, a $Mg\text{-}SiO_2$ biocompatible and environment-friendly nanocomposite was synthesized by using the Disintegrated Melt Deposition method followed by hot extrusion. Subsequently, it was subjected to four different sub-zero temperatures (−20 °C, −50 °C, −80 °C, and −196 °C). The results reveal the best densification at −80 °C, marginally improved ignition resistance at 50 °C, the best damping response at −80 °C, the best microhardness at −50 °C, and the best compressive response at −20 °C. The results clearly indicate that the cryogenic temperature should be carefully chosen depending on the property that needs to be particularly enhanced governed by the principal requirement of the end application.

Keywords: disintegrated metal deposition; nanocomposite; cryogenic treatment; silica; mechanical properties; magnesium; microstructure; biocompatible materials

1. Introduction

In recent times, magnesium-based materials have gained increasing traction in structural and biomedical contexts with green potential; this is by virtue of several desirable properties such as its light weight, high specific strength and stiffness, Young's modulus comparable with human bone, and biocompatibility owing to its non-toxic nature [1,2]. Furthermore, magnesium is sustainable as a resource, with worldwide production near 1 million tons annually in past years and having reserves assessed by the United States Geological Survey as sufficient to meet the needs of today and the future, positioning it as a metal expected to play a key and strategic role in the future [3].

Considering its low density of 1.74 g/cm^3 (distinctly less than that of aluminum and titanium) and subsequent light weight, its ease of processing and machining, as well as its suitability for recycling, which comes with significant energy savings (35 kWh/kg for primary production vs. 3 kWh/kg for remelting), magnesium is a very suitable material for multiple applications [4,5]. Combined with abundant reserves, as mentioned earlier, this allows for a green (sustainable, energy-saving) structural metal, critical in the resource- and energy-intensive landscape of today and for the future.

Magnesium-based materials are currently being actively explored as resorbable temporary implants and as fixation implants in orthopedic, craniofacial, and cardiovascular contexts [6]. These applications are currently being serviced with conventional metallic materials, such as those based on Ti, Fe, and Co, which require a secondary removal surgery.

Moreover, these conventional materials possess shortcomings, namely their high density, biological incompatibility arising from alloying elements (Ni, Co, Cr) [7,8], and stress-shielding effects. As magnesium is a bioresorbable material, it serves as a temporary fracture recovery support while also avoiding the aforementioned secondary/revision surgery, reducing the risk of complications [9,10]. Further, more recent engineering uses of magnesium targeted to reduce greenhouse gas emissions and increase human comfort include the automobile, rail, marine, aerospace, space, sports, electronics, and defense sectors [11,12].

SiO_2 (Silica) is attractive as a ceramic reinforcement; possessing strong covalent bonds, it is thermally stable up to 1300 °C [13] and also has a very high hardness (7.3 GPa) [14]. In the context of Mg materials, the use of such ceramic reinforcements (especially at the nanoscale) can be an inexpensive and enticing alternative to the use of alloys for improvements in mechanical properties via dispersion strengthening and the activation of non-basal slip systems [15,16]. Its usefulness also lies outside of structural and engineering contexts; also known as bioglass, it exhibits high biocompatibility [17] and bioactivity, encouraging bone growth and aiding in bone reconstruction due to its osteoinductive properties [18–20]. Furthermore, it also has positive effects on material properties, examples of which include improvements in the compressive strength, corrosion resistance, and thermal stability of SiO_2-reinforced magnesium composites [13,21,22].

Further improvements in material properties can be achieved via cryogenic treatment, a less pursued area in the case of magnesium nanocomposites. This involves exposing the material to sub-zero temperatures, which historically has been conducted for ferrous materials such as steel, with the aim of improving the reliability and lifespan of components based on these materials [23,24], but research on non-ferrous (including Mg-based) materials is still relatively sparse [25]. Preliminary research on Mg-based alloys and nanocomposites has been conducted [26–28], exhibiting positive outcomes such as improved mechanical properties and microstructure. However, these studies were conducted at temperatures of −20 °C (refrigeration) and/or −196 °C (liquid nitrogen treatment); limited information exists on the outcomes of cryogenic treatment of Mg-based materials at other sub-zero temperatures.

Accordingly, this work seeks to study and determine the effects of different sub-zero temperatures (including those between −20 °C and −196 °C) and correlate them with the microstructural, thermal, and compressive responses of a Mg-based nanocomposite containing nano-sized SiO_2.

2. Materials and Methods

2.1. Synthesis

A nanocomposite ingot was synthesized by using pure Mg turnings (99.9% purity, Acros Organics, Morris Plains, NJ, USA) and 2 wt%. SiO_2 nanoparticles (10–20 nm in size, Sigma Aldrich, Alfa Aesar GmbH & Co. KG, Haverhill, MA, USA) were synthesized using the DMD (Disintegrated Metal Deposition) process with a target superheat temperature of 750 °C. Using a conventional lathe, billets of 35.5 mm and 45 mm in length were machined from this ingot. These were then subjected to hot extrusion with an extrusion ratio of 20.25 at a temperature of 350 °C, using an extrusion die of 8 mm in diameter and with prior soaking at 400 °C for 1 h.

Materials from the resulting extruded rods were then subjected to cryogenic treatment, with the conditions summarized in Table 1.

Table 1. Material designations and treatments performed on Mg-2SiO_2 nanocomposites studied in this work.

Material Designation	Treatment Performed
AE	As-extruded (no treatment)
RF20	Refrigeration at −20 °C for 24 h
RF50	Refrigeration at −50 °C for 24 h
RF80	Refrigeration at −80 °C for 24 h
LN	Liquid nitrogen exposure at −196 °C for 24 h

Exposure for 24 h to different cryogenic temperatures was selected as previous studies on the cryogenic treatment of metallic materials have demonstrated that this duration results in distinct, noticeable effects on material properties such as strength and hardness [23,24,26] and is thus appropriate for this study given the variation in exposure temperatures.

Where applicable, property results of cryogenically-treated materials will be compared against pre-treated counterparts. Otherwise, they will be compared against the AE counterpart.

2.2. Density and Porosity

The Archimedes principle was used to determine the density and porosity of the materials. A Density Determination Kit (AND Company, Limited, Tokyo, Japan) mounted on a GH-252 weighing scale (AND Company, Limited, Tokyo, Japan) was used to obtain mass readings of the samples, with five samples characterized for each treatment case. The density and the porosity values of the cryogenically treated samples were also compared with those of the same samples prior to treatment.

2.3. Microstructural Characterization

Samples were ground with 4000-grit emery paper and fine-finished using 1 micron-sized alumina suspension for microstructural characterization. Microstructure images were taken using the JEOL JSM-6010PLUS/LV Scanning Electron Microscope (SEM, JEOL USA Inc., Peabody, MA, USA) equipped with Energy Dispersive Spectroscopy (EDS) at an accelerating voltage of 20 kV, 2000× magnification, and a working distance of 10 mm.

X-ray Diffraction (XRD) analysis of the materials was conducted using a Shimadzu XRD-6000 automatic spectrometer (Shimadzu Corporation, Kyoto, Japan). Longitudinal surfaces of the extruded materials were exposed to Cu Kα X-rays at a wavelength λ of 1.5418 Å, with a scanning speed of 2°/min and a scanning range from 25° to 80°.

2.4. Thermal Characterization

A Shimadzu DTG-60H thermogravimetric analyzer (TGA, Shimadzu Corporation, Kyoto, Japan) was used to determine the ignition response and a Shimadzu DSC-60 digital scanning calorimeter (DSC, Shimadzu Corporation, Kyoto, Japan) was used to determine the thermal response of materials in this study. These characterizations were conducted on samples with approximate dimensions of 2 mm × 2 mm × 2 mm, each subjected to a temperature range from 30 °C to 1400 °C at a rate of 10 °C/min in purified air at a flow rate of 50 mL/min for TGA, and from 30 °C to 600 °C at a rate of 5 °C/min in argon flowing at 25 mL/min for DSC.

A sample of 5 mm in height of each material was exposed to a temperature range from 50 °C to 400 °C at a rate of 5 °C/min in argon gas at a flow rate of 0.1 L/min to determine the coefficient of thermal expansion (CTE) using a TMA PT1000 Thermomechanical Analyzer (Linseis Messgeraete GmbH, Selb, Germany).

2.5. Mechanical Characterization

The damping/vibration response, including Young's modulus, was characterized by subjecting samples with a length of 50 mm to impulse excitation. The recorded vibration signals were analyzed using the Resonance Frequency Damping Analyzer (RFDA) software (RFDA version 8.1.2, IMCE, Genk, Belgium) to glean insight into the materials' damping response. One sample per material was analyzed.

Microhardness was characterized using an HMV 2 microhardness tester (Shimadzu Corporation, Kyoto, Japan) with an indenter force of 245.2 mN and a dwell time of 15 s, in accordance with procedures outlined in the standard ASTM E-384. A minimum of 15 hardness readings on one representative surface were taken per material for this test.

Flat and parallel samples with an aspect ratio of 1 were subject to quasi-static compressive loading using an MTS-E44 hydraulic tester (MTS Systems, Eden Prairie, MN, USA) at a strain rate of 0.000083 s^{-1} until failure. This analysis was conducted in accordance with the standard ASTM E9-09 [29]. Three representative samples were tested per material.

3. Results and Discussion

3.1. Density and Porosity

The density and porosity results of the Mg-2SiO$_2$ samples subjected to different cryogenic temperatures are shown in Table 2. Porosity was determined by using the Archimedes method with the following formula (assuming that any voids were filled with air):

$$\text{Porosity}: \frac{\rho_{theoretical} - \rho_{experimental}}{\rho_{theoretical} - \rho_{air}}$$

where $\rho_{\text{theoretical}}$, $\rho_{\text{experimental}}$, and ρ_{air} are material theoretical density, experimental density, and air density, respectively. Cryogenically treated materials exhibited, in general, elevated experimental densities and subsequently decreased porosity.

Table 2. Density and porosity results of Mg-2SiO$_2$ materials studied in this work.

Treatment	Average Experimental Density (g/cm^3)		Porosity Reduction (%)
	Before Treatment	After Treatment	
RF20	1.794 ± 0.006	1.794 ± 0.002	0
RF50	1.793 ± 0.002	1.796 ± 0.003 (↑0.167%)	7.0
RF80	1.787 ± 0.001	1.800 ± 0.004 (↑0.727%)	33.6
LN	1.789 ± 0.012	1.800 ± 0.006 (↑0.615%)	26.3

While no change in density was observed after exposure to −20 °C, lower exposure temperatures resulted in the densification of the materials, as indicated by the reduction in porosity. This was also observed in a study on CeO$_2$-reinforced Mg which exposed materials to −20–196 °C temperatures (the latter of which was achieved by immersion in LN, similar to in this study) [26]. This observation can be attributed to two mechanisms, namely the internal compressive stresses acting on the materials as a result of sub-zero environments [23], as well as the sinking of dislocations into existing pores [30]. The best results were achieved at an exposure temperature of −80 °C.

3.2. Microstructure

Figure 1 shows the general microstructural morphology, while Table 3 shows the EDS results of selected locations within the microstructure of the Mg-2SiO$_2$ materials in this work. Figure 2 shows one of the EDS plots/readings belonging to Mg-2SiO$_2$ AE, showing the normalized nature of the results (i.e., the table only includes results above a certain threshold).

The X-ray diffractograms for all Mg-2SiO$_2$ materials are shown in Figure 3. The peaks were verified against JCPDS cards (Mg: 00-004-0770, MgO: 00-084-0829, Mg$_2$Si: 00-035-0073) present in the Powder Diffraction File (PDF-4+ 2023) [31], showing that the detected peaks were predominantly Mg, with some low-intensity peaks corresponding to MgO and Mg$_2$Si. Furthermore, the relative intensities of the prismatic, basal, and pyramidal planes of the materials are also summarized in Table 4.

Cryogenic treatment did not appear to result in any discernible differences in the microstructure. The microstructure of all materials appeared to be predominantly a Mg matrix with small amounts of bright particles containing O; these are likely MgO and/or agglomerated SiO$_2$ particles. Mg$_2$Si was not visually detected despite some Si being detected in the Mg matrix regions. This lack of change in the general microstructural morphology and elemental composition is consistent with other Mg nanocomposites undergoing CT, such as those reinforced with CeO_2 nanoparticles [26].

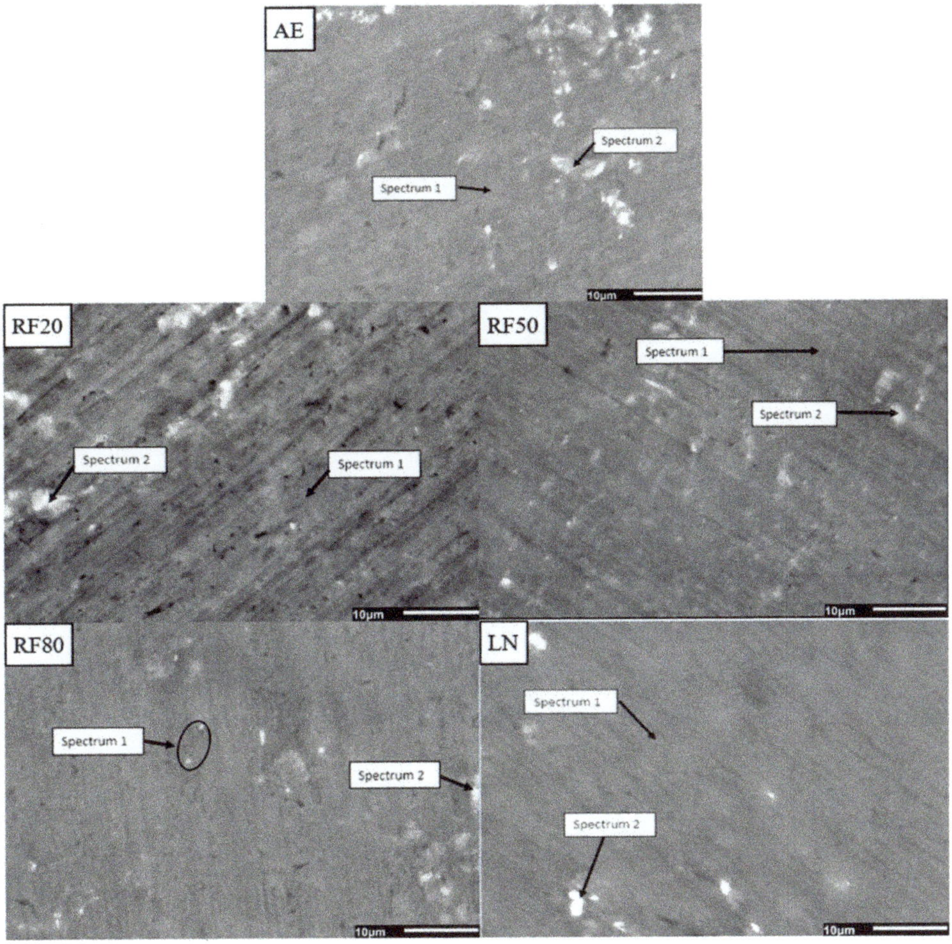

Figure 1. Scanning electron micrographs of Mg-2SiO$_2$ materials in this work, with selected regions used for the EDS studies.

Figure 2. EDS results/chart of Mg-2SiO$_2$ AE: matrix region.

Table 3. The EDS results of Mg-2SiO_2 materials in this work.

Treatment	Spectrum	Detected Element (wt.%)		
		Mg	Si	O
AE	1	97.7	2.3	-
	2	88.6	2.4	9.0
RF20	1	100	-	-
	2	88.0	2.3	9.7
RF50	1	97.4	1.1	1.5
	2	87.1	7.0	5.9
RF80	1	96.0	1.8	2.2
	2	92.9	1.8	5.4
LN	1	100	-	-
	2	95.5	1.5	3.1

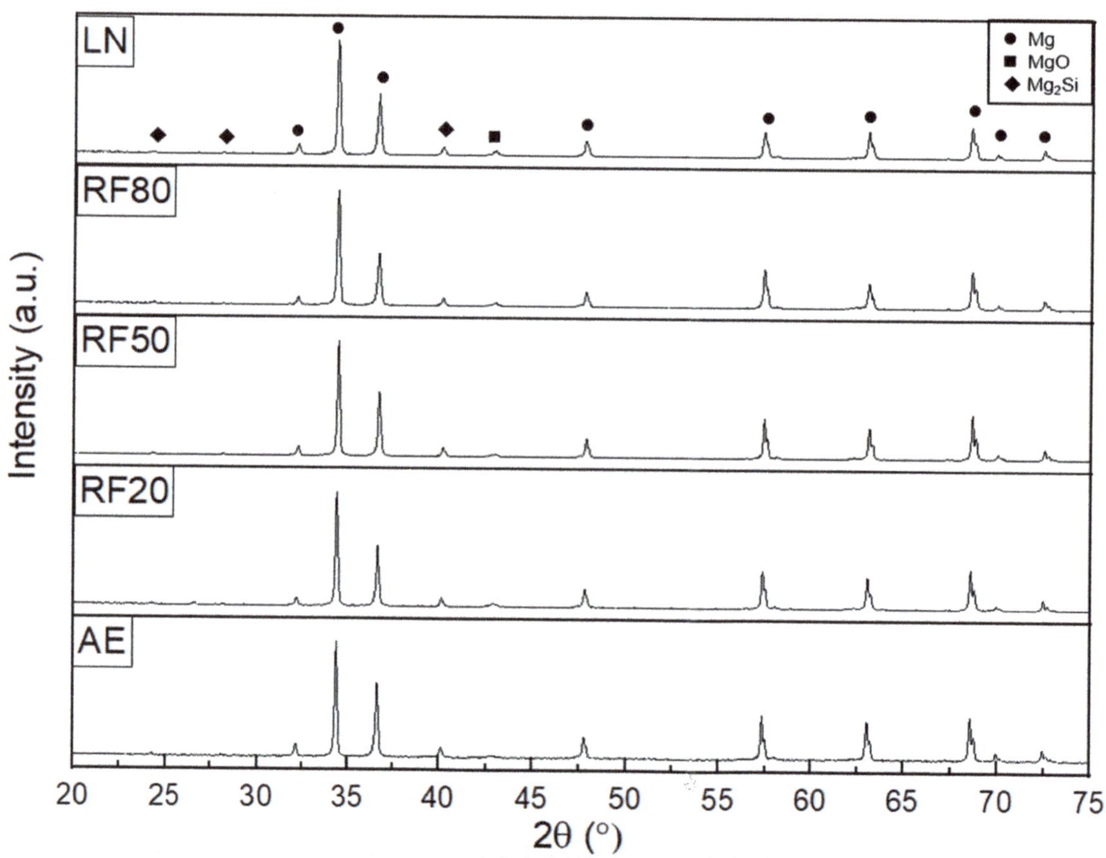

Figure 3. X-ray diffractograms of Mg-2SiO_2 materials in this work.

Table 4. XRD results of Mg crystallographic planes in this work.

Treatment	Plane	I/I_{max}
AE	10-10 prism	0.1181
	0002 basal	1
	10-11 pyramidal	0.6450
RF20	10-10 prism	0.0854
	0002 basal	1
	10-11 pyramidal	0.5373
RF50	10-10 prism	0.0898
	0002 basal	1
	10-11 pyramidal	0.5670
RF80	10-10 prism	0.0825
	0002 basal	1
	10-11 pyramidal	0.4610
LN	10-10 prism	0.0965
	0002 basal	1
	10-11 pyramidal	0.5326

All materials exposed to sub-zero temperatures exhibited a stronger basal texture relative to their as-extruded counterparts, shown by the lower relative intensities of the prismatic and pyramidal peaks. This is in line with previous studies concerning Mg nanocomposites containing CeO_2 where cryogenic treatment strengthened existing textures [26,32], such that the most dominant basal texture initially present in the extruded nanocomposites was further strengthened.

3.3. Thermal Response

The CTE and ignition temperatures of the materials studied in this work are summarized in Table 5. Only marginal changes in both properties were observed, indicating that cryogenic treatment does not result in significant alterations in the thermal properties of metallic composites.

Table 5. CTE ignition temperatures of Mg-2SiO_2 materials in this work.

Treatment	Average CTE ($\times 10^{-6}$/K)	Ignition Temperature (°C)
AE	22.95	602
RF20	23.63	604
RF50	23.07	606
RF80	22.79	604
LN	24.00	603

Figure 4 shows the ignition responses of the materials, with the ignition temperature determined at the point just before the spike and subsequent recovery in temperature, as seen in the magnified inset. The results shown in Table 5 indicate only a marginal increase (+2 °C) in ignition temperature following cryogenic exposure at −20 °C and −80 °C. For LN exposure, the ignition temperature was even more marginally affected (1 °C increase). The most significant increase occurred after exposure to −50 °C (4 °C increase).

The DSC results in Figure 5 indicate that cryogenic treatment does not result in noticeable changes in thermal response as there are no discernable differences in the trends.

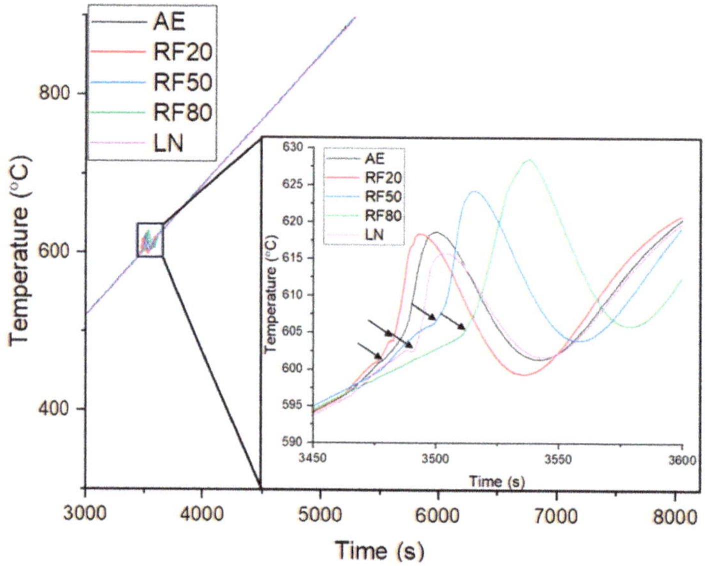

Figure 4. The ignition responses of Mg-2SiO_2 materials in this work, showing the locations where ignition temperatures were evaluated.

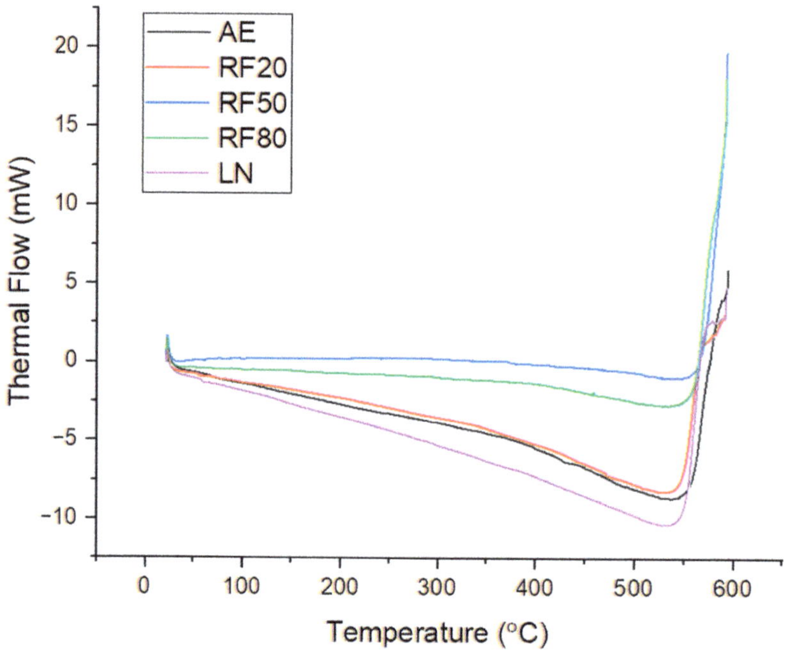

Figure 5. Thermal response of Mg-2SiO_2 materials in this work.

3.4. Mechanical Response

Cryogenic treatment was shown to variably change the damping capabilities of the materials, with a minimal effect on Young's modulus (Table 6).

Table 6. Results of damping properties and modulus.

Treatment	Attenuation Coefficient	Damping Capacity	E-Modulus (GPa)
AE	41.8	0.001238	47.54 ± 0.28
RF20 (pre-treatment)	48.42	0.001433	48.26 ± 0.28
RF20 (post-treatment)	49.31 (↑1.8%)	0.001292 (↓9.6%)	49.06 ± 0.29 (↑1.7%)
RF50 (pre-treatment)	38.73	0.001288	48.15 ± 0.28
RF50 (post-treatment)	35.19 (↓9.1%)	0.001174 (↓8.9%)	48.23 ± 0.38 (↑0.2%)
RF80 (pre-treatment)	38.09	0.001104	48.23 ± 0.28
RF80 (post-treatment)	40.94 (↑7.5%)	0.001322 (↑19.8%)	48.23 ± 0.38 (no change)
LN (pre-treatment)	44.54	0.001344	48.71 ± 0.29
LN (post-treatment)	44.88 (↑0.8%)	0.0001352 (↑0.6%)	48.67 ± 0.29 (↓0.1%)

Typically, the reduction in the damping efficacy of cryogenically treated materials can be attributed to the resulting densification; this would result in reductions in porosity and, correspondingly, in air pockets and voids, which have been shown to enhance damping capabilities in aluminum and Mg-based materials [33,34]. However, a notable exception at −80 °C exposure was observed when both the attenuation coefficient and damping capacity improved. The underlying mechanisms behind such observations are not clear and require a more detailed study that may include dislocation density measurements by TEM experts. Nevertheless, improvements in the damping capacity and attenuation coefficient indicate an increase in the ability to dissipate vibrational energy, crucial in end applications where such loading is expected to occur (e.g., in components of vehicles).

Microhardness increased after cryogenic treatment, as seen in Table 7, with exposure to −50 °C conferring the greatest increase in hardness. The hardness results suggest that hardness variation is not solely governed by the extent of porosity (Table 2), but by other subtle changes in microstructural features, such as variations in texture and dislocation density. These results open the doors for a more detailed study by experts in this area.

Table 7. Microhardness values of Mg-2SiO_2 materials in this work.

Treatment	Average Microhardness (H_V)
AE	85 ± 7
RF20	86 ± 6 (↑1.2%)
RF50	112 ± 8 (↑31.8%)
RF80	104 ± 5 (↑22.4%)
LN	86 ± 5 (↑1.2%)

Overall, improvements in compressive properties (combination of strength and failure strain) were observed for cryogenically treated materials (Table 8), most notably for RF20 and LN-treated materials which experienced more significant yield strength improvements. The variation in yield strength and failure strain with decreases in cryogenic temperature did not show a trend, while the average ultimate compressive strength increased after exposure to all cryogenic temperatures. The results reveal that the intrinsic microstructural factors that govern the friction stress of the matrix and subsequently the initiation of dislocation motion, affecting yield strength, evolve differently at different cryogenic temperatures. More detailed research is required in this area.

Table 8. Compressive properties of Mg-2SiO$_2$ materials in this work.

Treatment	Average 0.2% Yield Strength (MPa)	Average Ultimate Compressive Strength (MPa)	Average Failure Strain (%)	Average Work of Fracture (MJ/m^3)
AE	158 ± 2	380 ± 11	27.6 ± 0.4	60.8 ± 1.8
RF20	171 ± 1 (↑8.2%)	381 ± 12 (↑0.3%)	28.4 ± 1.2 (↑2.9%)	64.0 ± 3.7 (↑5.3%)
RF50	152 ± 3 (↓3.8%)	389 ± 11 (↑2.4%)	27.6 ± 1.2	63.2 ± 3.1 (↑4.0%)
RF80	154 ± 7 (↓2.5%)	385 ± 10 (↑1.3%)	27.5 ± 1.3 (↓0.4%)	60.9 ± 4.9 (↑0.2%)
LN	164 ± 1 (↑3.8%)	383 ± 1.4 (↑0.8%)	28.4 ± 1.04 (↑2.9%)	63.9 ± 4.3 (↑5.1%)

While improvements in compressive properties (especially work of fracture) have been explored with Mg materials subjected to sub-zero temperatures [26] in the past, relatively greater improvements in compressive properties (assessed through work of fracture—Table 8) were observed for cryogenically treated materials, attributable to a reduction in porosity which assists in the material's ability to bear load and reduces the potential for the development of crack initiation sites [35,36]. Specifically, RF20 and LN materials showed the greatest improvement. The results suggest that for strength-based design, exposure at −20 °C is the best way forward. Considering that this temperature can be achieved through refrigeration, further energy or resource savings can be attained by forgoing the use of liquid nitrogen for the treatment of Mg materials in the future.

Macro-scale images of the fractured samples can be seen in Figure 6, with an approximate 45° angle of fracture. Figure 7 shows that microcracks were not present for all materials, with just the characteristic shear bands resulting from compressive loading apparent. No effect of cryogenic treatment was observed on the fracture surfaces of differently cryogenically treated samples which displayed similar fractographic characteristics.

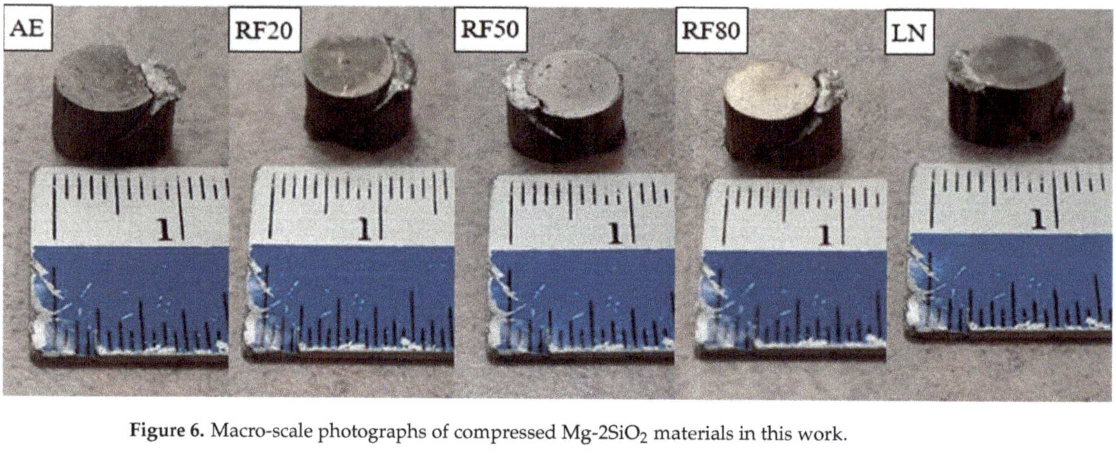

Figure 6. Macro-scale photographs of compressed Mg-2SiO$_2$ materials in this work.

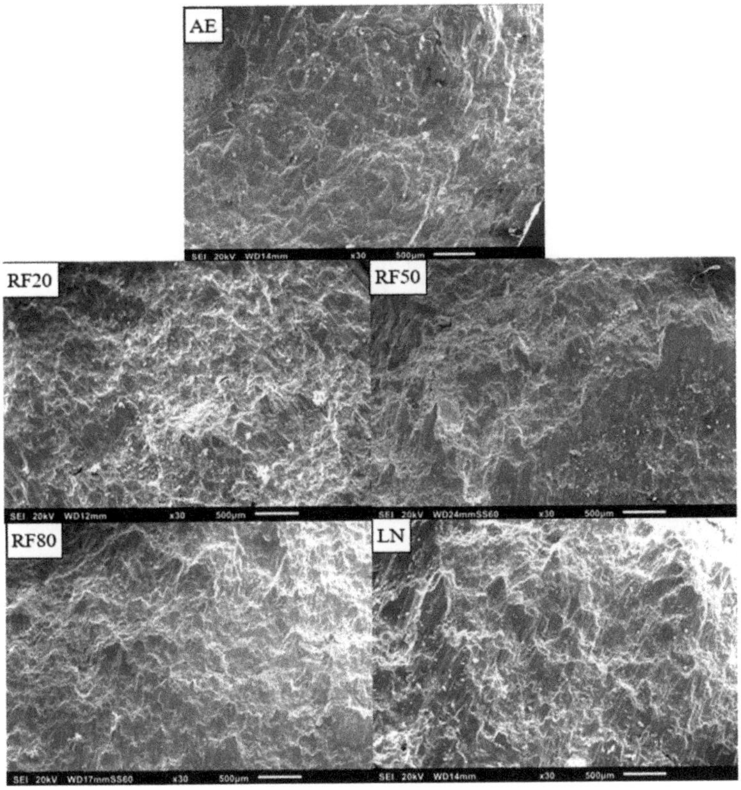

Figure 7. Fractographs of Mg-2SiO_2 materials in this work.

4. Conclusions

Following the cryogenic treatment of Mg-2SiO_2 nanocomposites exposed to different sub-zero temperatures, the following observations have been noted and conclusions made:

1. The greatest densification (33.6% reduction in porosity) was achieved with an exposure temperature of −80 °C.
2. Cryogenic treatment reinforced existing textures by increasing the relative intensity of the already dominant basal texture.
3. Exposure to −50 °C resulted in slight improvements in ignition resistance, with a 4 °C increase in ignition temperature.
4. The damping properties of Mg-2SiO_2 were most improved through exposure to −80 °C, with the attenuation coefficient improving by 7.5% and the damping capacity by 19.8%.
5. The hardness of Mg-2SiO_2 was most improved (31.8%) through exposure to −50 °C.
6. The greatest improvements in compressive response were achieved through exposure to −20 °C, with improvements in yield strength (8.2% increase), fracture strain (2.9% increase), and energy absorbed (5.3% increase).

The findings of the present work clearly indicate that exposure to differing cryogenic temperatures results in varying positive effects on the individual properties of Mg-2SiO_2 nanocomposites. Thus, the cryogenic treatment temperature should be determined based on the requirements of the end application. The results presented in this work also emphasize that the cause of variations in properties resulting from cryogenic exposure requires more detailed study by experts (typically in-depth microstructural characterization using TEM and EBSD) and opens multiple pathways for future work.

Author Contributions: Conceptualization, M.G.; methodology, M.G.; validation, M.J. and M.G; formal analysis, M.J. and S.M.; investigation, M.J., S.M. and V.V.; resources, M.G.; data curation, M.J.; writing—original draft preparation, M.J.; writing—review and editing, M.J. and M.G.; visualization, M.J., S.M. and V.V.; supervision, M.J. and M.G.; project administration, M.G. All authors have read and agreed to the published version of the manuscript.

Funding: This research received no external funding.

Data Availability Statement: The original contributions presented in the study are included in the article, further inquiries can be directed to the corresponding author.

Acknowledgments: The authors would like to acknowledge Juraimi Bin Madon for the extrusion work and Ng Hong Wei for their assistance with DSC, CTE, and TGA testing.

Conflicts of Interest: The authors declare no conflict of interest.

References

1. Bhat Panemangalore, D.; Shabadi, R.; Tingaud, D.; Touzin, M.; Ji, G. Biocompatible silica-based magnesium composites. *J. Alloys Compd.* **2019**, *772*, 49–57. [CrossRef]
2. Liu, C.; Ren, Z.; Xu, Y.; Pang, S.; Zhao, X.; Zhao, Y. Biodegradable Magnesium Alloys Developed as Bone Repair Materials: A Review. *Scanning* **2018**, *2018*, 9216314. [CrossRef] [PubMed]
3. Mineral Commodity Summaries 2024. In *Mineral Commodity Summaries*; Report; USGS: Reston, VA, USA, 2024. Available online: https://pubs.usgs.gov/publication/mcs2024 (accessed on 12 April 2024).
4. Ditze, A.; Scharf, C. Recycling of Magnesium Alloys. In *Magnesium—Alloys and Technology*; Wiley Online Books: Hoboken, NJ, USA, 2003; pp. 254–278. [CrossRef]
5. Lapovok, R.Y.; Thomson, P.F. Production of dense rod from magnesium swarf for re-melting. *Magnes. Technol.* **2004**, *1*, 149–154.
6. Li, L.; Zhang, M.; Li, Y.; Zhao, J.; Qin, L.; Lai, Y. Corrosion and biocompatibility improvement of magnesium-based alloys as bone implant materials: A review. *Regen. Biomater.* **2017**, *4*, 129–137. [CrossRef]
7. Pacheco, K.A. Allergy to Surgical Implants. *Clin. Rev. Allergy Immunol.* **2019**, *56*, 72–85. [CrossRef] [PubMed]
8. Jin, W.; Chu, P.K. Orthopedic Implants. In *Encyclopedia of Biomedical Engineering*; Narayan, R., Ed.; Elsevier: Oxford, UK, 2019; pp. 425–439. [CrossRef]
9. Selvaraj, D.; Raja, J.; Prasath, S. Interdisciplinary approach for bilateral maxillary canine: First premolar transposition with complex problems in an adult patient. *J. Pharm. Bioallied Sci.* **2013**, *5*, S190–S194. [CrossRef] [PubMed]
10. Ibrahim, H.; Moghaddam, N.S.; Elahinia, M. Mechanical and In Vitro Corrosion Properties of a Heat-Treated Mg-Zn-Ca-Mn Alloy as a Potential Bioresorbable Material. *Adv. Metall. Mater. Eng.* **2017**, *1*, 1–7. [CrossRef]
11. Ravi Kumar, N.V.; Blandin, J.J.; Suéry, M.; Grosjean, E. Effect of alloying elements on the ignition resistance of magnesium alloys. *Scr. Mater.* **2003**, *49*, 225–230. [CrossRef]
12. Tekumalla, S.; Joo Yuan, N.; Haghshenas, M.; Gupta, M. Enhancing Properties of Aerospace Alloy Elektron 21 Using Boron Carbide Nanoparticles as Reinforcement. *Appl. Sci.* **2019**, *9*, 5470. [CrossRef]
13. Wan, Y.; Cui, T.; Li, W.; Li, C.; Xiao, J.; Zhu, Y.; Ji, D.; Xiong, G.; Luo, H. Mechanical and biological properties of bioglass/magnesium composites prepared via microwave sintering route. *Mater. Des.* **2016**, *99*, 521–527. [CrossRef]
14. Michel, M.D.; Serbena, F.C.; Lepienski, C.M. Effect of temperature on hardness and indentation cracking of fused silica. *J. Non-Cryst. Solids* **2006**, *352*, 3550–3555. [CrossRef]
15. Umeda, J.; Kawakami, M.; Kondoh, K.; Ayman, E.L.S.; Imai, H. Microstructural and mechanical properties of titanium particulate reinforced magnesium composite materials. *Mater. Chem. Phys.* **2010**, *123*, 649–657. [CrossRef]
16. Goh, C.S.; Gupta, M.; Wei, J.; Lee, L.C. The Cyclic Deformation Behavior of Mg—Y_2O_3 Nanocomposites. *J. Compos. Mater.* **2008**, *42*, 2039–2050. [CrossRef]
17. Jaganathan, H.; Godin, B. Biocompatibility assessment of Si-based nano- and micro-particles. *Adv. Drug Deliv. Rev.* **2012**, *64*, 1800–1819. [CrossRef] [PubMed]
18. Maquet, V.; Boccaccini, A.R.; Pravata, L.; Notingher, I.; Jérôme, R. Porous poly(α-hydroxyacid)/Bioglass® composite scaffolds for bone tissue engineering. I: Preparation and in vitro characterisation. *Biomaterials* **2004**, *25*, 4185–4194. [CrossRef]
19. Roether, J.A.; Boccaccini, A.R.; Hench, L.L.; Maquet, V.; Gautier, S.; Jérôme, R. Development and in vitro characterisation of novel bioresorbable and bioactive composite materials based on polylactide foams and Bioglass® for tissue engineering applications. *Biomaterials* **2002**, *23*, 3871–3878. [CrossRef] [PubMed]
20. Vergnol, G.; Ginsac, N.; Rivory, P.; Meille, S.; Chenal, J.-M.; Balvay, S.; Chevalier, J.; Hartmann, D.J. In vitro and in vivo evaluation of a polylactic acid-bioactive glass composite for bone fixation devices. *J. Biomed. Mater. Res. Part B Appl. Biomater.* **2016**, *104*, 180–191. [CrossRef] [PubMed]
21. Bruckmann, F.D.; Nunes, F.B.; Salles, T.D.; Franco, C.; Cadoná, F.C.; Bohn Rhoden, C.R. Biological Applications of Silica-Based Nanoparticles. *Magnetochemistry* **2022**, *8*, 131. [CrossRef]
22. Johanes, M.; Gupta, M. An Investigation into the Potential of Turning Induced Deformation Technique for Developing Porous Magnesium and Mg-SiO_2 Nanocomposite. *Materials* **2023**, *16*, 2463. [CrossRef]

23. Sonar, T.; Lomte, S.; Gogte, C. Cryogenic Treatment of Metal—A Review. *Mater. Today Proc.* **2018**, *5*, 25219–25228. [CrossRef]
24. Zurecki, Z. Cryogenic Quenching of Steel Revisited. In *Heat Treating, Proceedings of the 23rd ASM Heat Treating Society Conference, Pittsburgh, PA, USA, 25–28 September 2005*; ASM: Almere, The Netherlands, 2006.
25. Baldiserra, P.; Delprete, C. Deep Cryogenic Treatment: A Bibliographic Review. *Open Mech. Eng. J.* **2008**, *2008*, 1–11. [CrossRef]
26. Gupta, S.; Parande, G.; Gupta, M. Comparison of Shallow (−20 °C) and Deep Cryogenic Treatment (−196 °C) to Enhance the Properties of a Mg/2wt.%CeO_2 Nanocomposite. *Technologies* **2024**, *12*, 14. [CrossRef]
27. Gupta, S.; Parande, G.; Tun, K.S.; Gupta, M. Enhancing the Physical, Thermal, and Mechanical Responses of a Mg/2wt.%CeO_2 Nanocomposite Using Deep Cryogenic Treatment. *Metals* **2023**, *13*, 660. [CrossRef]
28. Dong, N.; Sun, L.; Ma, H.; Jin, P. Effects of cryogenic treatment on microstructures and mechanical properties of Mg-2Nd-4Zn alloy. *Mater. Lett.* **2021**, *305*, 130699. [CrossRef]
29. *ASTM E9-09*; Standard Test Methods of Compression Testing of Metallic Materials at Room Temperature. ASTM International: West Conshohocken, PA, USA, 2018.
30. Xie, S.; Lv, Q.; Zhang, W.; Qu, Y.; Qi, H.; Yu, B.; Li, R.; Li, G.; Yang, F. Effect of Cryogenic Treatment on Microstructure and Mechanical Properties of $Al_{0.6}CrFe_2Ni_2$ Dual-Phase High-Entropy Alloy. *Metals* **2023**, *13*, 195. [CrossRef]
31. Gates-Rector, S.; Blanton, T. The Powder Diffraction File: A quality materials characterization database. *Powder Diffr.* **2019**, *34*, 352–360 (PDF-4+ 2023). [CrossRef]
32. Gupta, S.; Johanes, M.; Parande, G.; Gupta, M. An Investigation into the Effect of Length Scale of Reinforcement on the Cryogenic Response of a Mg/2wt.%CeO_2 Composite. *Micro* **2024**, *4*, 170–184. [CrossRef]
33. Colakoglu, M. Factors effecting internal damping in aluminum. *J. Theor. Appl. Mech.* **2004**, *42*, 95–105.
34. Xie, Z.-k.; Tane, M.; Hyun, S.-k.; Okuda, Y.; Nakajima, H. Vibration–damping capacity of lotus-type porous magnesium. *Mater. Sci. Eng. A* **2006**, *417*, 129–133. [CrossRef]
35. Kováčik, J. Correlation between Young's modulus and porosity in porous materials. *J. Mater. Sci. Lett.* **1999**, *18*, 1007–1010. [CrossRef]
36. Rivera-Salinas, J.E.; Gregorio-Jáuregui, K.M.; Romero-Serrano, J.A.; Cruz-Ramírez, A.; Hernández-Hernández, E.; Miranda-Pérez, A.; Gutierréz-Pérez, V.H. Simulation on the Effect of Porosity in the Elastic Modulus of SiC Particle Reinforced Al Matrix Composites. *Metals* **2020**, *10*, 391. [CrossRef]

metals

MDPI

Article

Production and Characterization of Hybrid Al6061 Nanocomposites

Beatriz Monteiro [1,2] and Sónia Simões [1,2,*]

[1] Department of Metallurgical and Materials Engineering, Faculty of Engineering, University of Porto, Rua Doutor Roberto Frias, 4200-465 Porto, Portugal; up201806755@edu.fe.up.pt
[2] LAETA/INEGI-Institute of Science and Innovation in Mechanical and Industrial Engineering, Rua Doutor Roberto Frias, 4200-465 Porto, Portugal
* Correspondence: ssimoes@fe.up.pt; Tel.: +351-220413113

Abstract: Aluminum-based hybrid nanocomposites, namely the Al6061 alloy, have gained prominence in the scientific community due to their unique properties, such as high strength, low density, and good corrosion resistance. The production of these nanocomposites involves incorporating reinforcing nanoparticles into the matrix to improve its mechanical and thermal properties. The Al6061 hybrid nanocomposites were manufactured by conventional powder metallurgy (cold pressing and sintering). Ceramic silicon carbide (SiC) nanoparticles and carbon nanotubes (CNTs) were used as reinforcements. The nanocomposites were produced using different reinforcement amounts (0.50, 0.75, 1.00, and 1.50 wt.%) and sintered from 540 to 620 °C for 120 min. The characterization of the Al6061 hybrid nanocomposites involved the analysis of their mechanical properties, such as hardness and tensile strength, as well as their micro- and nanometric structures. Techniques such as optical microscopy (OM) and scanning electron microscopy (SEM) with electron backscatter diffraction (EBSD) were used to study the distribution of nanoparticles, the grain size of the microstructure, and the presence of defects in the matrix. The microstructural evaluation revealed significant grain refinement and greater homogeneity in the hybrid nanocomposites reinforced with 0.75 wt.% of SiC and CNTs, resulting in better mechanical performance. Tensile tests showed that the Al6061/CNT/SiC hybrid composite had the highest tensile strength of 104 MPa, compared to 63 MPa for the unreinforced Al6061 matrix. The results showed that adding 0.75% SiC nanoparticles and CNTs can significantly improve the properties of Al6061 (65% in the tensile strength). However, some nanoparticle agglomeration remains one of the challenges in manufacturing these nanocomposites; therefore, the expected increase in mechanical properties is not observed.

Keywords: hybrid nanocomposites; powder metallurgy; SiC; CNTs; microstructure; mechanical properties

1. Introduction

The investigation of hybrid metal matrix nanocomposites (HMMNCs) has garnered significant attention in recent years due to their potential to develop material performance across various applications, particularly in the aerospace sector. HMMNCs combine the advantageous properties of metal matrix composites (MMCs) with the superior characteristics of nanomaterials, leading to enhanced mechanical, thermal, and wear-resistant properties. The incorporation of nanoparticles, such as silicon carbide (SiC), alumina (Al_2O_3), and carbon nanotubes (CNTs), into a metal matrix such as aluminum allows for the tailoring of material properties to meet the demanding requirements of aerospace applications, where weight reduction and performance optimization are critical [1,2].

Recent advancements in fabrication techniques, including stir casting, powder metallurgy, and additive manufacturing, have facilitated the production of HMMNCs with improved microstructural characteristics and mechanical performance. For instance, ultrasonic-assisted stir casting has been shown to enhance the dispersion of reinforcements within the matrix, resulting in composites with superior tensile strength and fatigue resistance [2,3]. The hybridization of different types of nanoparticles has also demonstrated synergistic effects, leading to composites that exhibit enhanced mechanical and tribological properties compared to conventional materials [2,4].

Hybrid nanocomposites based on the aluminum alloy Al6061 have emerged as a promising class of materials due to their ability to combine superior mechanical and functional properties. The Al6061 alloy is widely used in structural applications for its excellent strength-to-weight ratio, corrosion resistance, and good machinability. However, to meet growing industry demands for materials with even more advanced performance, the addition of reinforcing nanoparticles, such as ceramics (SiC, Al_2O_3, TiO_2) and carbon-based materials (graphene, carbon nanotubes), has been explored to produce hybrid nanocomposites. The advantages of Al6061 hybrid nanocomposites include significant improvements in mechanical properties, such as hardness and wear resistance, due to reinforcements that increase the aluminum matrix's yield strength and fracture resistance. In addition, incorporating nanoparticles can increase thermal stability and improve tribological properties, making these materials ideal for applications in the automotive, aerospace, and defense sectors, where the combination of lightness and high performance is crucial.

Recent advancements in fabrication techniques, such as stir casting and powder metallurgy, have facilitated the production of Al6061 nanocomposites with improved microstructural characteristics and mechanical performance. For instance, studies have demonstrated that the ultrasonic stir casting method can achieve a uniform distribution of reinforcements within the matrix, resulting in composites with enhanced tensile strength and fatigue resistance [5,6]. Additionally, using hybrid reinforcements—combining different types of nanoparticles—has yielded synergistic effects, leading to composites that exhibit superior mechanical and tribological properties compared to unreinforced materials [7,8].

Despite the clear advantages of producing Al6061 hybrid nanocomposites by stir casting and powder metallurgy, which are the processes that show the best results, these processes present significant challenges, such as dispersing the reinforcements in the matrix and obtaining a good interface between the matrix and the reinforcement, which can compromise the expected mechanical properties of the final nanocomposite. Obtaining a uniform distribution of nanoparticles in the aluminum matrix is complex due to the tendency of nanoparticles to agglomerate, which can lead to a heterogeneous distribution and, consequently, a reduction in the desired mechanical properties. For instance, Kumar et al. [9] demonstrated that the wear characteristics of Al6061 reinforced with multiwalled carbon nanotubes (MWCNTs) and graphene were significantly influenced by the dispersion of these fillers, which was enhanced through a ball milling process. Similarly, the work by Soni and Thomas [10] highlighted the importance of nanoparticle dispersion in influencing the mechanical properties of Al6061 matrix composites, emphasizing that poor dispersion can lead to weak interfacial bonding and reduced mechanical performance. This challenge is compounded by the high viscosity of the aluminum matrix during processing, which can hinder the adequate mixing of the nanofillers [11]. Reddy et al. [12] utilized ultrasonically assisted stir casting to fabricate Al6061 composites reinforced with silicon carbide and graphite, achieving improved mechanical properties. However, each method has its limitations; for instance, powder metallurgy can lead to difficulties in achieving the desired density and homogeneity of the final product [13]. Moreover, the processing parameters, including temperature and time, must be carefully controlled to prevent nanofiller degradation, mainly when using thermally sensitive materials such as CNTs and graphene [14].

Additional challenges include optimizing the interface between the matrix and the nanoparticles to maximize the load transfer and preventing microstructural defects that could compromise the integrity of the material. Recent studies have focused on functionalizing nanoparticle surfaces and modifying production techniques to improve the dispersion and adhesion of nanoparticles in the aluminum matrix [15,16].

The practical applications of these nanocomposites range from lightweight structural components in automotive and aerospace vehicles to sports equipment and consumer electronics, where superior mechanical performance and weight reduction are critical. Research to date has shown progress in improving the properties of nanocomposites but also highlights the need for further studies into the effects of different types of reinforcements and production methods. Continued progress in understanding the mechanisms governing the behavior of these hybrid materials is essential for the development of broader and more effective industrial applications.

In this sense, this work explores the possibility of producing hybrid nanocomposites of an Al6061 matrix reinforced with SiC and CNTs. An intense microstructural characterization was carried out to evaluate the effect of the percentage of reinforcement, sintering temperature, and type of reinforcement to obtain nanocomposites with the best balance between microstructure and mechanical properties with high potential for future implementation.

2. Materials and Methods

In the current study, the selected matrix metal was the 6061 aluminum alloy (Al6061) powder. Two different reinforcements were chosen to produce hybrid Al nanocomposites. These include multiwalled carbon nanotubes (MWCNTs) and silicon carbide (SiC) particles. The MWCNTs were obtained from Fibermax Nanocomposites Ltd. (London, UK), and the SiC particles were supplied by Merck (Darmstadt, Germany).

The as-received materials were characterized using optical microscopy (OM), scanning electron microscopy (SEM), and electron backscatter diffraction (EBSD). The equipment used for this characterization was an optical microscope, M 4000 M, with Leica Application Suite software (version 4.13.0, Leica Microsystems, Wetzlar, Germany) and a Thermo Fisher Scientific QUANTA 400 FEG SEM (Thermo Fisher Scientific, Hillsboro, OR, USA) with an EBSD TSL-EDAX detector unit (EDAX Inc. (Ametek), Mahwah, NJ, USA). The particle size, length, and width of the Al6061 powder were measured by Image J (version 1.51, Wayne Rasband, National Institutes of Health, Bethesda, MD, USA). The powder samples for EBSD observation were embedded in epoxy resin and prepared using the conventional metallographic preparation procedure. To ensure that the surface of the powders was free of significant plastic deformation and showed signs of EBSD, a colloidal silica finish was applied for 2 h without applying lower pressure. EBSD analysis was conducted using the TSL OIM Analysis 5.2 (EDAX Inc. (Ametek), Mahwah, NJ, USA). The grain size and the crystallographic orientation were evaluated using unique color and inverse pole figure (IPF) maps. The grain misorientations were evaluated by Kernel average misorientations (KAM) maps.

The CNTs were analyzed using transmission electron microscopy (TEM, FEI Company, Hillsboro, OR, USA) and high-resolution TEM (HRTEM, JEOL Ltd., Tokyo, Japan). To observe the CNTs, they were dispersed and placed on a TEM grid with a carbon film. The results of the HRTEM were analyzed using the DigitalMicrograph software GMS 3.6.1 (Gatan Ametek, Pleasanton, CA, USA). Raman spectroscopy (Jobin Yvon T64000, HORIBA Scientific, Kyoto, Japan) experiments were conducted with a laser wavelength of 514 nm.

The hybrid nanocomposites were produced via powder metallurgy schematically represented in Figure 1. The Al6061 powder was first mixed with different amounts of CNTs and SiC particles (0.50, 0.75, 1.00, and 1.75 wt.%) using an ultrasonication process in isopropanol for 15 min and 20.4 kHz. This method has proven effective and efficient in dispersing nanoparticles while minimizing agglomeration [17,18], thus ensuring a homogenous distribution of CNTs and SiC within the Al6061 powder. After ultrasonication, the mixture was dried and cold-pressed at 550 MPa to form cylindrical (∅ 10 mm) green compacts. Posteriorly, these compacts were sintered in a high-temperature tubular furnace under vacuum (better than 10^{-4} mbar), with sintering temperatures ranging from 540 to 620 °C, for 120 min with a 5 °C/min heating and cooling rate, which provokes diffusion and ensures enhanced bonding between the matrix and the reinforcements, thus optimizing the mechanical integrity of the hybrid nanocomposite. Nanocomposites reinforced only with CNTs and SiC and the Al matrix were also produced under the same condition for comparison. Figure 2 shows the dimensions of the sintered samples, and three samples of each condition were produced. A summary of the production conditions can be observed in Table 1.

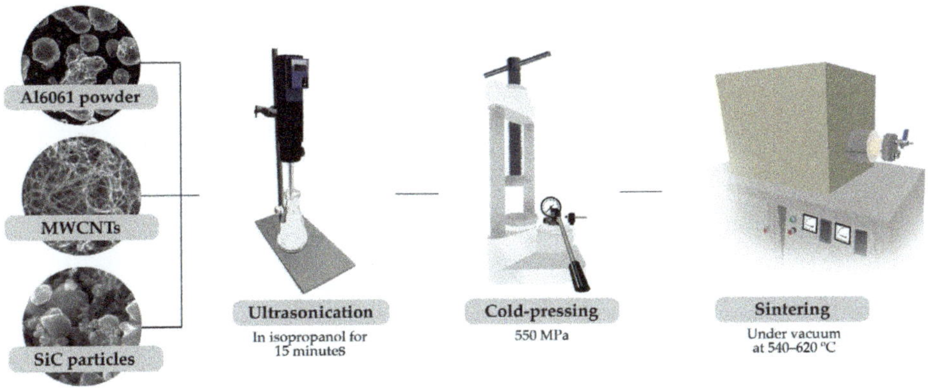

Figure 1. Schematic representation of the hybrid nanocomposite production process via powder metallurgy.

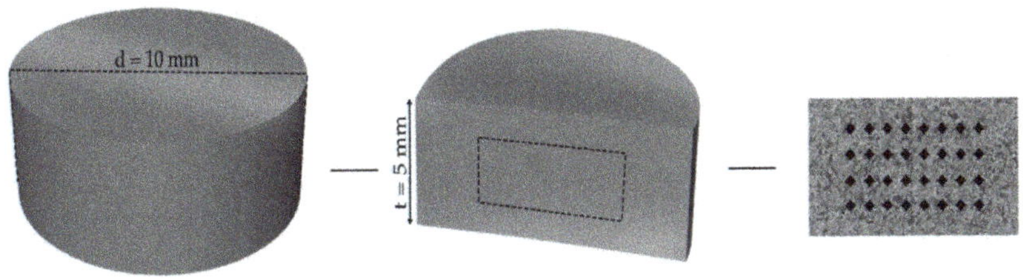

Figure 2. Schematic representation of the hardness indentation on polished cross-sections of the nanocomposites.

Table 1. Processing conditions of the samples.

Samples	Type of Reinforcement	Reinforcement (wt.%)	Ultrasonication (min/kHz)	Cold Pressing Pressure (MPa)	Temperature (°C)
Al 6061 matrix	–	–	15/20.4	550	540 560 580 600 620
Al 6061/CNTs	CNTs	0.50 0.75 1.00 1.50	15/20.4	550	540 560 580 600 620
Al 6061/SiC	SiC	0.50 0.75 1.00 1.50	15/20.4	550	540 560 580 600 620
Al 6061/SiC /CNTs	SiC CNTs	0.50 0.75 1.00 1.50	15/20.4	550	540 560 580 600 620

The approach used to evaluate the densification of the sintered composites resorted to determining both theoretical and experimental densities. The theoretical density was calculated using the rule of mixtures, accounting for the individual densities of Al6061, CNTs, and SiC. In contrast, the experimental density was determined by using the Archimedes principle. Posteriorly, the relative density of the nanocomposites was determined by comparing both densities. Four samples were used in the measurement.

A detailed microstructural analysis was conducted to understand the interaction between the aluminum matrix and reinforcements. Optical microscopy (OM), digital microscopy (DM), scanning electron microscopy (SEM), and electron backscatter diffraction (EBSD) were applied to polished cross-sections of the nanocomposites. The nanocomposite samples were embedded in epoxy resin and prepared using conventional metallographic preparation. For the EBSD analysis, a colloidal silica finish was applied for 2 h without lower pressure. EBSD analysis was conducted using the TSL OIM Analysis 5.2 (EDAX Inc. (Ametek), Mahwah, NJ, USA). This technique was used to study the grain size, the crystallographic orientation, and the texture of the samples.

The mechanical characterization of the sintered hybrid nanocomposites was performed through hardness and tensile tests. Thirty Vickers microhardness indentations were taken on polished cross-sections to determine the overall hardness of the samples as schematized in Figure 2. These were conducted with an application load of 196 mN for 10 s, using a FALCON 400 micro-/macro-Vickers hardness tester (INNOVATEST Europe BV, Maastricht, The Netherlands). The tensile tests were carried out using a Shimadzu EZ Test machine (Shimadzu Corporation, Kyoto, Japan) at a crosshead speed of 1 mm/s to understand the influence of both reinforcements on the overall strength, ductility, and failure mechanisms of the hybrid nanocomposite.

3. Results and Discussion

In this study, Al6061 hybrid nanocomposites reinforced with CNTs and SiC particles were produced and characterized to evaluate the effects of the reinforcements on the microstructure, phase composition, and mechanical properties of the resulting material. The initial materials have a significant influence on the final properties and microstructure. So, initially, a characterization of the reinforcements and as-received powders used to produce the nanocomposites was also carried out.

3.1. Powders and Reinforcements Characterization

The initial characterization of the as-received powders and reinforcements used in producing hybrid Al6061 nanocomposites is essential to understand their morphological, structural, and chemical properties before integration into the composite matrix.

The Al6061 powder, used as the matrix material, was characterized to ensure its suitability for the fabrication of nanocomposites. The particle size distribution was measured, and the results showed that the average particle size of the Al6061 powder was approximately 35 μm, with a narrow distribution, as depicted in Figure 3. The morphology of the Al6061 powder was examined using scanning electron microscopy (SEM), revealing predominantly spherical particles, which are preferred for uniform mixing and compaction. The spherical shape of the particles is advantageous for reducing the porosity of the sintered composite, thereby improving its mechanical properties.

The chemical composition of the Al6061 alloy powder was analyzed using energy-dispersive X-ray spectroscopy (EDS) to confirm the presence of primary alloying elements, such as magnesium (Mg), silicon (Si), and copper (Cu). The results, presented in Table 2, show that the elemental composition of the Al6061 powder follows the standard composition of the Al6061 alloy, ensuring that the matrix will exhibit the expected mechanical and thermal properties. Some oxygen is also detected on the surface of the particles.

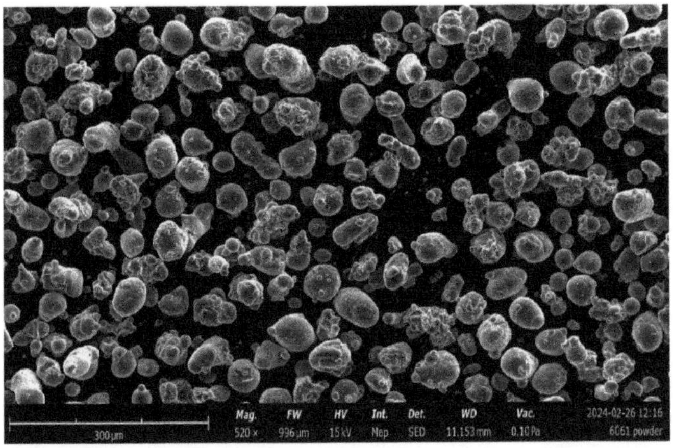

Figure 3. Scanning electron microscopy (SEM) image in secondary electron mode of the Al6061 as-received powders.

Table 2. Chemical composition (wt.%) of aluminum alloy 6061 as-received powder.

Al	Mg	Cu	Si	O
96.5	1.0	0.7	0.4	1.4

Electron backscatter diffraction (EBSD) was used to map crystallographic information, such as grain crystallographic orientation, phase distribution, and strain of the Al6061 particles. Figure 4 shows the results of the morphology and microstructural characterization of the Al6061 powder particle. Figure 4a shows the unique color grain map of the one powder particle. Each grain is colored in this map differently to represent different grain domains, and based on this map, the presence of grains of different sizes in each powder particle can be seen. The inverse pole figure (IPF) map in Figure 4b was obtained to show the crystallographic orientation of the individual grains in the powder. This helps to identify the grain structure, size distribution, and how the grains are oriented to each other. The map suggests that the particles are characterized by grains of different sizes, and the color variation indicates different crystallographic orientations. In the KAM map shown in Figure 4c, the colors represent the Kernel average misorientation from 0° to 5°. Regions with high disorientation may be associated with plastic deformation regions, such as grain or subgrain boundaries, where greater dislocation activity has occurred. Green or blue areas, more common in grain centers, indicate less misorientation, suggesting that these regions are less deformed or subject to significant internal stresses. In the Al6061 powder, these higher misorientation areas could indicate zones of intense processing, such as during milling, atomization, or rapid solidification, where deformation has accumulated due to internal stresses. Figure 4d is the image quality map, where the lighter areas typically correspond to well-ordered crystalline regions with high-quality diffraction patterns. The darker areas indicate areas of greater misorientation. This map helps to identify regions with stress or defects in the sample and to understand the quality of the EBSD results. The lines represent grain boundaries where the crystallographic orientation changes significantly. The blue lines may represent high-angle grain boundaries, while the red and green lines represent low-angle boundaries. The results agree with the KAM maps, which indicate that greater plastic deformation is observed on the particle surface where the low-angle boundaries are located, and there is a higher density of dislocations.

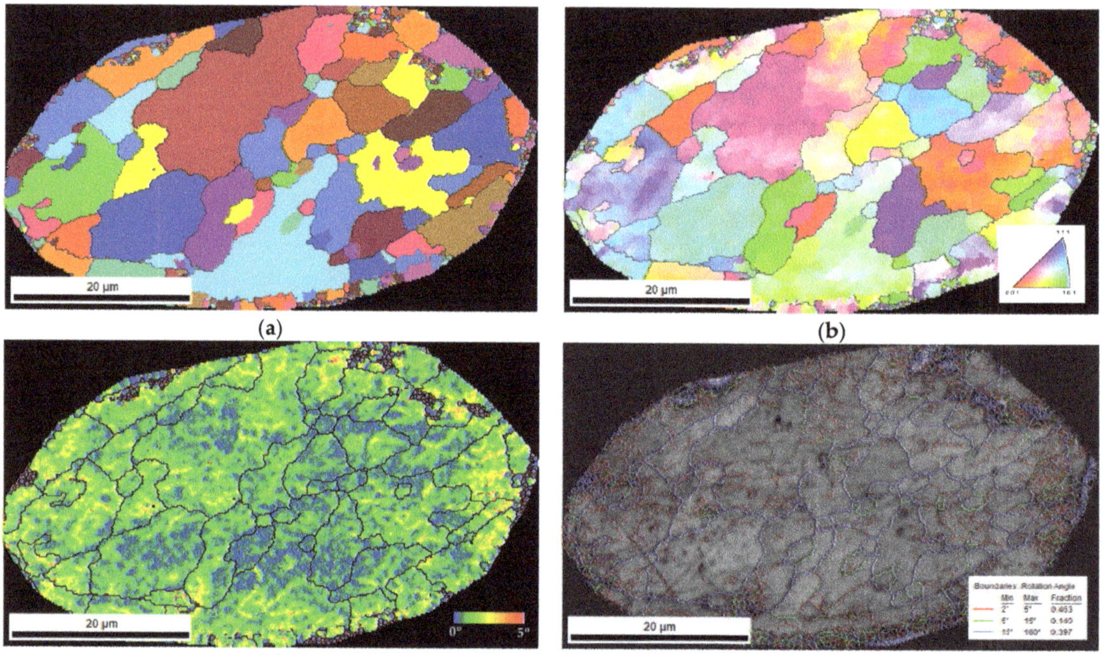

(**c**) (**d**)

Figure 4. EBSD results of an Al6061 powder particle (**a**) unique grain color map, (**b**) inverse pole figure map, (**c**) Kernel average misorientation map, and (**d**) image quality map with grain boundaries delimitated.

The CNTs, attending as the primary nanoscale reinforcement, were characterized to assess their purity, structure, and dispersion quality. Transmission electron microscopy (TEM) was used to examine the morphology and structural integrity of the CNTs. Figure 5 shows a TEM image of the CNTs, revealing a multiwalled structure with an average diameter of 10–20 nm and lengths in the micrometer range. The high aspect ratio of CNTs is crucial for improving the load transfer capability within the nanocomposite.

Raman spectroscopy was performed to confirm the structural quality of the CNTs. The Raman spectrum shows distinct D and G bands, corresponding to the disordered carbon (D band) and the graphitic structure (G band). The intensity ratio of these two bands (I_D/I_G) was calculated to be around 1.01. This value can be attributed to some defects in the CNTs, which is critical for their role as reinforcement in the composite.

The SiC particles were characterized to ensure their uniformity, size, and morphology before being used as the second reinforcement in the hybrid composite. The SiC powder was analyzed using SEM, and Figure 6 presents SEM images showing the angular morphology of the SiC particles, with an average particle size of approximately 5 μm. The sharp edges and irregular shape of the SiC particles are expected to enhance the mechanical interlocking within the matrix, contributing to improved load-bearing capacity.

In addition, the SiC particles were analyzed using EDS, which confirmed that the powder consists primarily of silicon and carbon, with no detectable impurities, as seen in Table 3. This high-purity SiC is necessary to prevent unwanted chemical reactions with the Al matrix, such as the formation of aluminum carbide (Al_4C_3), which can compromise the mechanical properties of the composite.

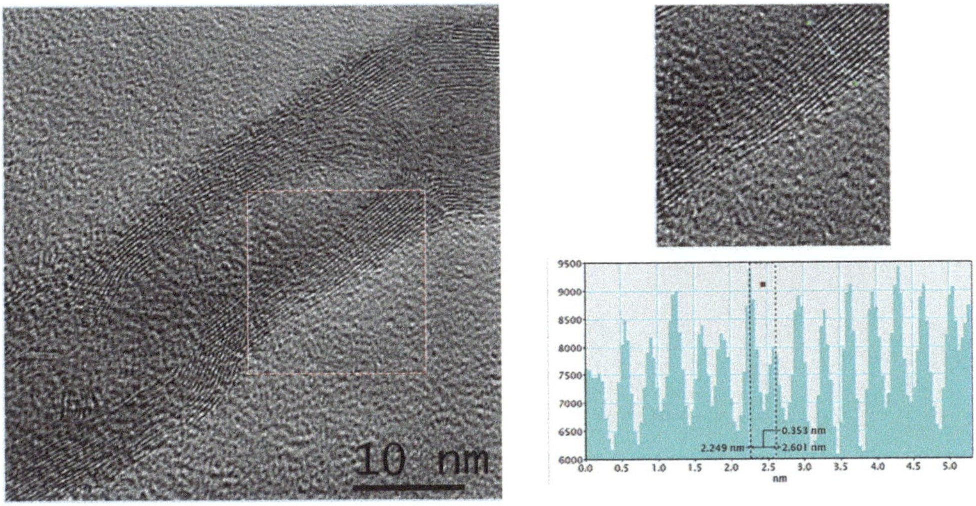

Figure 5. A high-resolution (HRTEM) image of an MWCNT shows a magnification of a region of the layers where we can see the image and measure the number of walls.

Figure 6. SEM image of the as-received SiC nanoparticles used as reinforcement material.

Table 3. Chemical composition (wt.%) of SiC powder obtained by EDS.

Si	C
13.2	86.8

The characterization of the Al6061 matrix powder, CNTs, and SiC particles demonstrated that these materials possess the necessary morphological and structural properties to produce hybrid nanocomposites. The spherical morphology of the Al6061 powder, the high aspect ratio and purity of the CNTs, and the angular, high-purity SiC particles contribute to efficient load transfer, grain refinement, and the overall improvement of the composite's mechanical properties. The uniform dispersion of CNTs and SiC within the matrix was achieved through ultrasonication to ensure homogeneity, minimizing potential defects in the final nanocomposite and reducing the observed agglomeration of the as-received nanoparticles and CNTs.

3.2. Nanocomposites Characterization

As already mentioned, to produce the hybrid nanocomposites, a study was first carried out on the effect of the amount of SiC and CNT reinforcement for producing the nanocomposites. Figure 7 shows the hardness evolution of Al6061 composites reinforced with different amounts (0.50, 0.75, 1.00, and 1.50 wt.%) of SiC and CNTs.

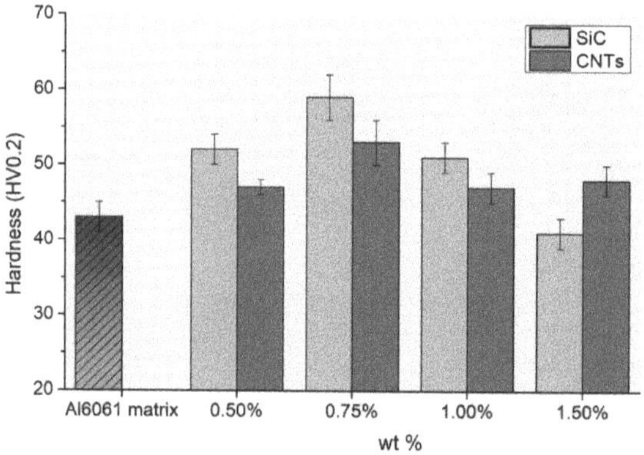

Figure 7. Evolution of hardness (HV0.2) of the nanocomposites with the different amounts of the reinforcement. The hardness of the Al6061 matrix is also shown for comparison purposes.

As the amount of SiC increases from 0.50 to 1.50 wt.%, a steady or rapid increase in hardness is expected. However, at very high SiC content (1.00 and 1.50 wt.%), the hardness tends to stabilize or slightly decrease due to poor particle distribution or clustering, which can cause defects or porosity. Adding CNTs to the aluminum matrix improves hardness until a value of 0.75 wt.% of CNTs is reached. However, this increase may be more gradual compared to SiC, due to the challenges of uniformly dispersing the CNTs within the matrix. As the CNT content increases, an improvement in hardness is expected. For both types of reinforcement, the maximum hardness value was observed for 0.75 wt.% reinforcement content.

A detailed microstructural characterization was performed in the nanocomposites reinforced with SiC to understand the hardness evolution with the reinforcement amount. Figure 8 shows the SEM images at a lower magnification of Al6061 nanocomposites reinforced with different amounts of SiC. Each image corresponds to a different SiC reinforcement weight percentage (0.50, 0.75, and 1.00 wt.% of SiC). The images display the distribution of SiC particles within the aluminum matrix. The particles appear as dark spots, while the aluminum matrix appears as a lighter background. Increasing the amount of SiC reinforcement changes the particles' distribution, agglomeration, and overall interaction within the matrix for the nanocomposite with 0.50 wt.% SiC, the image shows a relatively uniform distribution of SiC particles with some small clusters. The size of the SiC particles appear relatively uniform, with fewer large agglomerates. This amount may result in effective dispersion strengthening as the particles are well-distributed without significant clustering, enhancing the hardness and mechanical properties due to load transfer from the matrix to the particles. The concentration of particles increases for nanocomposites with 0.75 wt.% SiC, and there is a slight tendency for more clusters or agglomerations of SiC particles, especially in certain areas. Although there is still relatively good dispersion, the agglomeration of SiC particles could become more pronounced compared to the 0.50 wt.% sample. The presence of agglomerates may act as stress concentrators or weak points in the material, potentially affecting its overall mechanical properties. However, the hardness may continue to increase with the increase in SiC. For 1.00 wt.% SiC, the micrograph reveals a more pronounced agglomeration of SiC particles. The dark spots representing the SiC particles are more clustered and unevenly distributed. Agglomerates are visible across the image, suggesting that, at higher concentrations, it becomes more challenging to maintain a uniform distribution of the reinforcement within the matrix. This agglomeration could lead to reduced mechanical performance, as these clusters may act as points of weakness where failure can initiate. These areas could also introduce voids or other defects, negatively affecting the material's toughness and ductility.

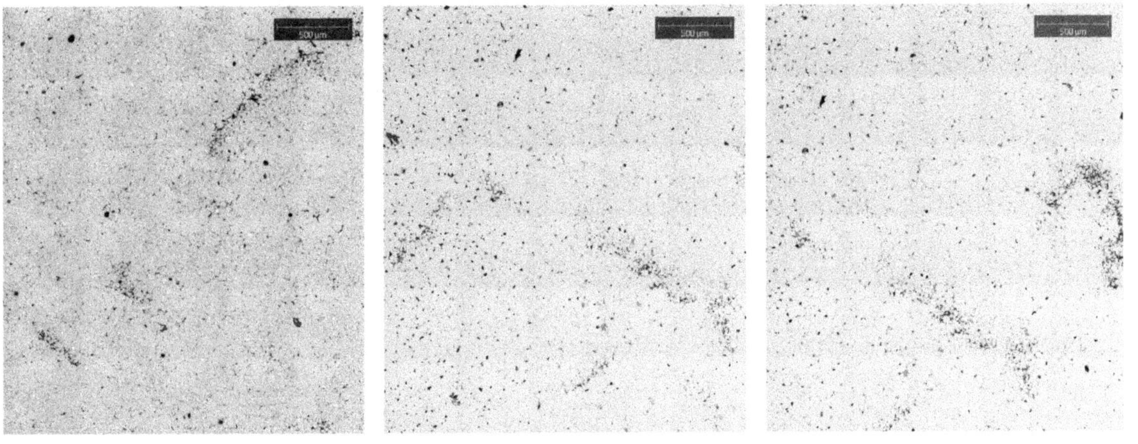

(a) (b) (c)

Figure 8. SEM images of the nanocomposites reinforced with SiC with (**a**) 0.50%, (**b**) 0.75%, and (**c**) 1.00% wt.

High-magnification SEM images of the nanocomposites reinforced with different amounts of SiC can be observed in Figure 9.

Based on these high-magnification images, which correspond to different compositions of SiC-reinforced Al6061 nanocomposites (0.50, 0.75, 1.00, and 1.50 wt.%), it is possible to confirm the evolution of the distribution and agglomeration of SiC particles within the aluminum matrix as the reinforcement concentration increases. As the amount of SiC increases from 0.50% to 1.50%, the images change from a uniform distribution to a highly agglomerated microstructure. At lower concentrations (0.50% and 0.75%), the uniform distribution of particles promotes effective strengthening by dispersion, improving the composite's mechanical properties without compromising its integrity. However, from 1.00% onwards, the formation of large clusters becomes significant, indicating that the reinforcement exceeds the matrix's dispersion capacity. At 1.50%, the large agglomerations result in a less homogeneous microstructure with the potential for deterioration in mechanical properties, especially fracture toughness and ductility. These observations suggest an ideal limit for incorporating SiC particles, probably around 0.75% to 1.00%, where there is a balance between increasing hardness and maintaining good mechanical properties. The formation of second-phase particles tends to increase with the concentration of SiC, and their presence strongly impacts the microstructure and mechanical properties of the nanocomposites. These particles are finely dispersed at lower reinforcement concentrations and act beneficially, increasing hardness and wear resistance. At higher amounts of reinforcement (1.00 and 1.50 wt.%), however, they group together and interact with the SiC agglomerates, introducing stress concentration points and defects that reduce the material's overall strength, especially under cyclic or impact loads. The quantity and distribution of these particles must be carefully controlled to avoid deteriorating mechanical properties at high reinforcement concentrations. Second-phase particles in aluminum composites often include compounds such as AlFeSi, AlMgSi, or other intermetallics arising from the reaction of alloying elements (such as Mg, Si, Fe) present in Al6061. Their formation is influenced by thermal processing, sintering time, and the nature of the reinforcement introduced. These phases are generally harder and more brittle than the pure aluminum matrix and can significantly impact the material's mechanical and tribological properties.

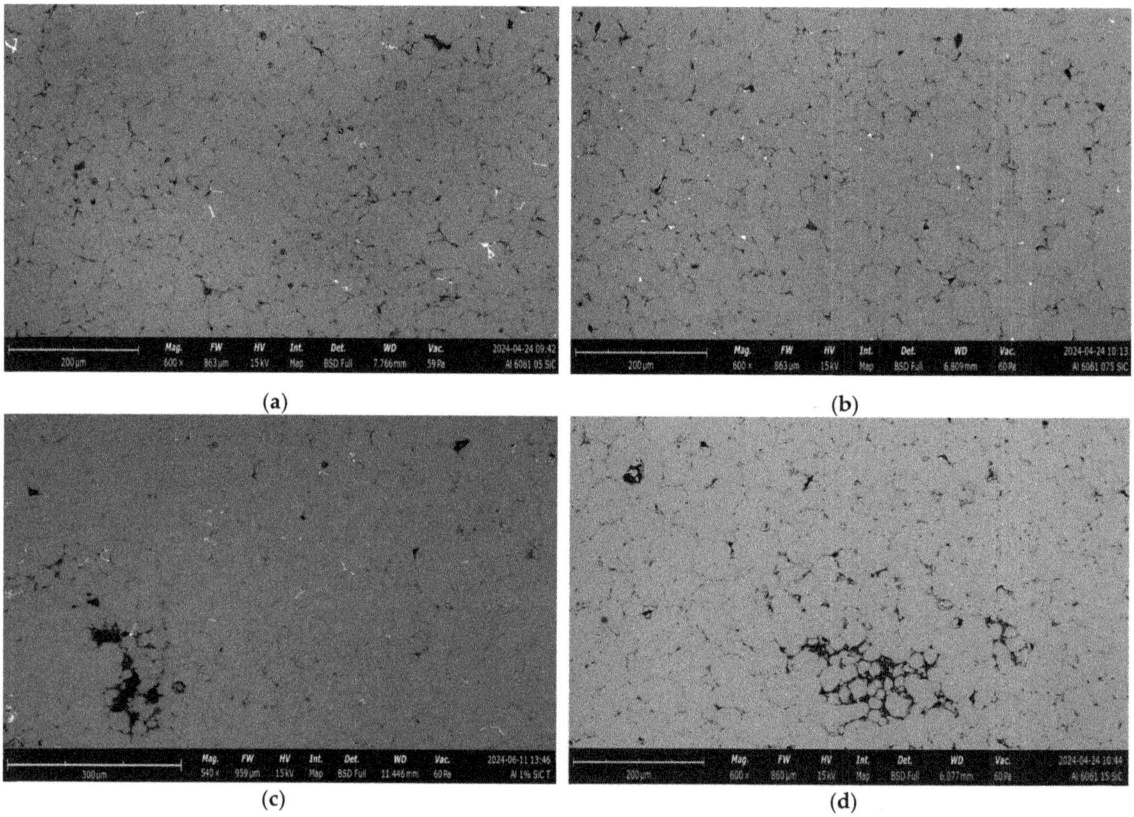

Figure 9. SEM images of the nanocomposites of the SiC reinforced with (**a**) 0.50 wt.%, (**b**) 0.75 wt.%, (**c**) 1.00 wt.%, and (**d**) 1.50 wt.%.

EDS analyses were carried out to predict the nature of these second phases by combining them with data from the Al-Mg-Si [19] and Al-Fe-Si [20] phase diagrams. Figure 10 shows the different zones observed in the nanocomposite samples. These regions were analyzed through EDS analysis, and the chemical composition was used to predict the phases through phase diagrams, then the most probable phase was indicated. Due to the interaction volume of the technique, and since some of the precipitates are smaller than this volume, there is an error associated with this technique. Based on this analysis, the lighter particles are precipitates that can be identified as $Al_{15}Fe_6Si_5$. Darker areas are characterized by (Si)ss and $M_{g2}Si$. It is also possible that a cluster of SiC particles has become concentrated in grain boundaries and pores in this image.

Figure 11 shows the EDS maps of the sample reinforced with 0.75 wt.% SiC. The Si element map shows that the element is more concentrated in the points corresponding to the dark and light particles in the SEM image. This confirms the point EDS analysis presented earlier. This distribution confirms that these dark regions are SiC particles used as reinforcement in the composite, corresponding to the (Si)ss, Mg_2Si, and $Al_{15}Fe_6Si_5$ phases. The Fe map reveals a significant concentration in the lighter second-phase particles. This suggests that the particle is composed of an Al-Fe intermetallic compound, possibly with Si. EDS analysis reveals the presence of a second-phase particle rich in Al, Fe, and Si, confirming the identification of the intermetallic compound of the Al-Fe-Si system formed during the solidification process. These intermetallic particles are generally harder and more brittle than the aluminum matrix and can act as stress concentration points and failure initiators under cyclic or impact loads. However, the formation of large intermetallic particles containing Mg or Si can compromise the material's overall strength, making it more susceptible to fracture.

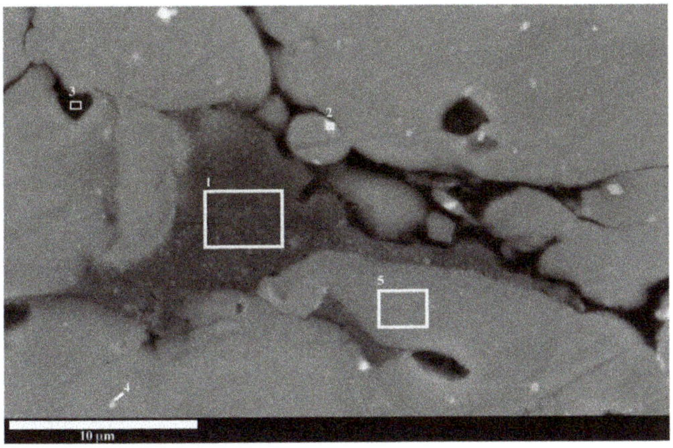

Figure 10. SEM image of Al6061 nanocomposites with the EDS analyzed regions marked and present in Table 4.

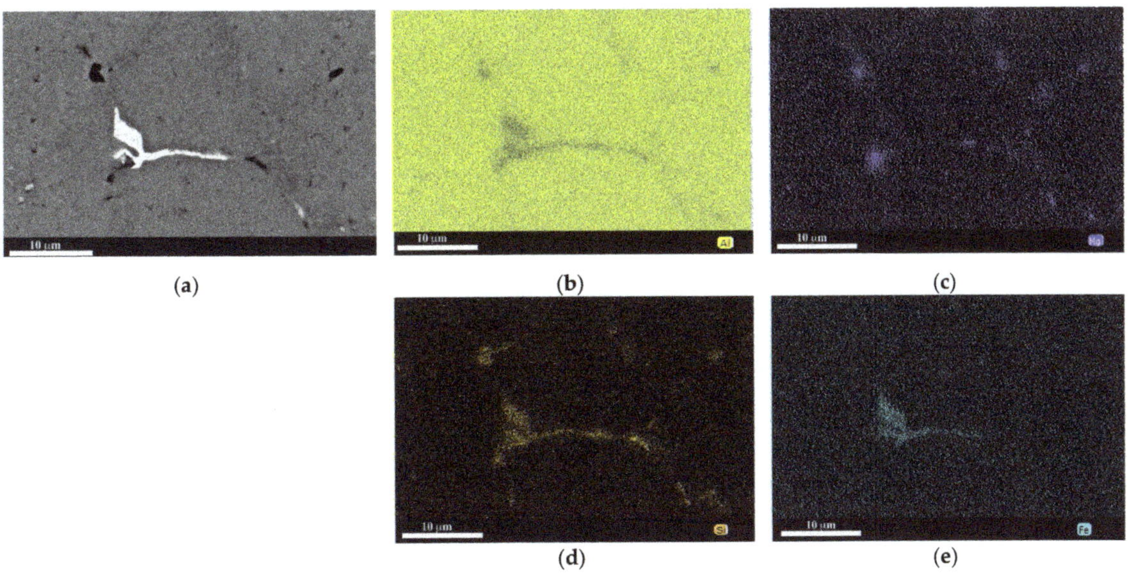

Figure 11. (**a**) SEM image of Al6061 matrix and EDS elemental distribution map of (**b**) Al, (**c**) Mg, (**d**) Si, and (**e**) Fe.

Table 5 shows the phase percentages of the samples for the different nanocomposites. The addition of reinforcement affects the formation of these second phases typical of the Al6061 alloy. Previous studies reported similar results when CNTs were added [17].

When CNTs are added to the Al6061 matrix, similar results are observed concerning dispersion and the formation of second phases. In this sense, a percentage of 0.75 wt.% was selected for both reinforcements to produce hybrid nanocomposites, considering all the microstructural and mechanical characterization results.

Table 4. EDS chemical composition in wt.% of the regions marked in Figure 10.

Zone	Al	Mg	Si	Fe	Cr	C	Possible Phases
1	58.9		10.1			31.0	SiC
2	93.2		4.7	2.1			$Al_{15}Fe_6Si_5$
3	59.6	15.5	14.9				α-Al + (Si)ss + Mg_2Si
4	95.6		3.0	1.3			$Al_{15}Fe_6Si_5$
5	100						α-Al

Table 5. Phase percentages and pores and SiC agglomerates of the Al6061 matrix and nanocomposites.

	$Al_{15}Fe_6Si_5$	(Si)ss + Mg_2Si	Pores + SiC Agglomerates
Al	0.5	1.3	1.3
Al0.5wt.%SiC	0.4	0.4	1.7
Al0.75wt.%SiC	0.8	0.7	1.6
Al1.00wt.%SiC	0.5	0.7	2.6
Al1.50wt.%SiC	0.3	0.7	3.2

3.3. *Hybrid Nanocomposites Characterization*

The hybrid nanocomposites were produced at different sintering temperatures. Figure 12 shows the density of the hybrid nanocomposites for the different sintering temperatures. The density for the matrix and the nanocomposites reinforced only with SiC and CNTs is also shown for comparison purposes. The sintering temperature where the greatest densification is obtained is 580 °C; for this reason, it was selected to produce the hybrid nanocomposites.

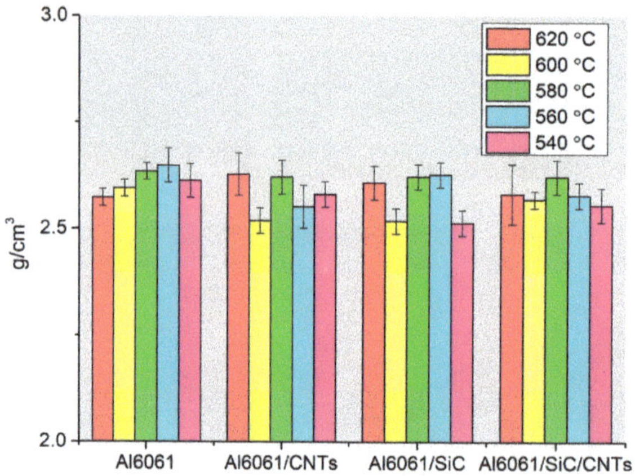

Figure 12. The density of the Al6061 matrix and Al6061 reinforced with CNTs, SiC, and SiC/CNTs for the different sintered temperatures.

The microstructure of the hybrid nanocomposites was analyzed using SEM, EDS, and EBSD. Figure 13 shows the SEM images of the composite at different magnifications.

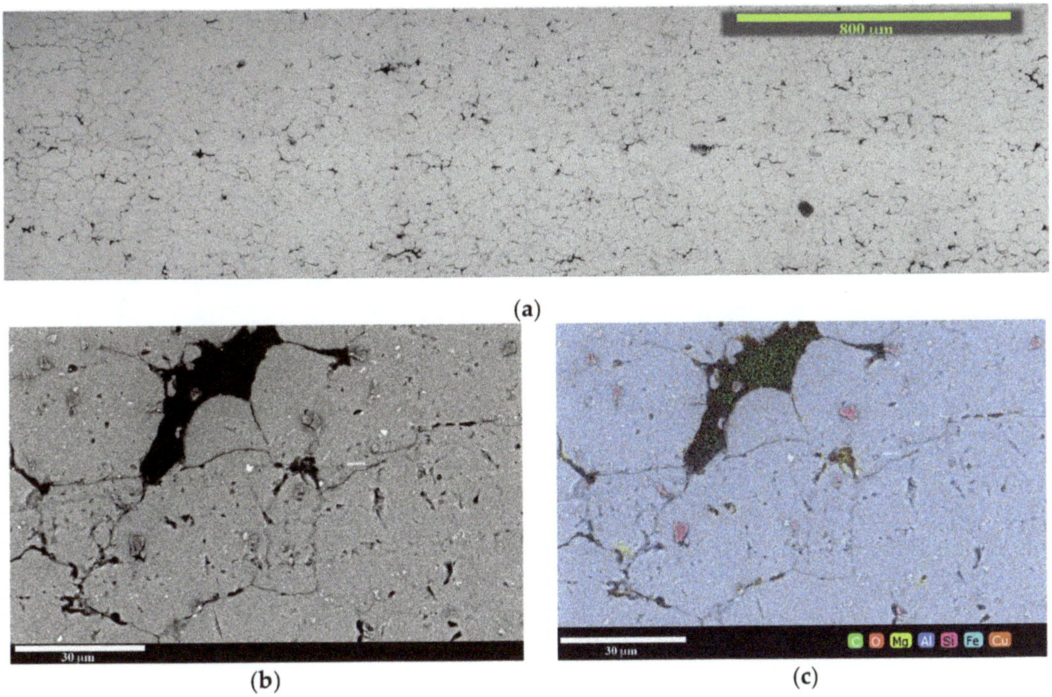

Figure 13. (**a**) Low magnification SEM image of Al6061/SiC/CNTs produced at 580 °C, (**b**) higher magnification SEM images, and (**c**) EDS elemental distribution map of Al, Fe, Mg, C, O, Si, and Cu.

It can be observed that the SiC particles are uniformly distributed within the matrix, while the CNTs form a network-like structure, reinforcing the matrix along the grain boundaries. No significant agglomeration of CNTs was detected, suggesting that the selected processing technique achieved effective dispersion. The higher magnification SEM image shows that dark spots and regions may correspond to the SiC or CNT clusters. The EDS map reveals the distribution of elements (Al, Fe, Mg, and Si) within the observed microstructure. Each color represents a different element, allowing for a visual correlation between the presence of specific elements and the microstructural features. For example, Si (silicon) might be concentrated around the SiC particles and combined with Fe in secondary phases.

In addition to the mechanical properties, the addition of reinforcements has a significant impact on the microstructure of the matrix. The microstructure of the nanocomposites was investigated in detail using EBSD. Figure 14 shows the unique color maps for Al6061, Al6061/CNTs, Al6061/SiC, and Al6061/SiC/CNTs. Using the unique color maps to represent grains helps visualize the impact of different reinforcements on grain refinement. The addition of SiC has a stronger effect on grain size reduction than CNTs, and the hybrid Al/SiC/CNT configuration shows the most refined microstructure. This refined grain structure enhances the resulting nanocomposites' mechanical performance, such as increased hardness, yield strength, and wear resistance.

Moreover, incorporating SiC particles resulted in grain refinement, as evidenced by the smaller grain sizes in the composite compared to the unreinforced Al6061 alloy. The fine dispersion of SiC likely acted as a barrier to grain growth during solidification, contributing to this observed refinement. CNTs may have further restricted grain boundary movement through a pinning effect, synergistically enhancing the microstructural refinement.

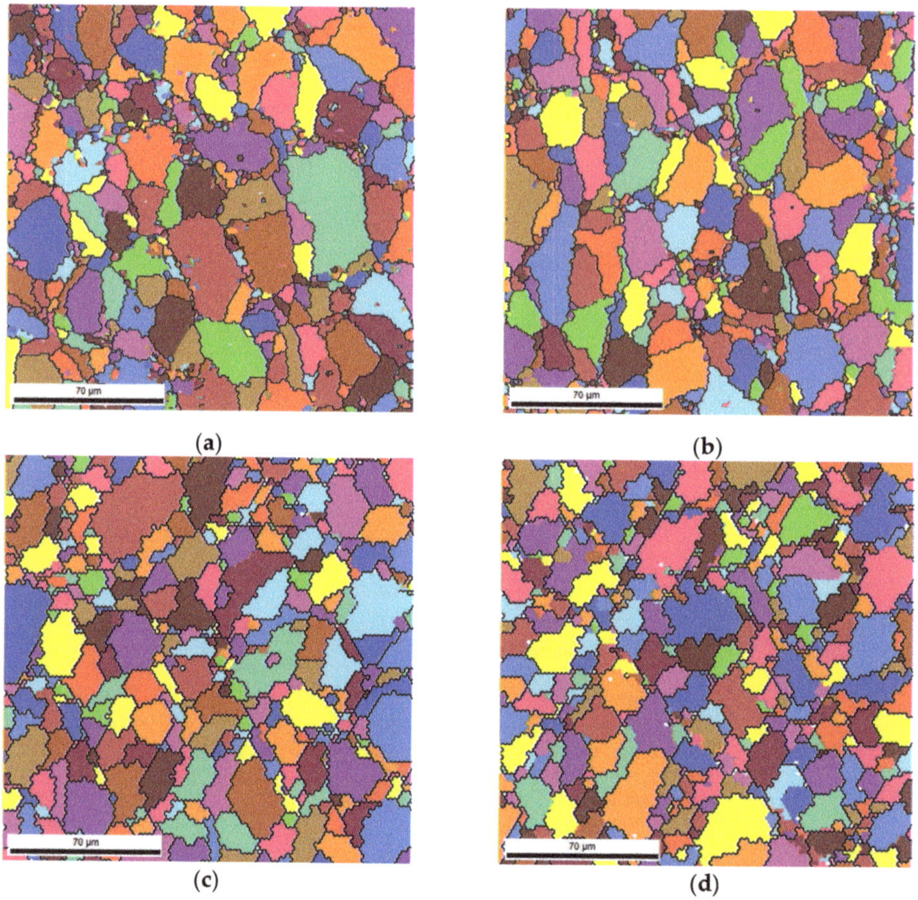

Figure 14. Unique grain color map of (**a**) Al6061 matrix, (**b**) Al6061/CNTs, (**c**) Al6061/SiC, and (**d**) Al6061/SiC/CNTs.

The inverse pole figure (IPF) maps the corresponding IPF orientation distribution function figures of the Al6061 matrix, and the different reinforced nanocomposites with CNTs, SiC, and hybrid reinforcements are presented in Figure 15. The IPF maps indicated the grain crystallographic orientation in the materials, while the figures show the intensity of specific crystallographic orientations within the material. Based on these results, the impact of different reinforcements on the grain orientation and microstructure of the Al6061 matrix is clear. Adding CNTs and SiC particles, individually and in combination, significantly refines the grain structure and modifies the crystallographic texture. The Al/SiC/CNT configuration exhibits the most refined microstructure and the strongest texture, suggesting the potential for superior mechanical properties, including increased hardness, strength, and possibly improved thermal stability and wear resistance. This influence in the crystallographic orientation of the metallic matrix grains can be explained by reinforcements, which create significant internal stresses in the matrix due to differences in mechanical properties, such as the modulus of elasticity and hardness. These internal stresses alter the distribution of forces within the grains, resulting in crystallographic reorientations to accommodate the local deformation.

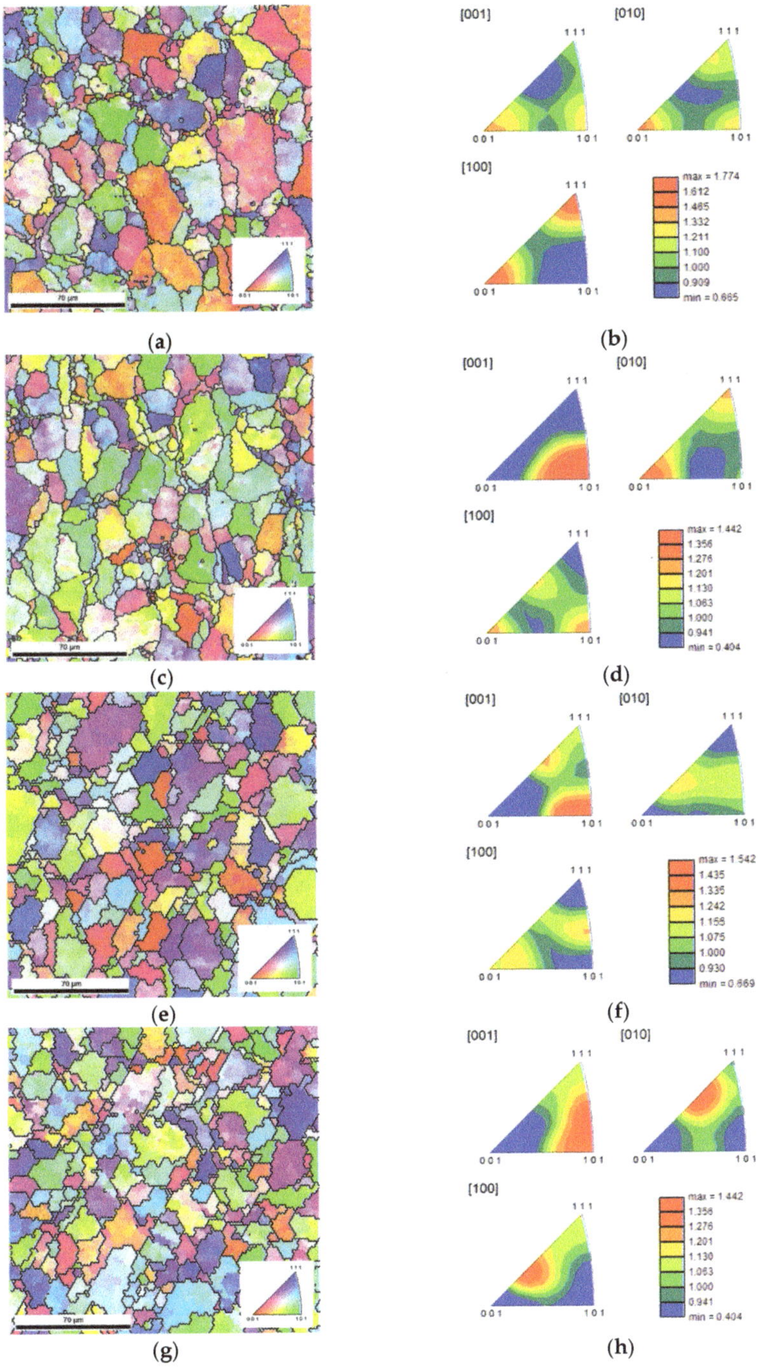

Figure 15. Inverse pole figures (IPF) maps and IPF figures of (**a**,**b**) Al6061 matrix, (**c**,**d**) Al6061/CNTs, (**e**,**f**) Al6061/SiC, and (**g**,**h**) Al6061/SiC/CNTs.

As a result, the grains near the reinforcements tend to realign, changing their orientations to balance the stress distribution. The high density of dislocations generated by the reinforcement particles can induce the formation of subgrains within the existing grains. These subgrains, which have slight orientation differences, arise to relieve the accumulated stress. Some subgrains may reorganize and reorient as the deformation continues, leading to new crystallographic orientation patterns. This reorientation is more pronounced when a combination of reinforcements, such as CNTs and SiC, synergistically affects the dislocation movement. The distribution of reinforcement particles influences the overall grain orientation in the material. The reinforcement particles can favor certain crystallographic orientations that minimize interfacial energy, promoting, for example, the formation of preferred textures. Therefore, adding reinforcements can lead to a stronger and more organized texture, especially in hybrid materials that combine different types of reinforcements, such as the Al/SiC/CNT nanocomposites. The reinforcements act as barriers to the movement of grain boundaries, restricting the ability of grains to reorganize under deformation freely. This restriction leads to the accumulation of deformation energy at the grain boundaries, forcing the grains to change their orientations to relieve this accumulated energy. The result is a significant modification in crystallographic orientation, with increased high-angle grain boundaries and greater structural complexity.

Process parameters have a predominant influence on the behavior of hybrid nanocomposites through internal shear stresses, which directly affect the texture and final properties of the material. Dispersoids alter the material's resistance to these stresses by acting as barriers to dislocation movement and either promoting or restricting grain reorganization. As a result, misorientation differences occur, forming new crystallographic orientations and preferred textures. In hybrid nanocomposites, these interactions can be even more complex due to the synergy between different types of reinforcements, such as CNTs and SiC, resulting in optimized properties and greater structural complexity.

The Kernel average misorientation (KAM) maps of four samples and a histogram showing the distribution of KAM misorientation angles are presented in Figure 16.

For the matrix, the KAM map shows the grain boundaries and internal misorientation, which are primarily green, indicating a relatively homogeneous microstructure. For the nanocomposites reinforced with CNTs, the distribution of colors suggests a slight change in the misorientation pattern, with more blue and red areas, indicating variations in strain or dislocation density compared to the pure Al6061 matrix. The Al6061 matrix reinforced with SiC revealed a KAM map with more prominent red and yellow areas, suggesting a higher misorientation in certain regions, which could indicate increased strain due to the presence of hard SiC particles. For the hybrid nanocomposites, the KAM map shows a more complex distribution of colors, with a mixture of high (red) and low (green) misorientations, indicating that the combination of reinforcements has a notable effect on the internal microstructure. The Al6061 matrix has the highest frequency at lower misorientation angles, while the Al6061/SiC/CNT hybrid composite has a broader distribution, indicating a more heterogeneous strain distribution. The trend shows how different reinforcements affect the internal strain and dislocation patterns of the Al6061 matrix.

Adding reinforcements in the matrix, such as SiC particles or carbon nanotubes (CNTs), causes a significant increase in dislocations and the average misorientation angle within the grains. This effect can be explained based on the interaction between the reinforcements and the metallic matrix during solidification and the subsequent plastic deformation processes. The reinforcements, especially the SiC particles, have much higher elasticity and hardness modulus than the aluminum matrix (Al6061). This generates incompatibility in the regions near the reinforcements, resulting in internal stresses and localized deformations. To accommodate these stresses, additional dislocations form, especially at the matrix–reinforcement interfaces. The reinforcement particles act as physical barriers to the movement of dislocations. When a dislocation encounters a rigid particle, it accumulates, increasing the density of dislocations near these regions. This effect leads to a local buildup of stress, which can be partially relieved by forming new dislocations in the remaining matrix, resulting in a more refined structure.

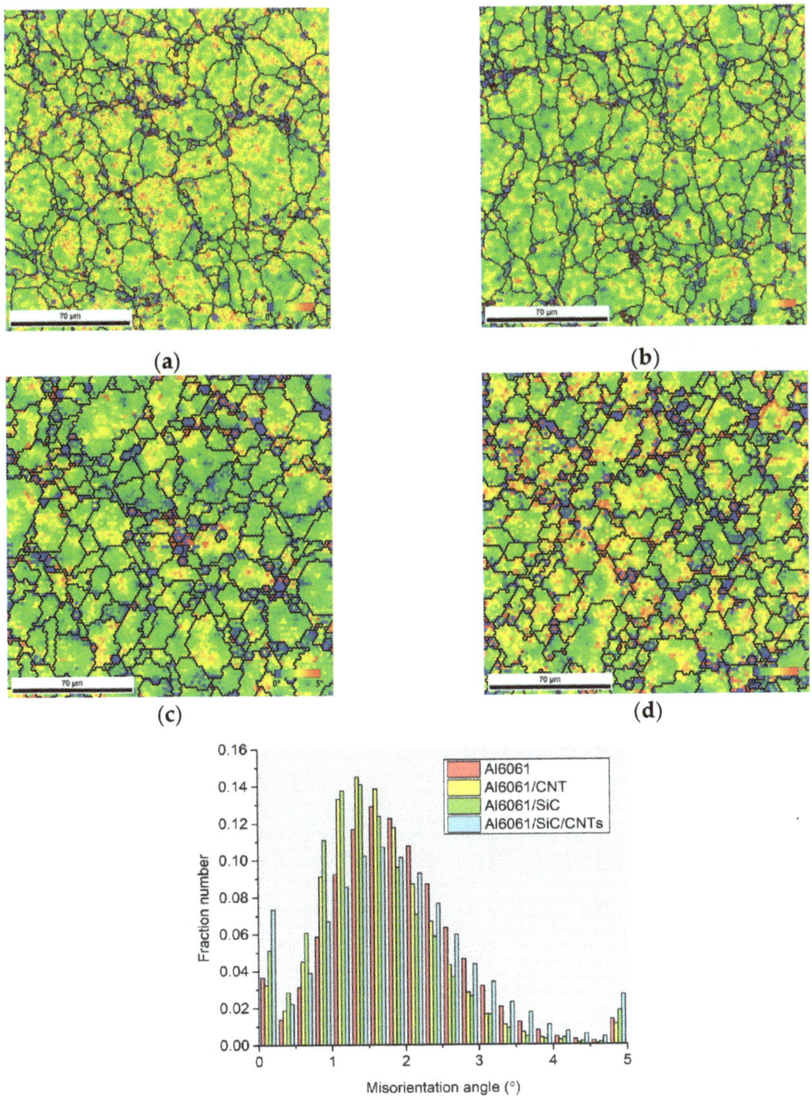

Figure 16. KAM maps of (**a**) Al6061 matrix, (**b**) Al6061/CNTs, (**c**) Al6061/SiC, (**d**) Al6061/SiC/CNTs, and (**e**) distribution of KAM misorientation angle for all samples.

The hardness evolution of Al6061-based composites sintered at different temperatures can be analyzed in Figure 17 by considering the reinforcing effects of CNTs and SiC particles on the Al6061 matrix.

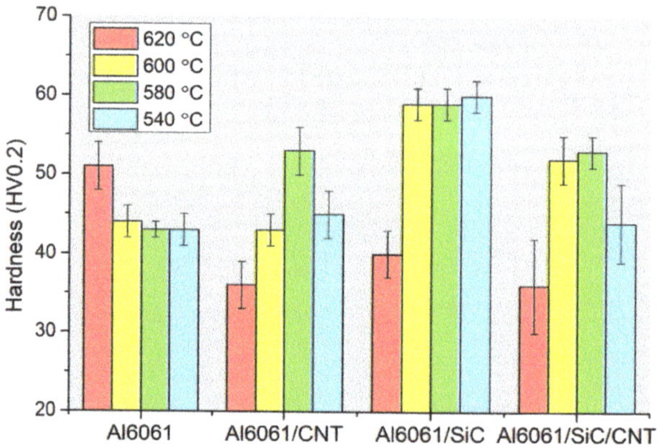

Figure 17. The evolution of hardness of the Al6061 matrix and Al6061 reinforced with CNTs, SiC, and SiC/CNTs for the different sintered temperatures.

Due to its relatively homogeneous microstructure, the Al6061 matrix typically shows the lowest hardness values across all sintering temperatures. As the sintering temperature increases, the hardness might initially rise due to densification but may decrease at higher temperatures due to grain growth. The addition of CNTs generally improves the hardness of the Al6061 matrix due to the strengthening effect of the nanoscale reinforcement and its ability to hinder dislocation motion. The CNTs can remain well-dispersed at lower sintering temperatures, reinforcing the matrix. However, CNT agglomeration or reaction with the matrix at higher sintering temperatures can reduce their strengthening effect, potentially decreasing hardness. SiC particles significantly enhance the hardness of the Al6061 matrix due to their high hardness and ability to act as barriers to dislocation movement. The improvement in hardness is more pronounced at all sintering temperatures than the CNT reinforcement. SiC particles can better maintain their reinforcing effect at high sintering temperatures than CNTs, as they are less prone to degradation or agglomeration. At lower sintering temperatures, hardness is primarily influenced by the matrix's distribution and dispersion of reinforcements. As the sintering temperature increases, hardness can increase due to improved bonding and matrix densification. At higher temperatures (ranging from 480 to 620 °C), grain growth and the potential degradation of reinforcements (e.g., CNTs) can decrease hardness.

Table 6 demonstrates the impact of different reinforcements on the mechanical properties of Al6061. The Al6061/CNT/SiC hybrid composite achieves the best overall strength but at the expense of ductility. Each reinforcement type (CNTs, SiC, and hybrid) improves tensile strength compared to the pure Al6061 matrix but reduces elongation, with the hybrid configuration providing a balance between the two properties. The Al6061 matrix has the lowest tensile strength, measuring 63 MPa. The addition of MWCNTs significantly enhances the tensile strength to 87 MPa, showing the strengthening effect of the carbon nanotubes. The Al6061/SiC composite exhibits an even higher tensile strength of 96 MPa, indicating that SiC particles are more effective at improving the load-bearing capacity of the matrix. The hybrid composite (Al6061/CNTs/SiC) shows the highest tensile strength of 104 MPa, suggesting a synergistic effect between CNTs and SiC in strengthening the matrix. The Al6061 matrix shows the highest elongation value of 8.2%, indicating good ductility. Adding MWCNTs decreases the elongation to 4.4%, reflecting the typical trade-off between strength and ductility as CNTs increase tensile strength but reduce flexibility. The Al6061/SiC composite has an even lower elongation of 3.2%, due to the inherent brittleness of SiC particles. The hybrid configuration (Al6061/CNTs/SiC) slightly improved elongation to 3.4% compared to the SiC-only composite, possibly due to the reinforcing effect of CNTs mitigating some of the brittleness introduced by SiC.

Table 6. Mechanical properties of matrix and nanocomposites obtained by tensile tests for samples sintered at 580 °C.

	Tensile Strength (MPa)	Elongation (%)
Al 6061	63	8.2
Al 6061/MWCNTs	87	4.4
Al6061/SiC	96	3.2
Al6061/CNTs/SiC	104	3.4

4. Conclusions

The production and characterization of hybrid composites based on the Al6061 alloy reinforced with carbon nanotubes (CNTs) and silicon carbide (SiC) particles have demonstrated that incorporating multiple reinforcements leads to significant enhancements in the material's mechanical properties. The production of hybrid nanocomposites with 0.75 wt.% SiC and CNTs by cold compaction and vacuum sintering demonstrated the possibility of obtaining an increase of 65% in mechanical properties compared to the matrix produced in the same conditions. Adding individual reinforcements, such as CNTs or SiC up to 0.75 wt.%, contributed to increased tensile strength and hardness due to the combined effects of grain refinement, load transfer, and barriers to dislocation movement. However, the hybrid configuration (Al6061/CNTs/SiC) exhibited the most pronounced improvements, showcasing superior tensile strength and a more refined microstructure, indicating a synergistic effect between the two reinforcements.

While the increase in strength is evident, a reduction in ductility was observed for all reinforced composites compared to the unreinforced Al6061 matrix, with the hybrid composites displaying a moderate compromise between strength (65% for tensile strength) and elongation (58% for the hybrid nanocomposites). This behavior is attributed to the high stiffness and hardness of SiC particles, coupled with the enhanced load transfer capability of the CNTs. The results suggest that the hybrid composites are particularly suitable for applications where high strength and wear resistance are critical, but moderate ductility can be tolerated.

Overall, the successful production of Al6061/CNT/SiC hybrid composites using powder metallurgy and subsequent characterization indicates that these materials possess great potential for advanced engineering applications. Future work should optimize the dispersion and bonding of reinforcements within the matrix to further enhance mechanical performance and explore the impact of different sintering conditions and reinforcement ratios on the microstructural evolution and overall properties of the composites.

Author Contributions: Conceptualization, S.S.; methodology, S.S.; validation, S.S.; formal analysis, B.M.; investigation, B.M.; writing—original draft preparation, B.M.; writing—review and editing, S.S.; visualization, B.M.; supervision, S.S. All authors have read and agreed to the published version of the manuscript.

Funding: This research received no external funding.

Data Availability Statement: Data can be made available upon request to the authors.

Acknowledgments: The authors are grateful to CEMUP—Centro de Materiais da Universidade do Porto—for the expert assistance with SEM.

Conflicts of Interest: The authors declare no conflicts of interest.

References

1. Mattli, M.; Reddy, M.; Khan, A.; Abdelatty, R.; Yusuf, M.; Ashraf, A.; Kotalo, R.; Shakoor, A. Study of microstructural and mechanical properties of Al/SiC/TiO_2 hybrid nanocomposites developed by microwave sintering. *Crystals* **2021**, *11*, 1078. [CrossRef]
2. Moustafa, E.; Taha, M. The effect of mono and hybrid additives of ceramic nanoparticles on the tribological behavior and mechanical characteristics of an Al-based composite matrix produced by friction stir processing. *Nanomaterials* **2023**, *13*, 2148. [CrossRef] [PubMed]

3. Anitha, P.; Rao, M. Influence of titanium diboride and graphite on microstructure characteristics and mechanical behaviors of aluminium hybrid nanocomposites. *Proc. Inst. Mech. Eng. Part E* **2022**, *236*, 2071–2081. [CrossRef]
4. Džunić, D.; Pantić, M.; Mitrović, S.; Babic, M.; Savić, S.; Đorđević, A.; Arsić, A. Assessment of tribological behaviour of ZA-27 zincaluminium alloy based nanocomposite. *Proc. Eng. Sci.* **2019**, *1*, 145–153. [CrossRef]
5. Singh, A.; Yadav, G.; Gupta, P. Correlation of structural and mechanical properties for ultrasonically stir casted al6061 reinforced SiC-AlN hybrid metal matrix composites. *ECS J. Solid State Sci. Technol.* **2023**, *12*, 057008. [CrossRef]
6. Iyengar, S.; Sethuram, D.; Shobha, R.; Koppad, P. Microstructure, microhardness, and tensile properties of hot-rolled al6061/TiB_2/CeO_2 hybrid composites. *J. S. Afr. Inst. Min. Metall.* **2021**, *121*, 543–548. [CrossRef]
7. Gireesh, C.; Prasad, K.; Ramji, K. Experimental investigation on mechanical properties of an Al6061 hybrid metal matrix composite. *J. Compos. Sci.* **2018**, *2*, 49. [CrossRef]
8. Kumar, R. Optimizing Al6061-based hybrid metal matrix composites: Unveiling microstructural transformations and enhancing mechanical properties through Ni and Cr reinforcements. *Evergreen* **2023**, *10*, 2161–2172. [CrossRef]
9. Kumar, V.; Nagegowda, K.; Boppana, S.; Sengottuvelu, R.; Palanikumar, K. Wear behavior of aluminium 6061 alloy reinforced with coated/uncoated multiwalled carbon nanotube and graphene. *J. Met. Mater. Miner.* **2021**, *31*, 17–24. [CrossRef]
10. Soni, S.; Thomas, B. Influence of TiO_2 and MWCNT nanoparticles dispersion on microstructure and mechanical properties of Al6061 matrix hybrid nanocomposites. *Mater. Res. Express* **2020**, *6*, 1265f3. [CrossRef]
11. Kinloch, I.; Suhr, J.; Lou, J.; Young, R.; Ajayan, P. Composites with carbon nanotubes and graphene: An outlook. *Science* **2018**, *362*, 547–553. [CrossRef] [PubMed]
12. Reddy, A.; Krishna, P.; Rao, R. Tribological behaviour of Al6061–2SiC-xGr hybrid metal matrix nanocomposites fabricated through ultrasonically assisted stir casting technique. *Silicon* **2019**, *11*, 2853–2871. [CrossRef]
13. Meignanamoorthy, M.; Ravichandran, M.; Mohanavel, V.; Afzal, A.; Sathish, T.; Alamri, S.; Khan, S.; Saleel, C. Microstructure, mechanical properties, and corrosion behavior of boron carbide reinforced aluminum alloy (Al-Fe-Si-Zn-Cu) matrix composites produced via powder metallurgy route. *Materials* **2021**, *14*, 4315. [CrossRef] [PubMed]
14. Rashid, A. Nanotechnology-enhanced fiber-reinforced polymer composites: Recent advancements on processing techniques and applications. *Heliyon* **2024**, *10*, e24692. [CrossRef] [PubMed]
15. Basak, A.K.; Pramanik, A.; Islam, M.N.; Anandakrishnan, V. Challenges and recent developments on nanoparticle-reinforced metal matrix composites. In *Fillers and Reinforcements for Advanced Nanocomposites*; Dong, Y., Umer, R., Lau, A.K.-T., Eds.; Composites Science and Engineering; Woodhead Publishing: Sawston, UK, 2015; pp. 349–367, ISBN 9780081000793.
16. Wieszczycka, K.; Staszak, K.; Woźniak-Budych, M.J.; Litowczenko, J.; Maciejewska, B.M.; Jurga, S. Surface functionalization—The way for advanced applications of smart materials. *Coord. Chem. Rev.* **2021**, *436*, 213846. [CrossRef]
17. Monteiro, B.; Simões, S. Optimization of Al6061 Nanocomposites Production Reinforced with Multiwalled Carbon Nanotubes. *J. Compos. Sci.* **2024**, *8*, 381. [CrossRef]
18. Carneiro, Í.; Monteiro, B.; Ribeiro, B.; Fernandes, J.V.; Simões, S. Production and Characterization of Cu/CNT Nanocomposites. *Appl. Sci.* **2023**, *13*, 3378. [CrossRef]
19. Raghavan, V. Al-Mg-Si (Aluminum-Magnesium-Silicon). *J. Phase Equilibria Diffus.* **2007**, *28*, 2. [CrossRef]
20. Raghavan, V. Al-Fe-Si (Aluminum-Iron-Silicon). *J. Phase Equilibria Diffus.* **2002**, *23*, 4.

metals MDPI

Article

Microstructure Development of Powder-Based Cu Composite During High Shear Strain Processing

Lenka Kunčická *, Josef Walek and Radim Kocich

Department of Metallurgical Technologies, Faculty of Materials Science and Technology, VŠB Technical University of Ostrava, 17. listopadu 2172-15, 708 00 Ostrava, Czech Republic

* Correspondence: lenka.kuncicka@vsb.cz; Tel.: +420-596-991-265

Abstract: Commercially pure Cu features excellent electric conductivity but low mechanical properties. In order to improve the mechanical properties of Cu, strengthening elements can be added to prepare alloys or composites featuring enhanced performances. This study focuses on the detailed characterization of the microstructure of a Cu composite strengthened with Al_2O_3 particles during high shear strain processing. The Cu-Al_2O_3 mixture was prepared by powder metallurgy and directly consolidated by the intensive plastic deformation method of hot rotary swaging. Samples cut from the consolidated piece were further processed by the severe plastic deformation method of high pressure torsion (HPT). The primary aim was to investigate the effects of varying degrees of the imposed shear strain, i.e., the number of HPT revolutions, microstructure development (grain size and morphology, texture, grain misorientations, etc.) of the consolidated composite; the microstructure observations were supplemented with measurements of Vickers microhardness. The results showed that the added oxide particles effectively hindered the movement of dislocations and aggravated grain fragmentation, which also led to the relatively high presence of grain misorientations pointing to the occurrence of residual stress within the microstructure. The high shear strain imposed into (the peripheral region of) the sample subjected to four HPT revolutions imparted equiaxed ultra-fine grains and an average Vickers microhardness of more than 130 HV0.1.

Keywords: copper; composite; high-pressure torsion; microstructure

1. Introduction

Highly pure Cu, typically prepared with the use of electrolytic refining, exhibits excellent electric conductivity but very low mechanical properties (ultimate tensile strength of only ~200 MPa) [1]. Adding alloying elements and introducing precipitates or second phases usually increases the mechanical properties of Cu but also deteriorates its electric conductivity by aggravating the movement of free electrons [2,3]. Similar effects can be introduced by deformation processing, favorably in combination with heat treatments [4,5]. Nevertheless, modern technological processes enable the fabrication of alloys and composites benefiting from the strengthening effects of introduced reinforcing particles and elements, while maintaining favorable electric conductivity provided by Cu (e.g., [6–9]). The introduced strengthening elements can have various forms, from individual layers (such as in lightweight Cu-Al laminated composite wires [10,11], or thermally stable Cu-Nb laminates [12]), through various oxide particles [13,14] to carbides [15–17], and more.

Cu-based composites featuring strengthening particles can advantageously be prepared by the means of powder metallurgy and/or additive manufacturing [18–20]. For example, Li et al. [21] prepared composite films with ultrahigh electric conductivity from Cu and carbon nanotubes by pulse reverse electrodeposition, and Rogachev et al. [22] documented the effects of mechanical activation of a Cu-Cr powder mixture before being processed by spark plasma sintering. Probably the first ever documented thorough scientific work on Cu-based alloys/composites featuring oxide particles was published by

Schilling and Grant [23], and the field has experienced rapid development ever since. As for the contemporary works, e.g., Khaloobagheri et al. [24] investigated the correlation of mechanical and electric properties for compacted and hot sintered Cu-ZrO_2 powders, Rajkovic et al. [25] prepared a Cu-Al_2O_3 composite featuring enhanced mechanical properties and electroconductivity of up to 88% by compacting mechanically alloyed mixtures of powders, and Mukhtar et al. [26] used Al_2O_3 nano-powder to fabricate an ultra-fine grained Cu-Al_2O_3 composite using warm powder compacting. Cu-Al_2O_3 composites can advantageously be used not only for electroconductive applications [27], due to their high hardness particularly for sliding contacts experiencing wear [28], but also for components working at elevated temperatures such as heat sinks, engine components, etc. [29,30], as their mechanical properties are relatively stable [31]. The compacted or sintered pieces can subsequently be subjected to additional (thermo)mechanical processing to optimize their microstructures and further improve the properties as, among the solutions, how to enhance the performance of electroconductive Cu and Cu-based alloys and composites is introducing nano-sized structural features, such as twins [32] and precipitates [33–36].

Nano-sized structural features can advantageously be introduced by imparting (high) shear strain via methods of intensive plastic deformation, such as rotary swaging [37–39], and severe plastic deformation (SPD), such as equal channel angular pressing (ECAP) and related methods [40–42], or high pressure torsion (HPT) [43–45]. Such methods can advantageously impart strengthening to the processed materials as they introduce generation and the accumulation of dislocations, the formation of dislocation cells, and the subsequent polygonization, i.e., the formation of subgrains, finally developing into full (refined) grains [46]. Rotary swaging is an industrially applicable method [47]. It was previously shown to be advantageous for the production of Cu-based electroconductive wires with refined structural features, e.g., [10,32]. ECAP-based methods' primary advantage is that they can be applied to process bulk workpieces of relatively large volumes (among the SPD methods) [48–50]. The positive effects of ECAP on the mechanical properties and the related microstructure refinement of commercially pure Cu were reported, e.g., by Volokitina [51] and Hlaváč et al. [52]. Guo et al. [53] used multi-pass ECAP combined with aging treatment to fabricate a CuCrSi alloy with high electric conductivity and enhanced strength and plasticity, Huang et al. [54] used multi-pass ECAP + aging procedures to impart an ultra-fine grained structure and, simultaneously, enhanced electric conductivity within a CuSeTe alloy, while Dalan et al. [55] achieved a high-strength material with favorable electric conductivity by subjecting a CuCrZr alloy to ECAP (routes A and Bc) and aging. Nevertheless, HPT can be considered to be even more efficient than ECAP as regards the grain refinement and introduction of nano-sized structural features.

According to Zhilyaev and Langdon [56], HPT is among the first ever introduced SPD methods and has several basic modifications as regards the geometry of the dies and dimensions of the processed specimen. The principle of the method is in deforming a round specimen (of variable height and diameter, according to the particular die design) between two massive dies under very high pressures (typically to 5–6 GPa). The dominant deformation mechanism during HPT processing is simple shear [57,58], and it is highly advantageous for fabricating ultra-fine grained (UFG) microstructures [59,60]. The combination of shear with very high pressure is highly advantageous. Imposing high shear strain usually promotes the formation of shear bands within the deformed structure, which progressively increases deformation inhomogeneity and, thus, also structure inhomogeneity. Nevertheless, the (supporting) application of high pressure helps to decrease the inhomogeneity of the imposed strain. The combined effect of these influencing factors results in substantial grains' fragmentation and substructure development (depending on the imposed strain) throughout the HPT-processed sample. Several studies reporting the effects of HPT on the development of a (sub)structure for commercially pure Cu were published. For example, An et al. [61] documented the effects of stacking fault energy (and Al content) on the mechanical properties of HPT-processed Cu (and selected Cu-Al alloys), Kosinova et al. [62] investigated the interactions of twins and coincidence site lattices (CLSs) within highly pure Cu processed by HPT with different die geometries, and Huang et al. [63] and Král et al. [64] observed the microstructures of HPT-processed Cu after room temperature annealing for differing time periods. As regards Cu composites with strengthening oxide particles, Qi et al. [65] subjected a cast Cu-ZnO composite to HPT and reported, among others, positive effects of HPT processing on the fragmentation of ZnO agglomerates. Nevertheless, as far as the authors' knowledge reaches, studies investigating the effects of HPT processing on the microstructures of powder-based Cu composites with additions of strengthening oxide particles are very scarce, or non-existent.

In accordance with the above mentioned, the presented study aims to assess the effects of severe plastic deformation on the development of microstructures of a Cu–Al_2O_3 composite prepared by powder metallurgy. The powder-based composite was pre-compacted by hot rotary swaging (i.e., directly consolidated) and finally processed via HPT at room temperature. HPT was performed by a varying number of revolutions in order to observe and evaluate in detail the progressive development of the microstructure. Last but not least, supplementary measurements of Vickers microhardness across the processed samples were performed to document the effects on the mechanical properties.

2. Materials and Methods

The first processing step involved the preparation of the powder mixture consisting of a powder of commercially pure Cu with the addition of 1 wt.% of Al_2O_3 powder. A scanning electron microscopy secondary electron (SEM-SE) image of the original Cu powder is depicted in Figure 1a, while Figure 1b shows an SEM-SE image of the Al_2O_3 particles.

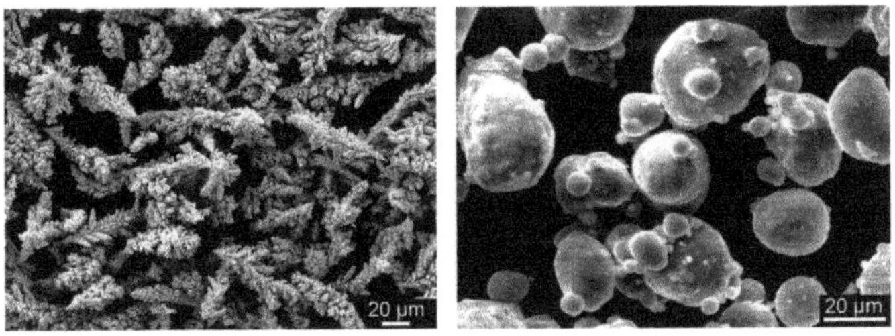

(a) (b)

Figure 1. SEM-SE images of used powders: Cu (**a**), and Al_2O_3 (**b**).

The powder mixture was homogenized using a TURBULA 3D shaker mixer (Willy A. Bachofen AG, Muttens, Switzerland) and subsequently filled into copper tube containers with a diameter of 35 mm, which were then vacuum sealed. The vacuum unit applied for the sealing was a D-35614 device (Pfeiffer Vacuum GmbH, Asslar, Germany). It consists of pre-pump, turbo molecular pump, angle valve, a pressure-measuring tube and vacuum-tight measuring cell and is able to achieve the vacuum of 5×10^{-4} Pa (used for the presented experiment). The sealed cans were further subjected to hot rotary swaging at a temperature of 600 °C for the powder mixture to directly consolidate. The swaging was carried out to the final diameter of 20 mm, to provide a workpiece from which 5 mm thick samples for final HPT processing could be cut. As the primary focus of the following HPT processing was to assess the effects of the imposed strain, i.e., the number of HPT revolutions, on the development of the consolidated microstructure, the samples were subjected to a single revolution (i.e., 360° rotation), double revolution (720° rotation), and 4 revolutions (1440° rotation) at the rotational speed of 1 RPM. The processed samples were then denoted at *HPT1*, *HPT2*, and *HPT4*, according to the total number of revolutions. HPT was performed under the pressure of 5 GPa. For each number of revolutions, three individual samples were processed to subsequently enable more reliable analyses of grain size and Vickers microhardness. A schematic depiction of the performed material processing is shown in Figure 2.

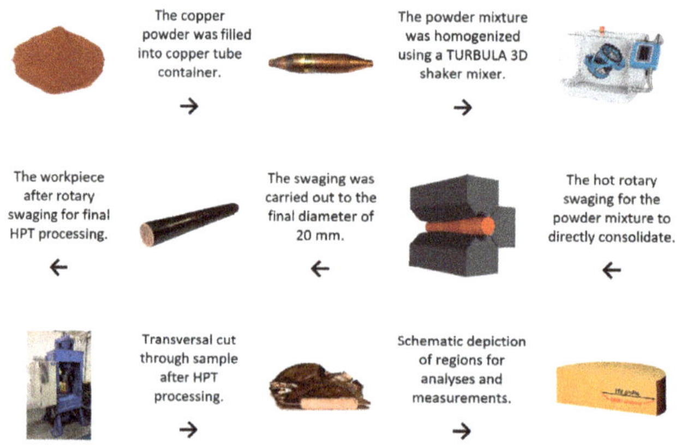

Figure 2. Schematic depiction of processing route.

All the processed samples were then subjected to detailed observations of microstructures via scanning electron microscopy (SEM) using the electron backscatter diffraction (EBSD) method (Tescan Fera device by Tescan Orsay Holding a.s., Brno, Czech Republic). The acquired microstructure scans were subsequently evaluated using the Atex software (Atex© 2018, version 1.32) by LEM3 laboratories [66]. To prepare the samples for the analyses, all the HPT-processed samples were at first transversally cut to expose their cross-sectional planes. See Figure 3a depicting a cut HPT specimen (*HPT1*), and Figure 3b showing a schematic depiction of regions of performed analyses and measurements. The cut samples were manually ground using SiC papers up to the coarseness of 1200 and subsequently polished using diamond solutions with particle dimensions of 3 μm, 1 μm, and, finally, 0.6 μm, all with the maximum load of 20 kN. Finally, the samples were electrolytically polished (solutions and equipment by Struers GmbH, Roztoky u Prahy, Czech Republic). The EBSD scans were acquired with an acceleration voltage of 20 keV. The scans of the directly consolidated composite were acquired with the dimensions of $1000 \times 1000\ \mu m^2$ and a scan step size of 0.25 μm, while the scans of the HPT-processed microstructures were acquired with the dimensions of $50 \times 50\ \mu m^2$ and a scan step size of 0.15 μm. The results were evaluated considering the following limits: LAGBs (low-angle grain boundaries) of 5–15°, HAGBs (high-angle grain boundaries) > 15°, and the maximum deviation considered during the analyses of ideal texture orientations was 10°.

(**a**) (**b**)

Figure 3. Transversal cut through sample *HPT1:* (**a**) schematic depiction of regions for analyses and measurements (**b**).

In order to assess the effects of the shear strain on the mechanical properties of the HPT-processed specimens, HV0.1 (load of 100 g) Vickers microhardness was measured along a horizontal line across the cross-sections of the cut samples (see Figure 3a depicting a cut sample, and Figure 3b schematically showing the analyzed locations). The acquired values were averaged from all three samples for each material state. Vickers microhardness was measured using FM ARS 900 equipment by Future Tech (Spectrographic Limited, Leeds, UK). The load time for each indent was 7 s.

3. Results

3.1. Direct Consolidation

At first, the microstructure of the directly consolidated workpiece was examined to assess the quality of consolidation and document the initial microstructure before HPT processing. Figure 4a shows the band contrast image of the consolidated microstructure and confirms the sufficient quality of consolidation, as the microstructure did not exhibit any visible voids or residual porosity. Figure 4a also documents that the Al_2O_3 particles were homogeneously distributed within the Cu grains, although they were relatively coarse; the Al_2O_3 particles are depicted as the black dots distributed within the Cu grains and marked with red circles in Figure 4a. The arithmetic mean Cu grain size value, evaluated via the parameter of equivalent circle diameter, was 47 μm. However, the grain size distribution was non-homogeneous, and the microstructure evidently also featured coarse Cu grains. The consolidated microstructure also featured numerous <111>60° twin boundaries, which can clearly be seen in the Orientation Image Map (OIM) in Figure 4b. The corresponding map of internal grain misorientations is then shown in Figure 4c—this image documents that the grain misorientations, pointing to a certain occurrence of residual stress within the microstructure (e.g., [67,68]), were primarily present in the vicinity of the occurring twins (depicted with red arrows in Figure 4a).

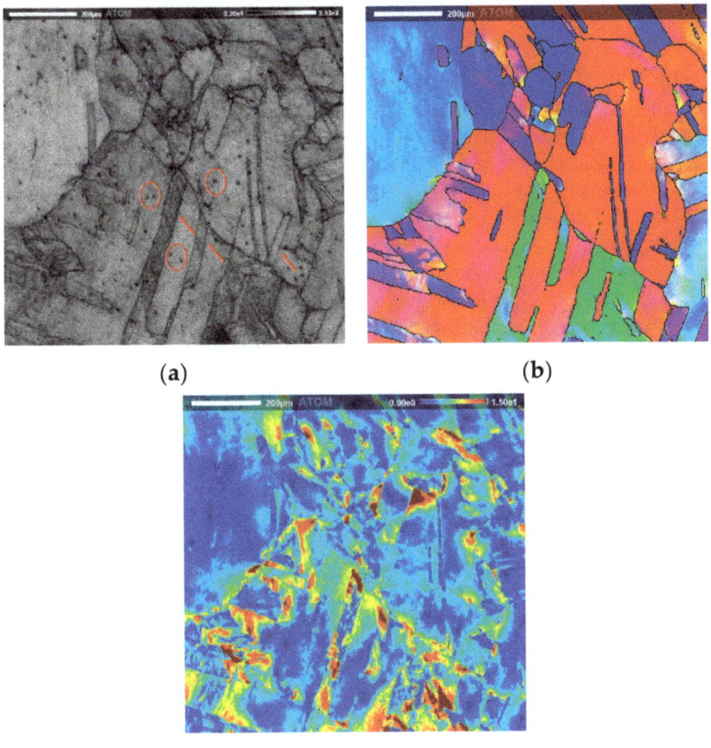

(a) (b) (c)

Figure 4. Microstructure of directly consolidated workpiece: band contrast image, red circles show oxides, and red arrows show twins (examples) (**a**); orientation image map (**b**); and internal grain misorientations (**c**).

3.2. *High Pressure Torsion*

3.2.1. Grain Size

The grain sizes in the individual regions of interest, according to Figure 3b, were evaluated via the parameter of equivalent circle diameter in μm for each HPT-processed sample. The arithmetic mean grain size values are summarized in Table 1. The table clearly shows that the grain size gradually decreased with the increasing number of HPT revolutions; this trend was evident especially in the peripheral regions of the samples. Also, comparing the individually analyzed regions within a sample, the mean grain size was larger in the axial region than at the periphery. This trend can be attributed to the character of the HPT process—as this is an axisymmetric process, the imposed strain gradually increases along the radius of a processed sample, from its axis towards the periphery.

Table 1. Arithmetic mean grain size with standard deviations in investigated regions of processed samples.

Sample	*HPT1*		*HPT2*		*HPT4*	
	Axis	**Periphery**	**Axis**	**Periphery**	**Axis**	**Periphery**
arithmetic mean equivalent circle diameter [μm]	3.75 ±0.22	2.65 ±0.24	2.80 ±0.25	2.30 ±0.16	2.40 ±0.09	0.95 ±0.03

All the examined samples exhibited highly refined grains compared to the directly consolidated state. However, the peripheral region of sample *HPT4* exhibited sub-micron grain size, i.e., a ultra-fine grained (UFG) microstructure (see Table 1).

3.2.2. Grain Analysis

The Cu grain orientations in the examined regions of the HPT-processed samples were evaluated via the Orientation Image Maps (OIMs). See Figure 5a–f depicting the individual results as follows: Figure 5a—the OIM of the sample *HPT1* axial region, Figure 5b—the OIM of the sample *HPT1* peripheral region, Figure 5c—the OIM of the sample *HPT2* axial region, Figure 5d—the OIM of the sample *HPT2* peripheral region, Figure 5e—the OIM of the sample *HPT4* axial region, and Figure 5f—the OIM of the sample *HPT4* peripheral region.

All the images confirm the significant effect of HPT processing on grain refinement, as documented in Table 1. As can be seen from Figure 5a–f, the microstructure of the *HPT1* sample featured the occurrence of numerous twins. Twins were evident especially in the axial region of the sample. The grains at the periphery of the sample primarily exhibited traces of severe shearing, although localized twins could also be observed there. After two HPT revolutions, the grains in the axial region of the *HPT2* sample exhibited visible flattening by the mutual effect of compression and the acting radial vector of shear strain. The peripheral region of this sample exhibited a bimodal microstructure consisting of heavily sheared original grains, as well as newly formed small equiaxed grains. The grains in the axial region of the *HPT4* sample then exhibited substantial flattening and shearing, and the occurrence of twinning in this region of the sample was negligible. The peripheral region of the *HPT4* sample featured heavily refined equiaxed ultra-fine grains with no evidence of larger sheared grains.

The presence of the Al_2O_3 particles within the microstructures of the examined regions of the HPT-processed samples is depicted via the phase maps shown in Figure 6a–f as follows: Figure 6a—the sample *HPT1* axial region, Figure 6b—the sample *HPT1* peripheral region, Figure 6c—the sample *HPT2* axial region, Figure 6d—the sample *HPT2* peripheral region, Figure 6e—the sample *HPT4* axial region, and Figure 6f—the sample *HPT4* peripheral region. As can be seen, the microstructures of the HPT-processed samples still featured the localized presence of larger clusters of the Al_2O_3 powder particles. Nevertheless, the majority of the particles were fragmented/refined and more or less homogeneously distributed within the examined microstructures.

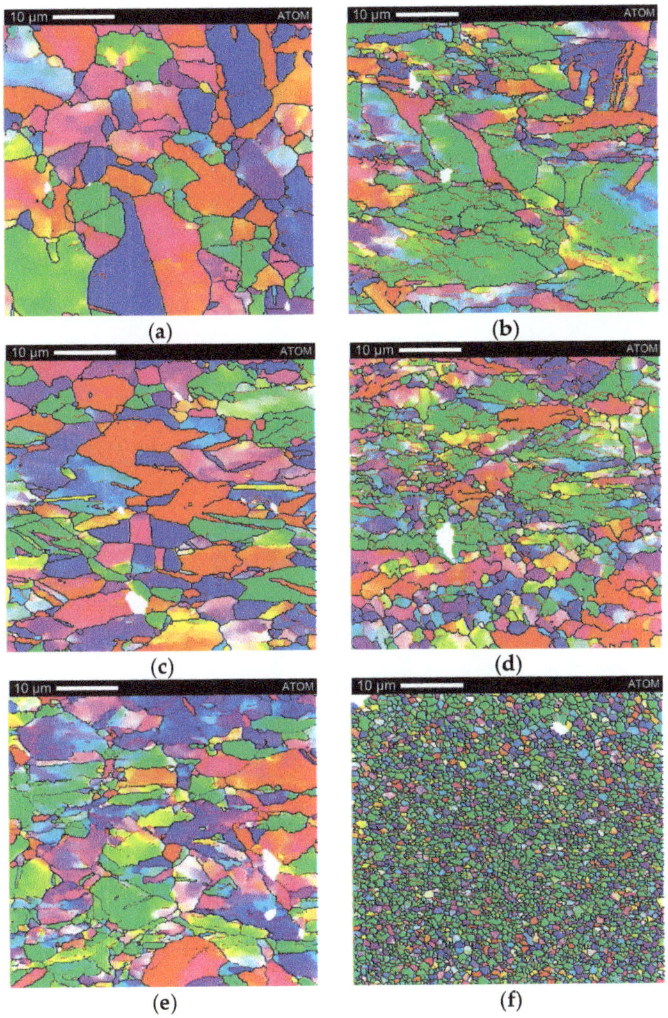

Figure 5. Orientation Image Maps (OIMs) for samples: *HPT1* axial region (**a**), *HPT1* peripheral region (**b**), *HPT2* axial region (**c**), *HPT2* peripheral region (**d**), *HPT4* axial region (**e**), and *HPT4* peripheral region (**f**). The grains coloring is according to conventional IPF triangle for FCC metals.

3.2.3. Texture

Figure 7a–f show Pole Figure (PF) texture plots for microstructures in the examined regions of the HPT-processed samples as follows: Figure 7a—PFs for the sample *HPT1* axial region, Figure 7b—PFs for the sample *HPT1* peripheral region, Figure 7c—PFs for the sample *HPT2* axial region, Figure 7d—PFs for the sample *HPT2* peripheral region, Figure 7e—PFs for the sample *HPT4* axial region, and Figure 7f—PFs for the sample *HPT4* peripheral region. The analyses of the dominant texture orientations were evaluated considering the ideal shear texture orientations according to Beyerlein and Tóth [69], who documented the characteristic ideal orientations within FCC metals subjected to high shear strain. The considered characteristic orientations were the following: *Ab* and *A* orientations of the {111}<110> fiber, constant Euler angles $\Phi = 35.26°$ and $\varphi 2 = 45°$; *A1* and *A2* orientations of the {111} fiber, constant Euler angles $\Phi = 45°$ and $\varphi 2 = 0°$; *Bb* and

B orientations of the <110> fiber, constant Euler angles $\Phi = 54.74°$ and $\varphi2 = 45°$; and *C* orientations of the <110> fiber, constant Euler angles of $\Phi = 90°$, and $\varphi2 = 45°$ [69,70].

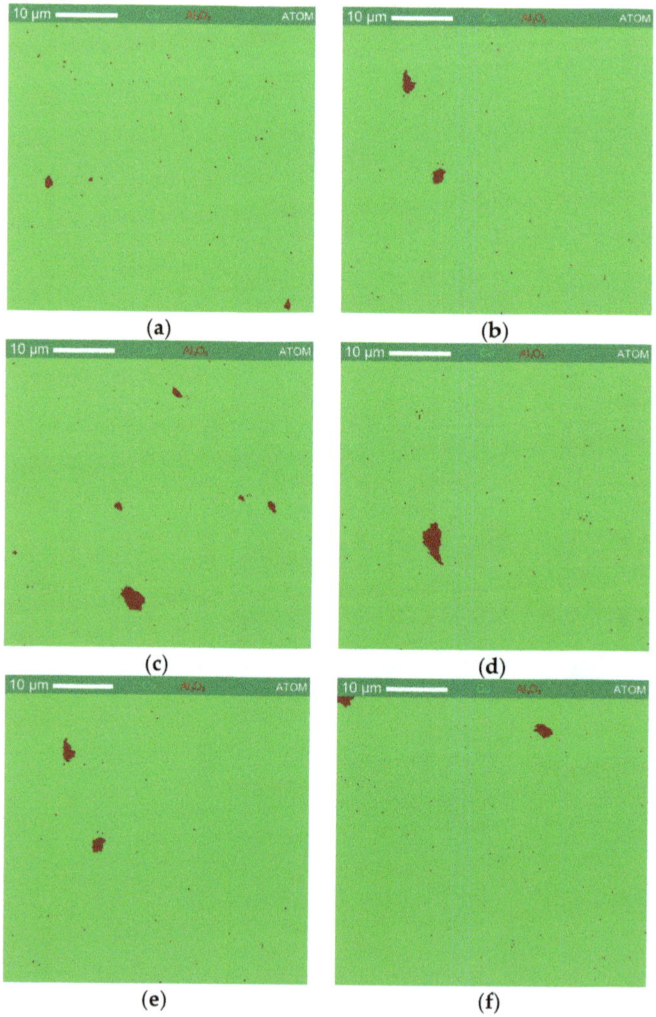

Figure 6. Phase maps for samples: *HPT1* axial region (**a**), *HPT1* peripheral region (**b**), *HPT2* axial region (**c**), *HPT2* peripheral region (**d**), *HPT4* axial region (**e**), and *HPT4* peripheral region (**f**).

Figure 7a,c,e documents the tendencies of alignment/orientation for the grains in the axial regions of the HPT-processed samples. Orientations close to the *A1* and *Bb* ones were observed within the axial region of the *HPT1* sample. The grains within the axial region of the *HPT2* sample exhibited the tendencies to align with the *A1* and C orientations, whereas the *HPT4* sample featured grains oriented close to the *A2* and *C* ideal orientations. The grains within the axial regions of all the examined samples thus featured the tendencies to orient along the {111} preferential fiber (*A1*, *A2* orientations), as well as along the <110> preferential fiber; the latter was represented by the *Bb* orientation within the samples featuring lower imposed strains (*HPT1* and *HPT2*), which then shifted to the *C* orientation for the sample deformed with four HPT revolutions.

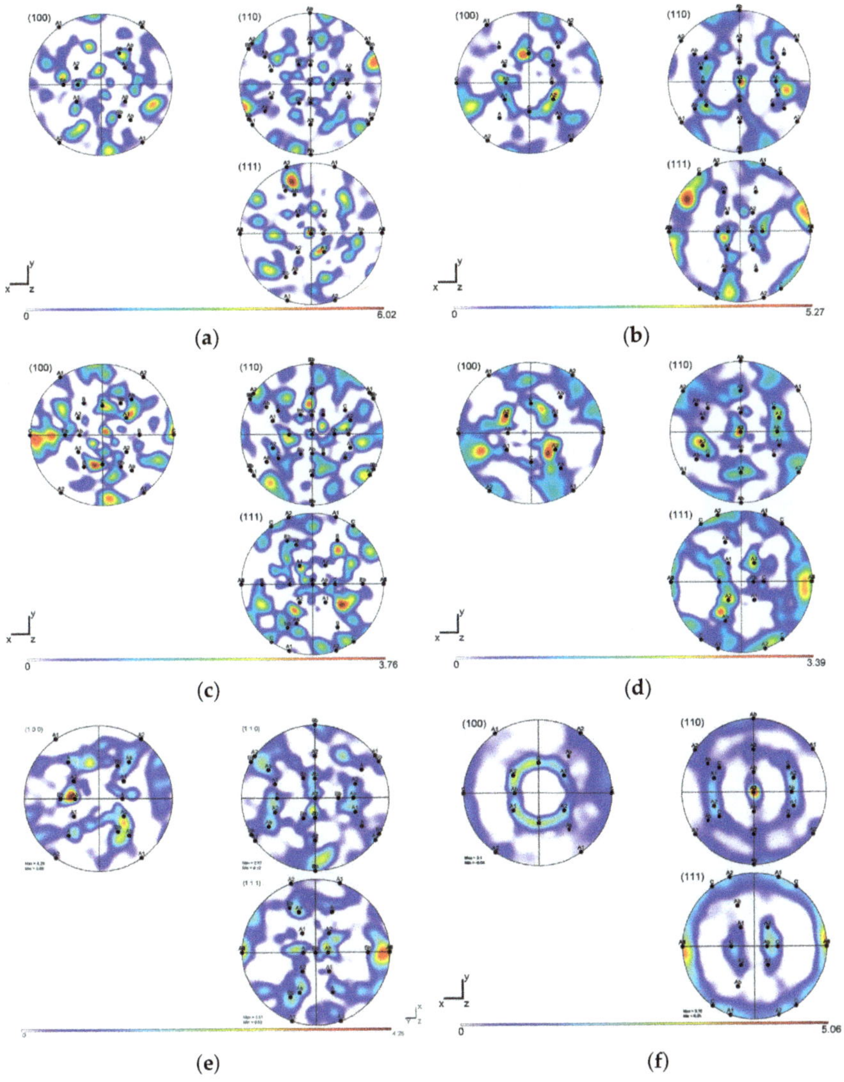

Figure 7. Pole figure plots for samples: *HPT1* axial region (**a**), *HPT1* peripheral region (**b**), *HPT2* axial region (**c**), *HPT2* peripheral region (**d**), *HPT4* axial region (**e**), and *HPT4* peripheral region (**f**).

The PF images in Figure 7b,d,f then show that the grains within the microstructure of the peripheral region of the *HPT1* sample featured the tendencies to align along the *A2* and *C* orientations. Increasing the imposed strain to two HPT revolutions then intensified the tendencies of the grains to form the {111} preferential fiber, as all the examined *A1*, *A2*, and *Ab* orientations, belonging to this fiber, were detected within the microstructure of the *HPT2* sample, together with the *C* orientation. The grains within the *HPT4* sample then featured a distinguishable tendency to form the {111} preferential fiber.

3.2.4. Grain Misorientations

The results of the analyses of internal grain misorientations in the scale from 0° to 15° are depicted in Figure 8a–f as follows: Figure 8a—misorientations for the sample *HPT1* axial region, Figure 8b—misorientations for the sample *HPT1* peripheral region, Figure 8c—misorientations for the sample *HPT2* axial region, Figure 8d—misorientations for the sample *HPT2* peripheral region, Figure 8e—misorientations for the sample *HPT4* axial region, and Figure 8f—misorientations for the sample *HPT4* peripheral region. The comparison of the misorientations occurring within the axial regions of the processed samples to those occurring within their peripheral regions shows that the increasing/decreasing trends with an increasing imposed shear strain in these regions were just the opposite. In the axial regions of the samples, the frequency of the occurrence of high misorientations within the microstructure increased with an increasing imposed shear strain. On the other hand, the microstructures in the peripheral regions of the samples exhibited a decreasing frequency of occurrence of high grain misorientations with an increasing imposed strain. The microstructure in the peripheral region of the *HPT4* sample, featuring refined equiaxed grains, exhibited a negligible presence of internal misorientations.

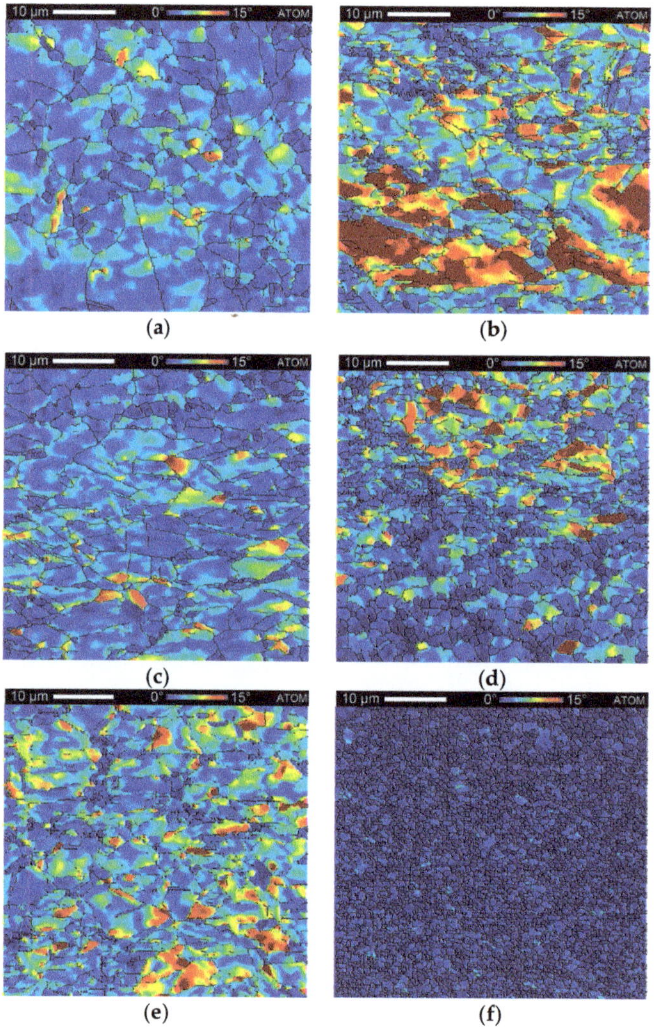

Figure 8. Internal grain misorientations for samples: *HPT1* axial region (**a**), *HPT1* peripheral region (**b**), *HPT2* axial region (**c**), *HPT2* peripheral region (**d**), *HPT4* axial region (**e**), and *HPT4* peripheral region (**f**).

3.3. Vickers Microhardness

Figure 9 shows the results of HV0.1 Vickers microhardness measurements performed along the cross-sectional line of HPT-processed samples, as depicted in Figure 3b. The initial hot consolidated Cu-Al_2O_3 composite sample featured the average microhardness value of 93.6 HV0.1, and the standard deviation from this value was negligible, only 3.2. As can be seen from Figure 9, the HV0.1 values increased (significantly) after processing via one, two, and especially four HPT revolutions. The average microhardness value for the *HPT1* sample was 116.9 HV0.1, and the standard deviation from the value was 9.7. The average microhardness value for the *HPT2* sample then increased to 123.2 HV0.1, and the standard deviation increased, too, to 11.2; this fact is evident also from Figure 9. Last but not least, the *HPT4* sample featured the highest observed average microhardness value of 133.0 HV0.1 with the lowest standard deviation (among the HPT-processed samples) of 4.0. Figure 9 depicts averaged values acquired for three samples from each material state, with standard error bars.

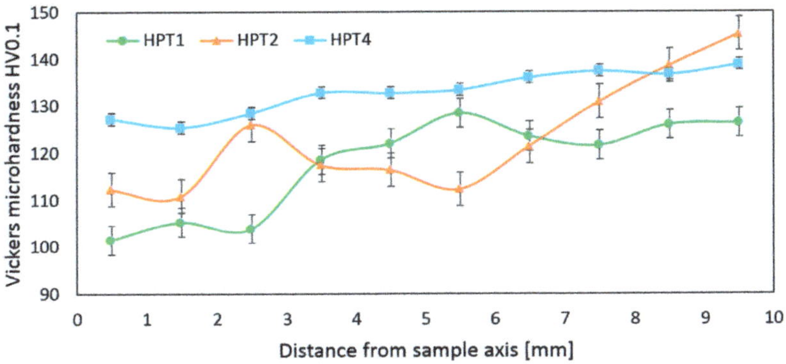

Figure 9. Vickers microhardness across HPT-processed samples.

4. Discussion

The results of the analyses showed that the hot consolidated workpiece featured relatively large grains with more or less homogeneously distributed Al_2O_3 particles (Figure 4a). The microstructure may seem comparable to a conventional commercially pure Cu. Nevertheless, the added value introduced by the powder metallurgy is in the homogenous distribution of the oxide particles within the consolidated microstructure. Due to the differences in the physical properties of copper and alumina (e.g., melting temperature, density, etc.), the conventional casting of composites consisting of these two phases features issues, which typically result in the non-homogeneous distribution of the oxides within the as-cast workpiece. Among the possible ways of the production of contemporary Cu-Al_2O_3 metal matrix composites are chemical routes [71], additive manufacturing [72], or methods of powder metallurgy; others previously documented that the most favorable way to produce (relatively) homogeneous Cu-Al_2O_3 composites is by powder metallurgy, i.e., the mixing of powders, followed by a selected/optimized processing technology [73,74].

Probably the most significant effect of HPT processing on the microstructure of the successfully consolidated workpiece was grain refinement, which was significant for all the samples. Samples subjected to HPT processing generally feature the minimum imposed strain along their axes, and the maximum imposed strain at their very peripheries. Considering the axial regions of the *HPT1*, *HPT2*, and *HPT4* samples, the grain sizes were more or less comparable, and increasing the number of revolutions had only a negligible effect on decreasing the grain size. In other words, when comparing the microstructures in the axial regions of the *HPT1* and *HPT4* samples, the decrease in the arithmetic mean grain size with an increasing imposed strain was 36%. Nevertheless, the peripheral regions of the samples were affected by the increasing imposed strain substantially, as the decrease in the arithmetic mean grain size, when compared to the microstructures in the peripheral regions of the *HPT1* and *HPT4* samples, was by 64%. Considering the high imposed shear strain, the grains within the microstructure of the Cu-Al_2O_3 composite after HPT processing were refined significantly less than typical for pure Cu processed by HPT (see, e.g., [64,75]). This phenomenon can be attributed to the effect of the Al_2O_3 particles. The high shear strain imposed during the HPT processing generally promotes the (repeating) development of the chain of the structure-forming phenomena of grain fragmentation, dislocation substructure development, and the nucleation of new (refined) grains. However, as the herein introduced composite featured the strengthening Al_2O_3 particles, the movement of dislocations and the transformation of grain boundaries were (significantly) hindered, as a result of which the structure (re)forming processes were aggravated [76].

The microstructure development went hand in hand with the development of internal misorientations pointing to substructure development, which typically goes hand in hand with a certain presence of residual stress [68,77]. The peripheral region of the *HPT1* sample evidently featured high internal misorientations, which points to a highly developed substructure; the highest misorientations were observed in the relatively larger grains, which also featured the presence of LAGBs (see Figures 5b and 8b). Increasing the imposed strain—the *HPT2* sample—caused fragmentation of the largest grains, but also imparted further development of the substructure, leading to further grain refinement (see Figures 5d and 8d). Finally, after four HPT revolutions, the imposed strain (i.e., energy) was sufficient for the grains to fragment and recrystallize to form a more or less relaxed microstructure consisting of equiaxed uniform UF grains (see Figure 5f). The microstructure in the peripheral region of the *HPT4* sample thus featured a negligible presence of internal misorientations (see Figure 8f). On the other hand, the micro and substructure development in the axial regions of the HPT-processed samples was hindered as the shear strain was not imposed with such an intensity in this region. Therefore, the *HPT1* sample featured a relatively low presence of internal misorientations, and the presence of misorientations in this region increased with an increasing number of HPT revolutions. Note that the arithmetic mean grain size in the axial region of the *HPT4* sample was comparable to that observed in the peripheral region of the *HPT2* sample, and so was the presence of internal misorientations. Such similarity between these two samples could also be seen in the textures. The PF images in Figure 7d,f showed that the grains within the peripheral region of the *HPT2* sample featured the tendency to align along the *A fiber* and *C* orientations, which is comparable to the tendencies of the grains in the axial region of the *HPT4* sample.

Last but not least, the observed microstructure development went hand in hand with the development of the mechanical properties, i.e., Vickers microhardness. When compared to the initial hot consolidated Cu-Al_2O_3 sample, the average HV0.1 microhardness value increased by ~25% after processing by a single HPT revolution. However, after processing by two revolutions, the average microhardness increased by 32% and then increased even by 43% after processing by four revolutions. The achieved values were again higher than previously achieved for pure Cu subjected to comparable processing [64].

5. Conclusions

The presented study investigated the effects of severe plastic deformation, performed by room-temperature high pressure torsion (HPT) on a directly consolidated Cu-Al_2O_3 composite. The preparation method consisting of the mixing of powders combined with direct consolidation via rotary swaging can be considered suitable for the preparation of Cu-Al_2O_3 composites as the consolidated microstructure featured the homogeneous distribution of Al_2O_3 particles with no evident presence of pores and uniform microhardness distribution. Subsequent HPT processing then revealed the effects of a (gradually) imposed shear strain on the (gradual) development of the consolidated microstructure, and its effects were primarily seen in the peripheral regions of the samples. One HPT revolution fragmented the grains and introduced substantial substructure development, as documented by a frequent presence of high internal grain misorientations. The sample processed by two HPT revolutions featured the most homogeneous distribution of both grain size and internal misorientations. However, after four HPT revolutions, the composite sample featured ultra-fine grains (mean grain size of 0.95 μm), the negligible presence of internal grain misorientations (pointing to the occurrence of substructure and residual stress), and an average Vickers microhardness of more than 130 HV0.1 (highly homogeneous distribution across the cross-section).

Author Contributions: Conceptualization, methodology, data curation, and writing—original draft preparation, L.K.; investigation, data curation, and writing—review and editing, J.W.; resources, investigation, project administration, and funding acquisition, R.K. All authors have read and agreed to the published version of the manuscript.

Funding: The authors acknowledge the support of project no. 22-11949S by the Czech Science Foundation.

Data Availability Statement: The raw data supporting the conclusions of this article will be made available by the authors on request.

Acknowledgments: We acknowledge the help of Jaromír Kopeček (Institute of Physics, CAS, CZ) with microstructure acquisition.

Conflicts of Interest: The authors declare no conflicts of interest. The funders had no role in the design of this study, in the collection, analyses, or the interpretation of data, in the writing of the manuscript, or in the decision to publish the results.

References

1. Russell, A.; Lee, K.L. *Structure-Property Relations in Nonferrous Metals*, 1st ed.; John Wiley & Sons, Inc.: Hoboken, NJ, USA, 2005; ISBN 978-0-471-64952-6.
2. Abyzov, A.M.; Kidalov, S.V.; Shakhov, F.M. High Thermal Conductivity Composite of Diamond Particles with Tungsten Coating in a Copper Matrix for Heat Sink Application. *Appl. Therm. Eng.* **2012**, *48*, 72–80. [CrossRef]
3. Champion, Y.; Couzine, J.P.; Tusseau Nenez, S.; Bréchet, Y.; Islamgaliev, R.K.; Valiev, R. High Strength and Electrical Conductivity of UFG Copper Alloys. *Mater. Sci. Forum* **2010**, *667–669*, 755–759. [CrossRef]
4. Solouki, H.; Jamaati, R.; Jamshidi Aval, H. High-Temperature Annealing Behavior of Cold-Rolled Electrolytic Tough-Pitch Copper. *Heliyon* **2024**, *10*, e33276. [CrossRef]
5. Strzępek, P.; Mamala, A.; Zasadzińska, M.; Kiesiewicz, G.; Knych, T.A. Shape Analysis of the Elastic Deformation Region throughout the Axi-Symmetric Wire Drawing Process of ETP Grade Copper. *Materials* **2021**, *14*, 4713. [CrossRef]
6. Tian, W.; Zhang, F.; Han, S.; Chen, X.; Gao, P.; Zheng, K. Analysis of Microstructure and Properties in Cold Rotary Swaged Copper-Clad Magnesium Wires. *Metals* **2023**, *13*, 467. [CrossRef]
7. Geng, Y.; Ban, Y.; Wang, B.; Li, X.; Song, K.; Zhang, Y.; Jia, Y.; Tian, B.; Liu, Y.; Volinsky, A.A. A Review of Microstructure and Texture Evolution with Nanoscale Precipitates for Copper Alloys. *J. Mater. Res. Technol.* **2020**, *9*, 11918–11934. [CrossRef]
8. Liu, B.; Sui, D.S.; Guo, Z.M.; Wang, Y.L.; Yang, X.L.; Hu, G.H.; Zhang, J.X.; Luo, C.Z. Microstructural Characteristics and Mechanical Properties of Bronze/Steel Bimetallic Laminated Composite Prepared by Protective Atmosphere Casting Process. *China Foundry* **2021**, *18*, 497–504. [CrossRef]
9. Banno, N.; Takeuchi, T. Enhancement of Electrical Conductivity of Copper/Carbon-Nanotube Composite Wire. *J. Japan Inst. Met.* **2009**, *73*, 651–658. [CrossRef]
10. Kocich, R.; Kunčická, L. Optimizing Structure and Properties of Al/Cu Laminated Conductors via Severe Shear Strain. *J. Alloys Compd.* **2023**, *953*, 170124. [CrossRef]
11. Kunčická, L.; Kocich, R. Effect of Activated Slip Systems on Dynamic Recrystallization during Rotary Swaging of Electro-Conductive Al-Cu Composites. *Mater. Lett.* **2022**, *321*, 10–13. [CrossRef]
12. Carpenter, J.S.; Miller, C.; Savage, D.J.; Coughlin, D.R.; Tegtmeier, E.L.; Winter, W.P. The Impact of Rolling at Temperature on Conductivity and Texture in Nanolamellar Cu/Nb Bimetallic Composites. *Metall. Mater. Trans. A* **2022**, *53*, 2208–2213. [CrossRef]
13. Saito, T.; Yu, H.; Gao, Z.; Inoue, K.; Kondo, S.; Kasada, R.; Nagai, Y.; Oba, Y.; Hiroi, K. Multiscale Observation of Nano-Oxide Particles in Oxide-Dispersion-Strengthened Cu Alloys. *J. Alloys Compd.* **2024**, *983*, 173852. [CrossRef]
14. Chen, Y.; Wang, Z.; Jin, G.; Zhu, S.; Liu, W. Achieving High Strengthening Efficiency and Good Balance between Strength and Ductility/Electrical Conductivity in the Laminated RGO/Copper Composites. *ACS Appl. Electron. Mater.* **2024**, *6*, 5903–5913. [CrossRef]
15. Wang, L.; Tong, Y.G.; Zhou, C.; Li, Y.J.; Liu, J.; Cai, Z.H.; Wang, H.D.; Hu, Y.L. Self-Lubricating Ti_3SiC_2 Strengthening Cu Composites with Electrical Conductivity and Wear Resistance Trade Off. *Tribol. Int.* **2023**, *187*, 108724. [CrossRef]
16. Azmi, K.; Al Bakri Abdullah, M.M.; Mohd Salleh, M.A.A. Fabrication of Cu-SiCp Composites via the Electroless Copper Coating Process for the Electronic Packaging Applications. *Adv. Mater. Res.* **2013**, *795*, 272–275. [CrossRef]

17. Wang, J.; Zuo, T.; Ru, Y.; Xue, J.; Wu, Y.; Liu, Y.; Gao, Z.; Xiao, L. Carbide Particle Strengthened Cu/Graphene Composites in-Situ Fabricated by Alloying. *Mater. Lett.* **2024**, *362*, 136164. [CrossRef]
18. Eremina, M.A.; Lomaeva, S.F.; Paranin, S.N.; Demakov, S.L.; Elsukov, E.P. Effect of Compaction Method on the Structure and Properties of Bulk Cu+Cr_3C_2 Composites. *Phys. Met. Metallogr.* **2016**, *117*, 510–517. [CrossRef]
19. Gorbatyuk, S.; Pashkov, A.; Chichenev, N. Improved Copper-Molybdenum Composite Material Production Technology. *Mater. Today Proc.* **2019**, *11*, 31–35. [CrossRef]
20. Zykova, A.; Panfilov, A.; Chumaevskii, A.; Vorontsov, A.; Moskvichev, E.; Nikonov, S.; Gurianov, D.; Savchenko, N.; Kolubaev, E.; Tarasov, S. In-Situ Dispersion Hardened Aluminum Bronze/Steel Composites Prepared Using a Double Wire Electron Beam Additive Manufacturing. *Prog. Addit. Manuf.* **2023**, *8*, 1067–1082. [CrossRef]
21. Li, D.; Xue, J.; Zuo, T.; Gao, Z.; Xiao, L.; Han, L.; Li, S.; Yang, Y. Copper/Functionalized-Carbon Nanotubes Composite Films with Ultrahigh Electrical Conductivity Prepared by Pulse Reverse Electrodeposition. *J. Mater. Sci. Mater. Electron.* **2020**, *31*, 14184–14191. [CrossRef]
22. Rogachev, A.S.; Kuskov, K.V.; Moskovskikh, D.O.; Usenko, A.A.; Orlov, A.O.; Shkodich, N.F.; Alymov, M.I.; Mukasyan, A.S. Effect of Mechanical Activation on Thermal and Electrical Conductivity of Sintered Cu, Cr, and Cu/Cr Composite Powders. *Dokl. Phys.* **2016**, *61*, 257–260. [CrossRef]
23. Schilling, W.F.; Grant, N.J. Oxide Dispersed Copper Alloys by Surface Oxidation. *Metall. Trans.* **1970**, *1*, 2205–2210. [CrossRef]
24. Khaloobagheri, M.; Janipour, B.; Askari, N.; Abad, E. Characterisation of Powder Metallurgy Cu-ZrO_2 Composites. *Adv. Prod. Eng. Manag.* **2013**, *8*, 242–248. [CrossRef]
25. Rajkovic, V.; Bozic, D.; Jovanovic, M.T. Properties of Copper Matrix Reinforced with Nano- and Micro-Sized Al_2O_3 Particles. *J. Alloys Compd.* **2008**, *459*, 177–184. [CrossRef]
26. Mukhtar, A.; Zhang, D.L.; Kong, C.; Munroe, P. Consolidation of Ultrafine-Grained Cu Powder and Nanostructured Cu–(2.5–10) Vol% Al_2O_3 Composite Powders by Powder Compact Forging. *J. Mater. Sci.* **2010**, *45*, 4594–4605. [CrossRef]
27. Liu, M.; Li, Y.; Luo, Y.; Bai, Y.; Wu, Y.; Li, Z. Uniformly Dispersed Nano-Al_2O_3 Particles Reinforced Copper Matrix by Chemical Coprecipitation Method. *Ferroelectrics* **2019**, *546*, 129–136. [CrossRef]
28. Wu, S.; Zhou, L.; Cai, X.; Shen, Q.; Cheng, Y.; Yang, C. Fabrication of Multiscale-Alumina Particles Reinforced Copper Matrix Composites with High-Energy Ball Milling and Hot Pressing. *Integr. Ferroelectr.* **2022**, *226*, 113–124. [CrossRef]
29. Lv, L.; Jiang, X.; Xiao, X.; Sun, H.; Shao, Z.; Luo, Z. Study on Corrosion Resistance of Copper Matrix Composites Reinforced by Al_2O_3 Whiskers. *Mater. Res. Express* **2020**, *7*, 026534. [CrossRef]
30. Tian, B.H.; Liu, P.; Han, S.L.; Song, K.X.; Dong, Q.M.; Cao, X.; Niu, L. Mechanical Behavior of Nano-Al_2O_3 Dispersion-Hardened Copper-Base Composite at Elevated Temperature. *Key Eng. Mater.* **2004**, *274–276*, 367–372. [CrossRef]
31. Lin, H.; Guo, X.; Song, K.; Feng, J.; Li, S.; Zhang, X. Synergistic Strengthening Mechanism of Copper Matrix Composite Reinforced with Nano-Al_2O_3 Particles and Micro-SiC Whiskers. *Nanotechnol. Rev.* **2021**, *10*, 62–72. [CrossRef]
32. Kocich, R.; Kunčická, L. Crossing the Limits of Electric Conductivity of Copper by Inducing Nanotwinning via Extreme Plastic Deformation at Cryogenic Conditions. *Mater. Charact.* **2024**, *207*, 113513. [CrossRef]
33. Zhou, M.; Geng, Y.; Zhang, Y.; Ban, Y.; Li, X.; Jia, Y.; Liang, S.; Tian, B.; Liu, Y.; Volinsky, A.A. Enhanced Mechanical Properties and High Electrical Conductivity of Copper Alloy via Dual-Nanoprecipitation. *Mater. Charact.* **2023**, *195*, 112494. [CrossRef]
34. Kunčická, L.; Kocich, R. Comprehensive Characterisation of a Newly Developed Mg-Dy-Al-Zn-Zr Alloy Structure. *Metals* **2018**, *8*, 73. [CrossRef]
35. Krinitcyn, M.; Toropkov, N. Structure, Phase Composition, and Properties of Ti_3AlC_2—Nano-Cu Powder Composites. *Coatings* **2022**, *12*, 1928. [CrossRef]
36. Druzhinin, A.V.; Ariosa, D.; Siol, S.; Ott, N.; Straumal, B.B.; Janczak-Rusch, J.; Jeurgens, L.P.H.; Cancellieri, C. Effect of the Individual Layer Thickness on the Transformation of Cu/W Nano-Multilayers into Nanocomposites. *Materialia* **2019**, *7*, 100400. [CrossRef]
37. Canelo-Yubero, D.; Kocich, R.; Hervoches, C.; Strunz, P.; Kunčická, L.; Krátká, L. Neutron Diffraction Study of Residual Stresses in a W–Ni–Co Heavy Alloy Processed by Rotary Swaging at Room and High Temperatures. *Met. Mater. Int.* **2022**, *28*, 919–930. [CrossRef]
38. Wang, Z.; Chen, J.; Kocich, R.; Tardif, S.; Dolbnya, I.P.; Kunčická, L.; Micha, J.-S.; Liogas, K.; Magdysyuk, O.V.; Szurman, I.; et al. Grain Structure Engineering of NiTi Shape Memory Alloys by Intensive Plastic Deformation. *ACS Appl. Mater. Interfaces* **2022**, *14*, 31396–31410. [CrossRef] [PubMed]
39. Martynenko, N.S.; Bochvar, N.R.; Rybalchenko, O.V.; Prosvirnin, D.V.; Rybalchenko, G.V.; Kolmakov, A.G.; Morozov, M.M.; Yusupov, V.S.; Dobatkin, S.V. Increase in the Strength and Electrical Conductivity of a Cu–0.8Hf Alloy after Rotary Swaging and Subsequent Aging. *Russ. Metall.* **2023**, *2023*, 466–474. [CrossRef]
40. Kocich, R.; Kunčická, L. Development of Structure and Properties in Bimetallic Al/Cu Sandwich Composite during Cumulative Severe Plastic Deformation. *J. Sandw. Struct. Mater.* **2021**, *23*, 4252–4275. [CrossRef]
41. Kunčická, L.; Kocich, R.; Ryukhtin, V.; Cullen, J.C.T.; Lavery, N.P. Study of Structure of Naturally Aged Aluminium after Twist Channel Angular Pressing. *Mater. Charact.* **2019**, *152*, 94–100. [CrossRef]
42. Kunčická, L.; Kocich, R.; Drápala, J.; Andreyachshenko, V.A. FEM Simulations and Comparison of the Ecap and ECAP-PBP Influence on Ti6Al4V Alloy's Deformation Behaviour. In Proceedings of the METAL 2013—22nd International Conference on Metallurgy and Materials, Conference Proceedings, Brno, Czech Republic, 15–17 May 2013; pp. 391–396.

43. Zhao, C.; Niu, R.; Xin, Y.; Brown, D.; McGuire, D.; Wang, E.; Han, K. Improvement of Properties in Cu–Ag Composites by Doping Induced Microstructural Refinement. *Mater. Sci. Eng. A* **2021**, *799*, 140091. [CrossRef]
44. Islamgaliev, R.K.; Sitdikov, V.D.; Nesterov, K.M.; Ganeev, A.V.; Bochkova, E.V. Microstructure and Crystallographic Texture of Titanium Subjected to Combined Severe Plastic Deformation Processing. *Russ. Metall.* **2014**, *2014*, 234–240. [CrossRef]
45. Kunčická, L.; Lowe, T.C.; Davis, C.F.; Kocich, R.; Pohludka, M. Synthesis of an Al/Al_2O_3 composite by Severe Plastic Deformation. *Mater. Sci. Eng. A* **2015**, *646*, 234–241. [CrossRef]
46. Verlinden, B.; Driver, J.; Samajdar, I.; Doherty, R.D. *Thermo-Mechanical Processing of Metallic Materials*; Elsevier: Amsterdam, The Netherlands, 2007; ISBN 9780080444970.
47. Wang, Z.; Zhang, Y.; Liogas, K.; Chen, J.; Vaughan, G.B.M.; Kocich, R.; Kunčická, L.; Uzun, F.; You, Z.; Korsunsky, A.M. In Situ Synchrotron X-Ray Diffraction Analysis of Two-Way Shape Memory Effect in Nitinol. *Mater. Sci. Eng. A* **2023**, *878*, 145226. [CrossRef]
48. Murav'ev, V.V.; Murav'ev, O.V.; Len'kov, S.V.; Vladykin, A.L.; Belosludtsev, K.Y. Acoustic Structuroscopy of Copper Samples after Equal-Channel Angular Pressing. *Russ. Metall.* **2023**, *2023*, 1542–1548. [CrossRef]
49. Zhu, S.; Jivkov, A.P.; Borodin, E.; Bodyakova, A. Triple Junction Disclinations in Severely Deformed Cu–0.4%Mg Alloys. *Acta Mater.* **2024**, *264*, 119600. [CrossRef]
50. Bodyakova, A.I.; Mishnev, R.V.; Kaibyshev, R.O. Microstructure and Properties of Cu–Cr–Zr Alloys After Plastic Deformation and Aging. *Met. Sci. Heat Treat.* **2024**, *66*, 343–354. [CrossRef]
51. Volokitina, I.E. Evolution of the Microstructure and Mechanical Properties of Copper under ECAP with Intense Cooling. *Met. Sci. Heat Treat.* **2020**, *62*, 253–258. [CrossRef]
52. Hlaváč, L.M.; Kocich, R.; Gembalová, L.; Jonšta, P.; Hlaváčová, I.M. AWJ Cutting of Copper Processed by ECAP. *Int. J. Adv. Manuf. Technol.* **2016**, *86*, 885–894. [CrossRef]
53. Guo, T.; Wang, J.; Wu, Y.; Tai, X.; Jia, Z.; Ding, Y. Fabricate of High-Strength and High-Conductivity Cu–Cr–Si Alloys through ECAP-Bc and Aging Heat Treatment. *Materials* **2020**, *13*, 1603. [CrossRef]
54. Huang, R.; Zhu, D.; Liao, X.; Yan, Q. Effect of ECAP Process and Subsequent Annealing on Microstructure and Properties of Cu-0.25Se-0.25Te Alloy. *J. Electron. Mater.* **2020**, *49*, 2617–2624. [CrossRef]
55. Dalan, F.C.; de Lima Andreani, G.F.; Travessa, D.N.; Faizova, S.; Faizov, I.A.; Cardoso, K.R. Effect of ECAP Processing on Hardness, Electrical Conductivity, and Precipitation Kinetics of the Cu-0.81Cr-0.07Zr Alloy. *J. Electron. Mater.* **2021**, *50*, 6171–6182. [CrossRef]
56. Zhilyaev, A.P.; Langdon, T.G. Using High-Pressure Torsion for Metal Processing: Fundamentals and Applications. *Prog. Mater. Sci.* **2008**, *53*, 893–979. [CrossRef]
57. Kulagin, R.; Beygelzimer, Y.; Ivanisenko, Y.; Mazilkin, A.; Hahn, H. Modelling of High Pressure Torsion Using FEM. *Procedia Eng.* **2017**, *207*, 1445–1450. [CrossRef]
58. Sevastyanov, G.M. Analytical Solution for High-Pressure Torsion in the Framework of Geometrically Nonlinear Non-Associative Plasticity. *Int. J. Solids Struct.* **2020**, *206*, 383–395. [CrossRef]
59. Valiev, R.Z.; Korznikov, A.V.; Mulyukov, R.R. Structure and Properties of Ultrafine-Grained Materials Produced by Severe Plastic Deformation. *Mater. Sci. Eng. A* **1993**, *168*, 141–148. [CrossRef]
60. Murashkin, M.Y.; Sabirov, I.; Sauvage, X.; Valiev, R.Z.; Sabirov, I.; Sauvage, X.; Valiev, R.Z.; Sabirov, I.; Sauvage, X.; Valiev, R.Z. Nanostructured Al and Cu Alloys with Superior Strength and Electrical Conductivity. *J. Mater. Sci.* **2016**, *51*, 33–49. [CrossRef]
61. An, X.H.; Lin, Q.Y.; Wu, S.D.; Zhang, Z.F.; Figueiredo, R.B.; Gao, N.; Langdon, T.G. The Influence of Stacking Fault Energy on the Mechanical Properties of Nanostructured Cu and Cu–Al Alloys Processed by High-Pressure Torsion. *Scr. Mater.* **2011**, *64*, 954–957. [CrossRef]
62. Kosinova, A.; Straumal, B.; Kilmametov, A.; Straumal, P.; Bulatov, M.; Rabkin, E. Faceting of Twin Grain Boundaries in High-Purity Copper Subjected to High Pressure Torsion. *Adv. Eng. Mater.* **2020**, *22*, 1900589. [CrossRef]
63. Huang, Y.; Sabbaghianrad, S.; Almazrouee, A.I.; Al-Fadhalah, K.J.; Alhajeri, S.N.; Langdon, T.G. The Significance of Self-Annealing at Room Temperature in High Purity Copper Processed by High-Pressure Torsion. *Mater. Sci. Eng. A* **2016**, *656*, 55–66. [CrossRef]
64. Král, P.; Staněk, J.; Kunčická, L.; Seitl, F.; Petrich, L.; Schmidt, V.; Beneš, V.; Sklenička, V. Microstructure Changes in HPT-Processed Copper Occurring at Room Temperature. *Mater. Charact.* **2019**, *151*, 602–611. [CrossRef]
65. Qi, Y.; Kosinova, A.; Kilmametov, A.R.; Straumal, B.B.; Rabkin, E. Plastic Flow and Microstructural Instabilities during High-Pressure Torsion of Cu/ZnO Composites. *Mater. Charact.* **2018**, *145*, 389–401. [CrossRef]
66. Beausir, B.; Fundenberger, J.J. Analysis Tools for Electron and X-Ray Diffraction, ATEX—Software. 2017. Available online: www.atex-software.eu (accessed on 9 August 2024).
67. Zhang, Y.; Yu, T.; Xu, R.; Thorborg, J.; Liu, W.; Tischler, J.; Godfrey, A.; Juul Jensen, D. Local Residual Stresses and Microstructure within Recrystallizing Grains in Iron. *Mater. Charact.* **2022**, *191*, 112113. [CrossRef]
68. Sanchez Chavez, G.; Farid Estefen, S.; Gurova, T.; Leontiev, A.; Silva Gomes, L.; Bottega Peripolli, S. Redistribution of Grain Boundary Misorientation and Residual Stresses of Thermomechanically Simulated Welding in an Intercritically Reheated Coarse Grained Heat Affected Zone. *Metals* **2021**, *11*, 1850. [CrossRef]
69. Beyerlein, I.J.; Tóth, L.S. Texture Evolution in Equal-Channel Angular Extrusion. *Prog. Mater. Sci.* **2009**, *54*, 427–510. [CrossRef]
70. Toth, L.S.; Gilorminis, P.; Jonas, J.J. Effect of Rate Sensitivity on the Stability of Torsion Textures. *Acta Metall.* **1988**, *36*, 3077–3091. [CrossRef]

71. Jena, P.; Brocchi, E.; Motta, M. In-Situ Formation of Cu–Al2O3 Nano-Scale Composites by Chemical Routes and Studies on Their Microstructures. *Mater. Sci. Eng. A* **2001**, *313*, 180–186. [CrossRef]
72. Cui, L.; Yang, W.; Zhang, L.; Hao, J.; Zou, J.; Zhang, W.; Jia, L.; Hao, E.; Zhu, J.; Liu, G. Preparation of High-Tensile-Ductility and High-Conductivity Alumina Dispersion-Strengthened Copper via a Cold Spray Additive Manufacturing-Friction Stir Processing Composite Process. *J. Alloys Compd.* **2024**, *1008*, 176467. [CrossRef]
73. Goswami, R.; Qadri, S.B.; Ryou, H. Correlation of Fine Scale Microstructure and Mechanical Properties of Copper-Alumina Nanocomposites. In *Metal-Matrix Composites: Advances in Analysis, Measurement, and Observations*; Springer International Publishing: Berlin/Heidelberg, Germany, 2021; pp. 133–142.
74. Oh, S.-T.; Sekino, T.; Niihara, K. Effect of Particle Size Distribution and Mixing Homogeneity on Microstructure and Strength of Alumina/Copper Composites. *Nanostructured Mater.* **1998**, *10*, 327–332. [CrossRef]
75. Jiang, H.; Zhu, Y.T.; Butt, D.P.; Alexandrov, I.V.; Lowe, T.C. Microstructural Evolution, Microhardness and Thermal Stability of HPT-Processed Cu. *Mater. Sci. Eng. A* **2000**, *290*, 128–138. [CrossRef]
76. Rollett, A.; Humphreys, F.; Rohrer, G.S.; Hatherly, M. *Recrystallization and Related Annealing Phenomena*, 2nd ed.; Elsevier Ltd.: Amsterdam, The Netherlands, 2004; ISBN 9780080441641.
77. Bandyopadhyay, R.; Stopka, K.S.; Sangid, M.D. Initializing Intragranular Residual Stresses within Statistically Equivalent Microstructures for Crystal Plasticity Simulations. *J. Mech. Phys. Solids* **2024**, *184*, 105529. [CrossRef]

metals

MDPI

Article

Microstructure, Mechanical, and Tribological Behaviour of Spark Plasma Sintered TiN, TiC, TiCN, TaN, and NbN Ceramic Coatings on Titanium Substrate

Ganesh Walunj [1,2,†], Amit Choudhari [1,†], Satyavan Digole [1,†], Anthony Bearden [1], Omar Kolt [1], Praful Bari [3] and Tushar Borkar [1,*]

1 Department of Mechanical Engineering, Cleveland State University, Cleveland, OH 44115, USA; a.choudhari@vikes.csuohio.edu (A.C.); s.digole@vikes.csuohio.edu (S.D.)

2 Department of Engineering Technology, SUNY Buffalo State University, Buffalo, NY 14222, USA

3 ZF Group, 800 Heath Street, Lafayette, IN 47904, USA; praful.bari@zf.com

* Correspondence: t.borkar@csuohio.edu

† These authors contributed equally to this work.

Abstract: Titanium (Ti) is widely used in structural, maritime, aerospace, and biomedical applications because of its outstanding strength-to-weight ratio, superior corrosion resistance, and excellent biocompatibility. However, the lower surface hardness and inferior wear resistance of the Ti and Ti alloys limit their industrial applications. Coating Ti surfaces can initiate new possibilities to give unique characteristics with significant improvement in the Ti component's functionality. The current research designed and synthesized titanium nitride (TiN), titanium carbide (TiC), titanium carbonitride (TiCN), tantalum nitride (TaN), and niobium nitride (NbN) ceramic coating layers (400 μm) over a Ti substrate using a spark plasma sintering process (SPS). The coatings on the Ti substrate were compact and consolidated at an SPS temperature of 1500 °C, pressure of 50 MPa, and 5 min of holding time in a controlled argon atmosphere. Microstructure investigation revealed a defect-less coating-substrate interface formation with a transition/diffusion zone ranging from 10 μm to 20 μm. Among all of the ceramic coatings, titanium carbide showed the highest improvement in surface hardness, equal to 1817 ± 25 HV, and the lowest coefficient of friction, equal to 0.28 for NbN.

Keywords: Ti substrate; ceramic coating; spark plasma sintering (SPS); microstructure; hardness; wear resistance

1. Introduction

The fascinating properties such as high strength-to-weight ratio, flame retardation, damage tolerance, and corrosion resistance make titanium (Ti) and its alloys most suitable for aerospace, marine, automotive, and chemical applications [1]. In the biomedical industry, the excellent biocompatibility of titanium makes it the first choice for orthopaedic implant applications [2]. Despite such valuable characteristics, Ti has inherent shortcomings like lower hardness. The hardness varies from 150 to 200 HV for pure Ti processed via different processes such as casting, powder metallurgy, additive manufacturing, etc. [3]. Different alloying elements have been added to improve the hardness, but the achieved hardness is still lower than 400 HV (e.g., as-cast Ti-Ni: 190 HV; as-cast Ti-25Pd-5Cr: 260 HV; as-cast Ti-20Cr-0.2Si: 320 HV and Ti-6Al-4V: 290–375 HV) [3]. Along with hardness, Ti and its alloy possess inferior wear resistance [4,5]. In biomedical applications, the implants are subject to intensive wear, which leads to the formation of wear debris responsible for toxic effects (cancer and Alzheimer's disease) in the human body [6]. The inferior wear resistance of titanium at high temperatures leads to accelerated surface degradation, reduced component lifespan, increased friction, and potential failure in high-performance applications [7]. To overcome the above shortcomings, various surface modification techniques have been employed over the past few decades [8]. For example, applying a ceramic coating on bare metal or alloy substrate significantly affects the hardness and wear resistance of the substrate [9]. Additionally, ceramics possess very low thermal conductivity, so they act as a thermal barrier and prevent the failure of the Ti component by reducing the heat effect [10,11]. In implant applications, the deposition of a biocompatible ceramic coating enhances antibacterial activity and cell adhesion along with improved hardness and superior wear resistance, sustaining the implant for a longer duration [12]. Various conventional surface modification techniques, such as spray coating, sol–gel coating, chemical vapor deposition (CVD), physical vapor deposition (PVD), and electrophoretic deposition (EPD), etc., have been discussed in the previous literature to improve the performance of Ti substrates [3]. Additionally, additive manufacturing and powder metallurgy have lately been published as surface modification techniques for Ti and its alloys [13,14].

The in situ formation of a titanium carbide (TiC) coating on a titanium substrate was generated through a CVD process, utilizing a gas mixture of H_2, CH_4, and $TiCl_4$ [15]. The resulting TiC layer exhibited a thickness of approximately 6 μm. The formation of a TiC coating significantly enhanced the hardness of the titanium substrate to 2000 HV, which is 15 times the hardness of the substrate. This substantial improvement in hardness is attributed to the successful formation of the hard TiC ceramic phase on the titanium surface. A laser cladding (LC) process was employed to fabricate an in situ TiC/Ti composite coating on a pure titanium substrate using 20% carbon nanotubes and titanium powder [16]. The final coating predominantly consists of α-Ti and dendritic TiC particles. The results show that the coating exhibits significant improvements in mechanical properties, with a hardness of 1125 HV and exceptional wear resistance, demonstrating the potential of LC for enhancing the surface performance of titanium-based materials.

Ti substrates were coated with titanium nitride (TiN) using PVD techniques [17]. The results demonstrated that the 3.3 μm thick TiN coating significantly improved the substrate's hardness, wear, and corrosion resistance. Specifically, the TiN coating enhanced the hardness by a factor of 7.5, increased corrosion resistance by a factor of 8, and improved wear resistance by a factor of 4 compared to the uncoated titanium substrate. Similar to TiC and TiN, other ceramics such as titanium carbonitride (TiCN), tantalum nitride (TaN), and niobium nitride (NbN) have also shown the potential to improve the mechanical characteristics of the substrate through conventional coating techniques [18–22].

While various coating technologies can improve the substrate's hardness and wear resistance, several disadvantages often limit their application. Techniques such as CVD and PVD are primarily effective for producing very thin coatings, which may not be suitable for heavy load applications since, under high load conditions, the relatively low toughness of these coatings can lead to premature failure [3]. The high laser energy density in additive manufacturing techniques such as laser cladding can lead to the melting of both the coating and the substrate, resulting in the formation of a heterogeneous microstructure [23]. This intense localized heating induces uneven cooling rates, which may promote phase segregation and microstructural inhomogeneity within the deposited material. The resulting microstructural variations can significantly affect the mechanical properties and performance of the LC components.

Moreover, the complexity of the process and the associated high operational costs significantly restrict their broader industrial implementation. Recently, attention has been diverted toward the spark plasm sintering (SPS) process to fabricate the coating over the substrate [24]. The in situ synthesis of $TiC/SiC/Ti_3SiC_2$ composite coatings of approximately 400 μm thickness was achieved on a Ti substrate using a spark plasma sintering process [25]. The results showed a three times increase in hardness and 100 times less wear loss for a coated Ti substrate than a pure Ti substrate.

In the SPS process, pulsed direct current and uniaxial pressure is applied to compact and consolidate powder materials [26]. The heat generated through Joule heating at the points of particle contact induces solid-state sintering, enabling the material to undergo densification within a short time frame. SPS enables the fabrication of thick coating layers with precise control over temperature, pressure, and processing time, ensuring controlled interactions between the coating and substrate while preserving the mechanical properties of the substrate, making it suitable for load-bearing applications. So, SPS was considered as the processing technique for this research. The role of interlayer adhesion in mechanical properties was studied by Liparoti et al. [27] for PLA material using fused deposition modelling. Future researchers can build upon this work to investigate the effect of the interference zone width on the mechanical properties of composite coatings on Ti and other substrates.

To overcome the limitations associated with the inherent properties of pure Ti surfaces, various ceramic coatings, including TiN, TiC, TiCN, TaN, and NbN, have been applied on Ti substrates using the SPS process. The influence of the SPS process and the different ceramic coatings on the microstructure, hardness, and wear resistance of the ceramic-coated Ti samples has been thoroughly investigated. This study aimed to provide a comprehensive understanding of the effects of these coatings on the performance characteristics of titanium, with a focus on enhancing surface durability and functionality.

2. Materials and Methods

To fabricate ceramic coatings over a Ti substrate, five different ceramic powders (TiN, TiC, TiCN, NbN, and TaN) were procured from Atlantic Equipment Engineers (Upper Saddle River, NJ, USA) and Sigma–Aldrich (St. Louis, MO, USA), respectively. Pure Ti discs with a diameter of 20 mm and a thickness of 5 mm were employed as substrates in this work. The Ti–ceramic sample fabrication process layout is shown in Figure 1. Each substrate disc was inserted into the graphite die, and a homogeneous layer of amorphous powder was applied to the substrate surface to generate a >400 μm thick coating. Before applying the coating, the substrate surfaces were extensively polished using 400 and 600 mesh abrasive papers, followed by ethanol cleaning to avoid contamination.

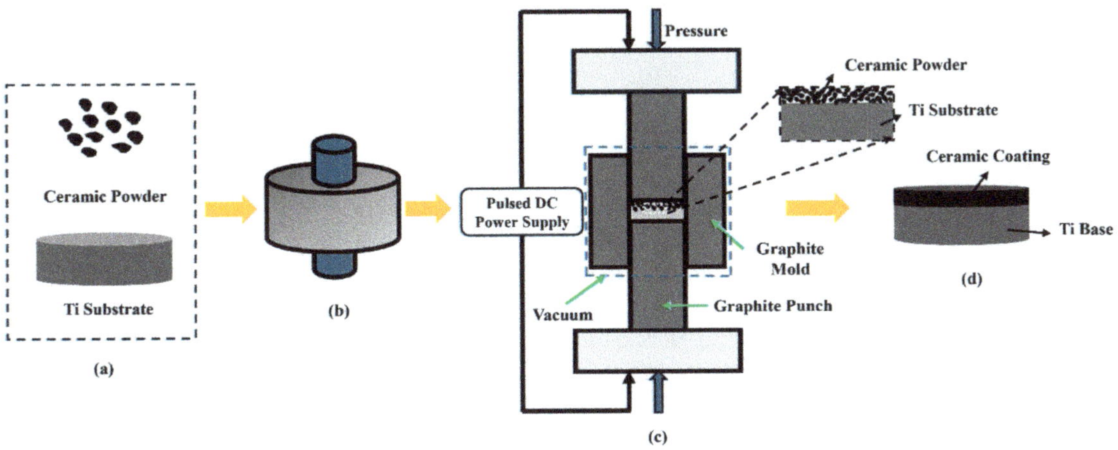

Figure 1. Ti–ceramic sample fabrication process. (**a**) Raw materials, (**b**) punch die assembly, (**c**) spark plasma sintering, and (**d**) coated Ti substrate.

The SPS technique was employed to compact and consolidate ceramic powder onto a titanium substrate. Initially, the ceramic powder was pre-compacted in a graphite die under a pressure of 5 MPa. The pre-compacted powders were then sintered using an SPS system (SPS 10-3, Thermal Technology LLC, Santa Rosa, CA, USA) at a temperature of 1500 °C, under a pressure of 65 MPa, with a dwell time of 5 min, all within a controlled argon atmosphere. The resulting sintered samples were cut in half for cross-section analysis. The samples were then mounted in graphite moulds before polishing. Following sintering and cutting, the samples were sequentially ground and polished using 240, 400, 600, 800, and 1200 grit SiC abrasive papers with a BUEHLER Grinder-Polisher (AutoMet 250, Buehler, Lake Bluff, IL, USA). The samples were further polished with colloidal silica to achieve a smooth and fine surface finish. Finally, the polished samples underwent ultrasonic cleaning in deionized water (DI), soap solution, and ethanol to remove residual silica from the surface.

The surface hardness of the coating and the coating–substrate interface were evaluated using a microhardness test conducted with a Wilson BUEHLER microhardness tester (Wilson, Chicago, IL, USA) equipped with a Vickers indenter. A test load of 5 N was applied for 10 s. To ensure result consistency, fifteen initial readings were taken on each sample. The reported hardness value of 230 HV corresponds to 230 HV0.5, where 0.5 kgf (5 N) was the applied load. To investigate the tribological performance of sample wear, tests were performed under dry conditions at room temperature using a ball-on-disc tribometer (Nanovea™, Irvine, CA, USA). A Si_3N_4 ball with a diameter of 6 mm was used as the counter body against the coating surface. The wear test was conducted with a 1 N applied load, a sliding velocity of 200 rpm, and a track diameter of 5 mm. Hardness and wear measurements in this study were performed using a statistically significant sample size ($p < 0.05$) and carefully controlled testing conditions to ensure reliability and reproducibility. The samples consisted of coatings synthesized with consistent SPS parameters and identical substrate geometries (20 mm diameter, 5 mm thickness). The reported average hardness values were derived from fifteen readings taken at various locations, with reproducibility further validated by independent testing with five readings taken at different locations. For COF calculations, four readings were taken at distinct locations on the samples. Phase identification and microstructural characterization of the Ti–ceramic samples were carried out using scanning electron microscopy (SEM) (Thermo Fisher Scientific, Hillsboro, OR, USA).

3. Results and Discussion

3.1. Microstructure Analysis

The coatings examined in this study, including TiC, TiN, Ti-TaN, Ti-CNT, and Ti-NbN, provide significant benefits when applied to titanium substrates, particularly in engineering applications that demand enhanced surface properties. While similar coatings have been widely investigated for steel substrates [28–32], titanium substrates, despite their higher cost, offer superior corrosion resistance, lower density, and excellent biocompatibility. These advantages make titanium ideal for applications where weight reduction, durability, and performance are critical, such as in biomedical devices (e.g., implants) and aerospace components (e.g., aircraft parts and engines). The results of this study highlight the feasibility of SPS as a scalable and cost-effective method for coating titanium, opening the door to its broader industrial adoption and potential for innovative applications in advanced engineering systems.

To enhance the hardness and wear performance of pure Ti surfaces, various ceramic coatings, including TiN, TiC, TiCN, TaN, and NbN, were applied to the Ti substrate through the SPS process. Figure 2 depicts a schematic representation of the coating on the substrate and the corresponding SEM image of a cross-sectional view of the processed sample along the vertical direction.

The bottom substrate consists of Ti, subjected to extensive cleaning and polishing to achieve a homogeneous surface texture. This meticulous surface preparation is crucial, as any contamination or non-uniformity in surface topography, such as voids, micropores, or overgrowth regions, can significantly compromise the mechanical interlocking between the substrate and the applied coating. Such imperfections can lead to weak adhesion, thereby increasing the susceptibility to coating detachment and failure, particularly under tribological conditions [33,34]. The Ti substrate without any significant porosity is identified in the SEM image shown in Figure 2. Furthermore, a clean and crack-less interaction line confirms the excellent bonding of the substrate and deposited ceramics. The upper part of the sample is the ceramic consolidated region, which accommodates significant porosity.

During SPS sintering of ceramic powder over a Ti base, a temperature higher than 1600 °C will melt the titanium base, and diffusion of ceramic elements will result in alterations of the substrate's mechanical characteristics, which is undesirable in several mechanical applications. Additionally, the higher temperature used in sintering will result in the diffusion of carbon from the punch and die, causing the formation of unnecessary carbide, which degrades the mechanical characteristics of the substrate and coating [26]. To avoid this, the SPS temperature during the sintering of Ti–ceramics coating samples was limited to 1500 °C. All of the ceramics have a very high melting point (>2500 °C), and sintering ceramics over a Ti plate at low temperature results in porosity, which can be seen in Figure 2.

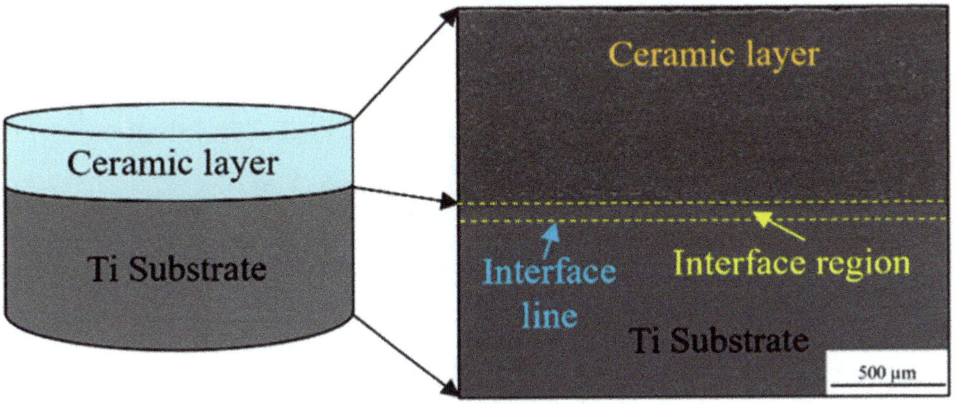

Figure 2. Schematic of ceramic coating over Ti substrate, and SEM image of a ceramic–Ti sample cross-section highlighting the ceramic layer, interface region, and substrate.

Figure 3a shows the SEM image of the vertical cross-section for TiN over the Ti substrate. The appearance of a defect-less interface line within the microstructure reflects the excellent bonding of Ti and TiN. Despite titanium nitride's higher melting point (~2930 °C) [35], at elevated temperatures, nitrogen from the TiN layer begins to diffuse into the underlying titanium substrate. Nitrogen has a relatively significant diffusion coefficient (10^{-11} m^2/sec) [36], which facilitates the formation of a transition zone approximately 20 µm in length. This zone is characterized by the diffusion of nitrogen into the titanium matrix, where nitrogen occupies interstitial sites, forming a solid solution (Ti(N)) [37]. The energy dispersive spectroscopy (SEM-EDS) mapping shown in Figure 3c–e confirms the presence of nitrogen in the transition zone, thereby supporting the formation of the Ti(N) solid solution. However, further analysis of the consolidated TiN layer reveals the presence of significant microcracking and porosity within the microstructure, indicating poor bonding and consolidation of TiN within the coating after sintering. The higher magnification SEM image shown in Figure 3b highlights the presence of voids within the TiN consolidated region, which can be attributed to insufficient sintering temperature. The presence of cracks within consolidated regions always leads to lower tribological characteristics, discussed in the succeeding section.

The Ti-TiC sample cross-section image is shown in Figure 4. The bottom portion of the image (Figure 4a) shows the uniform Ti substrate without substantial porosity, similar to the Ti-TiN sample. Further investigating the Ti and TiC ceramic coating interface, a transition zone approximately 10 µm thick is observed, notably thinner than the Ti-TiN transition region. The reduced thickness of the Ti-TiC transition zone can be attributed to the relatively low diffusivity (10^{-12} m^2/sec) of carbon in titanium [38]. As carbon diffuses into titanium, it reacts to form TiC, further restricting carbon diffusion and creating a shallower transition zone. Further examination of the SEM images in the TiC region reveals the presence of finer microcracks compared to the Ti-TiN sample, indicating improved bonding of TiC after sintering relative to TiN. However, despite the better bonding in the TiC region, the formation of cracks and voids (Figure 4a,b) still compromises the integrity and overall performance of the coating under operating conditions, potentially degrading its mechanical properties.

Figure 5a represents the SEM image of the Ti-TiCN sample section. A comparison of the Ti-TiCN microstructure with that of Ti-TiN and Ti-TiC reveals a similar, homogeneous titanium substrate and a crack-free interface. A careful examination of the transition zone reveals a thickness of approximately 10 μm, comparable to that of the Ti-TiC sample, likely due to the presence of carbon. However, significant porosity is observed within the transition zone, notably more extensive than in the Ti-TiN and Ti-TiC composites. Furthermore, many agglomerated voids due to incomplete fusion of particles were detected in the sintered TiCN region (Figure 5a,b), indicating ineffective bonding during the sintering process.

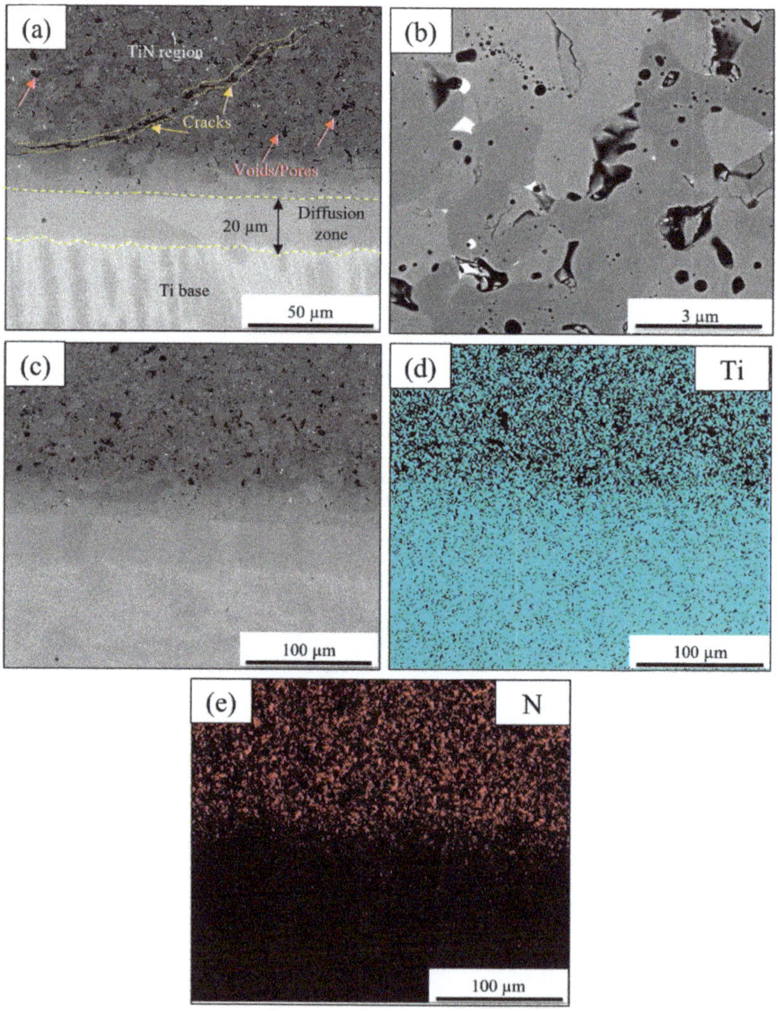

Figure 3. SEM images of (**a**) Ti-TiN vertical cross-section, (**b**) higher magnification image in the TiN region, (**c**) SEM image for EDS map, (**d**) Ti distribution, (**e**) N distribution.

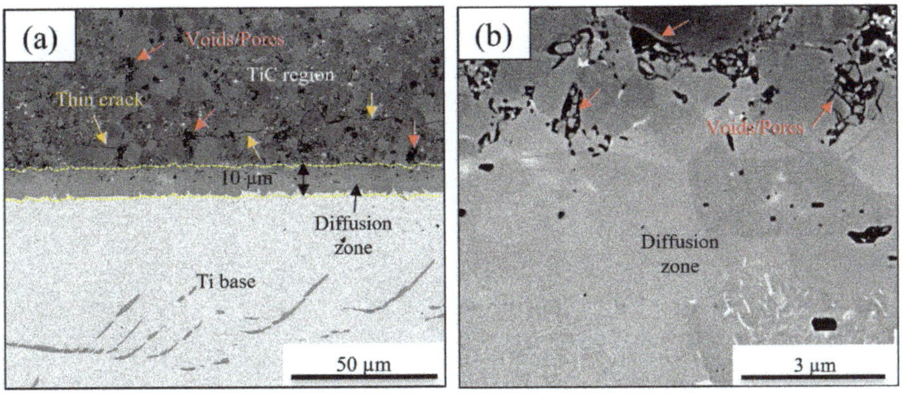

Figure 4. SEM images of (**a**) Ti-TiC vertical cross-section, (**b**) higher magnification image in the TiN region.

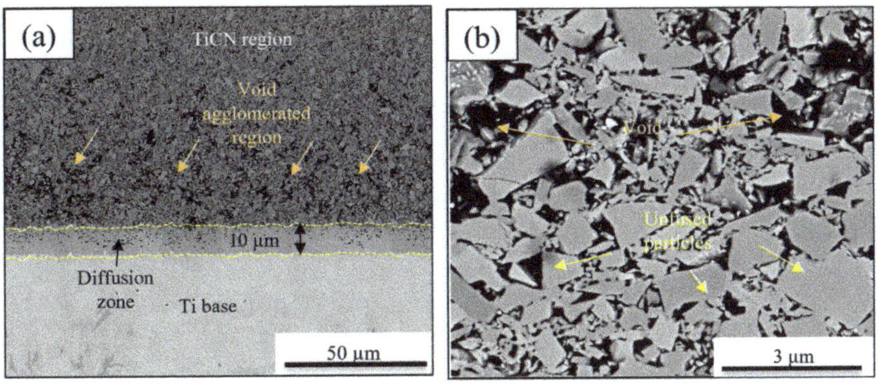

Figure 5. SEM images of (**a**) Ti-TiCN vertical cross-section, (**b**) higher magnification image in the TiCN region.

The microstructures of NbN and TaN coatings on Ti substrates are depicted in Figures 6 and 7, respectively. The Ti-NbN sample exhibits a transition zone of thickness of approximately 15 µm. In contrast, the Ti-TaN system shows a transition zone of roughly 12 µm, thinner than the Ti-NbN. This can be attributed to Nb's significantly higher diffusion coefficient than Ta [39]. Both NbN and TaN coatings display transition zones with no porosity, indicating strong interfacial bonding compared to other ceramic coatings. Some fine cracks were observed within the transition zone of the Ti-TaN sample, but no cracks were detected in the Ti-NbN transition zone. Overall, the consolidated regions of both NbN and TaN coatings exhibit superior bonding, with some porosities without cracks, suggesting excellent interfacial integrity compared to the other ceramic coatings.

Among all of the ceramic-coated samples, the Ti-NbN coating exhibits a superior transition zone with an intermediate thickness, devoid of porosity and cracks. In addition, the NbN coating layer is more uniform, exhibiting minimal porosity and no cracking compared to the TiN, TiC, TiCN, and TaN coatings. In general, an increase in porosity tends to lower the hardness and wear resistance of a material. The voids created by porosity reduce the overall amount of solid material available to resist deformation under stress. The hardness typically decreases as the volume of the pores increases, as the material becomes less dense and more prone to plastic deformation or indentation. The presence of voids can act as stress concentrators, leading to localized weakening and an increased tendency for material loss under tribological loading. The reduced thickness of the transition zone, along with the minimal porosity and no cracks in the Ti-NbN sample, are key factors that contribute to enhanced performance under tribological loading conditions.

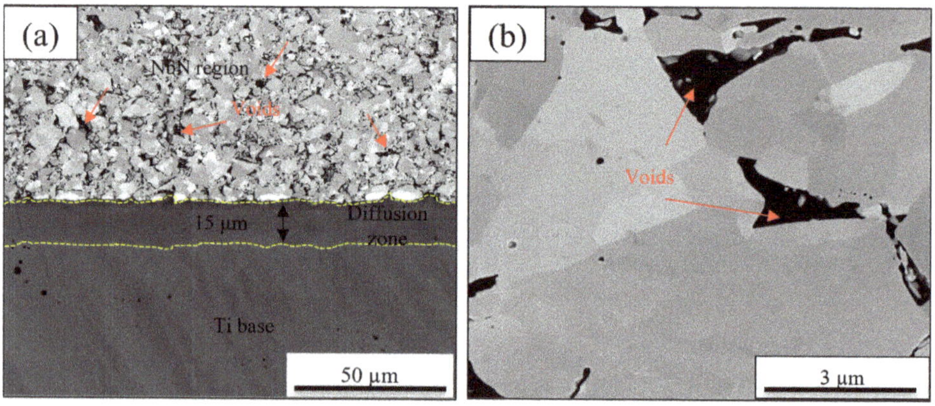

Figure 6. SEM images of (**a**) Ti-NbN vertical cross-section, (**b**) higher magnification image in the NbN region.

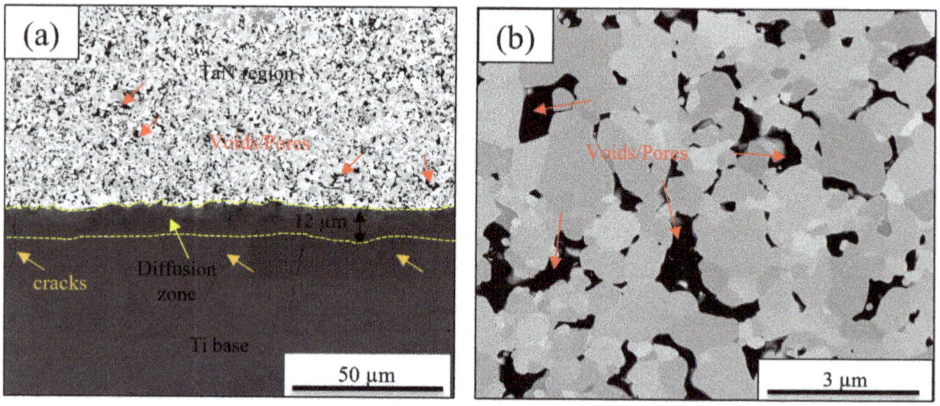

Figure 7. SEM images of (**a**) Ti-TaN vertical cross-section, (**b**) higher magnification image in the TaN region.

3.2. Microhardness

Figure 8 illustrates the average hardness variation across the Ti-plate, the interface zone, and the composite coating regions for various coatings. The baseline hardness of the Ti-plate was approximately 184 ± 20 HV. Upon deposition of the coatings, the hardness in the Ti-plate zone increased to a range of ~225–260 ± 20 HV, depending on the coating type. For the Ti-TiCN coating, the hardness progressively increased from 226 ± 20 HV near the Ti-plate to ~557 ± 20 HV in the interface zone and reached ~846 ± 20 HV in the composite coating zone. In the case of Ti-TaN, the hardness measured was 225 ± 20 HV in the Ti-plate, ~1048 ± 20 HV in the interface zone, and ~1160 ± 20 HV in the composite zone. Similarly, the Ti-TiN coating exhibited a hardness of 261 ± 20 HV in the Ti-plate zone, 1192 ± 20 HV in the interface zone, and ~760 ± 20 HV in the composite zone. The influence of carbon content on hardness is particularly noteworthy, with hardness increasing proportionally to the carbon percentage. The TiC coating demonstrated the highest hardness in the composite zone, with values reaching ~1817 ± 25 HV, while the Ti-plate and interface zones recorded ~230 ± 25 HV and ~466 ± 25 HV, respectively. In the case of Ti-NbN, the hardness observed was 256 ± 20 HV in the Ti-plate zone and ~1058 ± 20 HV in the interface zone, indicating the significant role of the coating material in enhancing surface hardness.

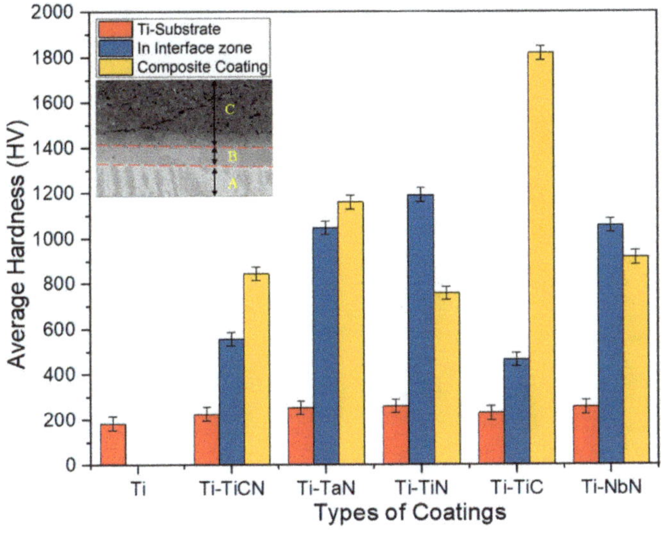

Figure 8. Average hardness across the Ti plate, interface zone, and composite coating regions for various coating types. Note: In the figure, "A" represents the Ti substrate, "B" indicates the interface zone, and "C" corresponds to the composite coating region.

The presence of cracks on the surface of a material directly influences its surface hardness, as cracks act as stress concentrators and reduce the material's resistance to localized deformation [40]. In the case of coatings such as TiC, TiN, and TiCN, the initiation and propagation of cracks during loading reduces the effective contact area that contributes to bearing the applied load. This leads to localized plastic deformation around the cracks, thereby lowering the apparent hardness values measured in those regions. For example, TiC coatings, known for their high inherent hardness (~1817 ± 25 HV in the composite zone), often display microcracks due to their brittleness and limited ductility. While these cracks can serve as stress relief points and limit catastrophic failure, they can also lead to a non-uniform hardness distribution across the surface. Microhardness measurements taken over regions with dense crack networks often exhibit slightly lower values than crack-free regions because the localized deformation is concentrated near the crack tips. In TiN and TiCN coatings, microstructural defects such as porosities and cracks further exacerbate this effect by reducing the cohesive strength of the coating [41]. The uneven stress distribution caused by cracks can lead to premature plastic deformation during hardness testing, diminishing the overall surface hardness.

Figure 9 illustrates the hardness variation across different zones, including the Ti-plate, the interface zone, and the composite coating region, for various coating types. The average hardness within the Ti-plate region was relatively consistent, ranging from approximately 185 ± 20 HV to 260 ± 20 HV. The interface zone exhibited a uniform width across most coating types, as evidenced by the convergence trend depicted in Figure 9. For coatings such as Ti-TiN and Ti-NbN, a decrease in hardness was observed within the composite coating zone compared to the interface zone. Conversely, coatings like Ti-TiCN, Ti-TaN, and Ti-TiC demonstrated an increasing trend in hardness within the composite coating zone, indicating variations in material properties and bonding characteristics influenced by the coating type. These observations highlight the role of coating composition and interface integrity in determining the hardness distribution across the coated structure.

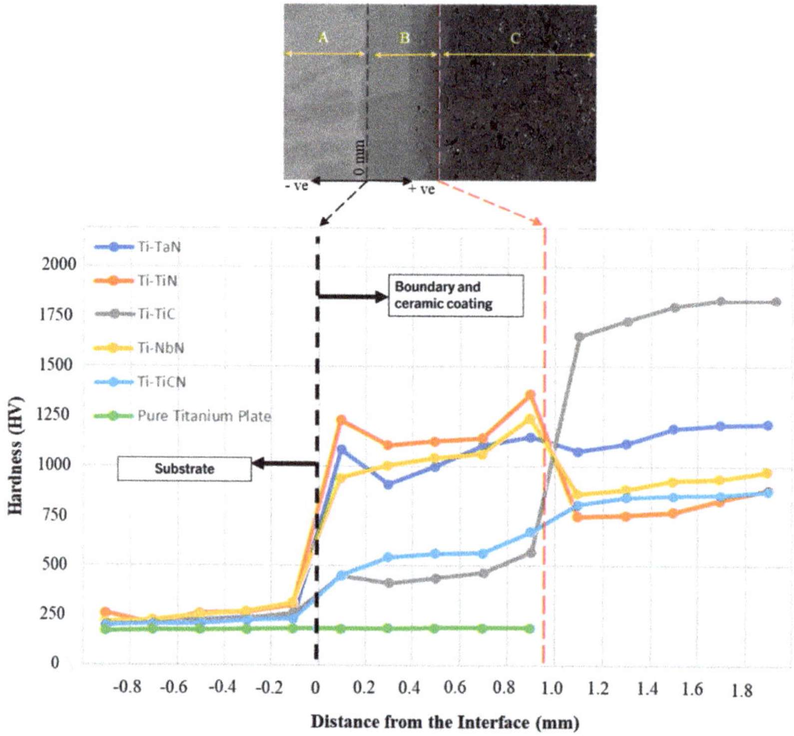

Figure 9. Hardness variation trends across the Ti substrate, interface zone, and composite coating regions for different coating types. Note: In the figure, "A" represents the Ti substrate, "B" indicates the interface zone, and "C" corresponds to the composite coating region. Positive (+ve) values represent the right side, while negative (−ve) values represent the left side of the Ti substrate and interface zone.

3.3. *Tribological Analysis*

3.3.1. Coefficient of Friction

Tribological tests for the coatings were conducted under controlled laboratory conditions simulating real-world environments, with a relative humidity of 50% and room temperature. These conditions facilitated the evaluation of the self-lubricating oxide layers, such as tantalum oxide (Ta_2O_5) and niobium oxide, formed during sliding. The results demonstrated the effectiveness of these oxide layers in reducing the COF and wear with the Ti-TaN and Ti-NbN coatings. This highlights their potential for applications in humid and moderate-temperature conditions, validating their tribological performance under practical environmental constraints.

The wear behaviour of coatings is predominantly influenced by factors such as morphology, toughness, surface roughness, and bonding strength, which are typically quantified through the coefficient of friction (COF) and wear weight loss [42]. Figure 10a presents the COF trends for a Ti substrate and five ceramic coatings fabricated via the SPS technique as a function of applied load and sliding distance. The Ti substrate exhibited an inconsistent COF within the range of ~0.75–0.80 ± 0.020, indicating non-uniform sliding behaviour during the wear test, which may result from surface roughness variations and uneven oxide layer formation

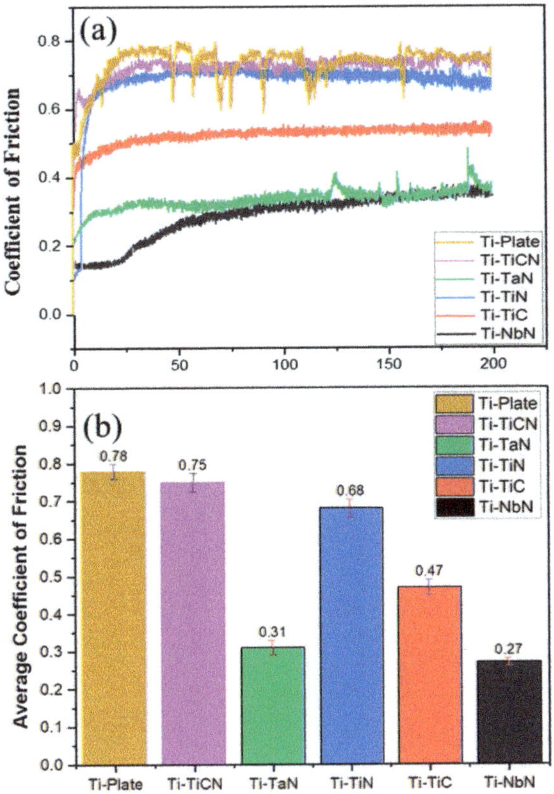

Figure 10. (**a**) Variation in the coefficient of friction over time and (**b**) the corresponding average coefficient of friction values for Ti plates and various ceramic coatings fabricated using SPS.

TiCN is emerging as a potential alternative to tungsten carbides due to its desirable properties, such as high hardness, wear resistance comparable to TiC, and toughness akin to TiN. However, the densification of TiCN requires extremely high sintering temperatures due to its elevated melting point, which poses challenges for its processing and limits its widespread application. The average COF of TiCN, as shown in Figure 10b, was measured to be approximately 0.75 ± 0.025, with consistent performance across the sliding distance, indicating good stability during tribological testing. Tantalum nitride (TaN) coatings demonstrated a COF in the range of ~0.32 ± 0.020, slightly higher than niobium nitride (NbN) coatings, which had a COF of ~0.27 ± 0.010. The lower COF in TaN coatings can be attributed to the formation of a self-lubricating tantalum oxide layer during sliding, enhancing its tribological performance. In comparison, NbN coatings benefit from a lower COF due to their reduced surface roughness and higher nitrogen diffusion into the Ti substrate, creating a robust interfacial bond. For TiC coatings, the COF was significantly lower, ranging from ~0.46 to 0.49, owing to their superior hardness and wear resistance. Conversely, Ti-TiN coatings exhibited an average COF of ~0.68 ± 0.025, which, while higher than TiC, demonstrates the coating's effective combination of hardness and toughness. Additionally, adhesive wear was observed in TiCN coatings, as evidenced by EDS mapping, which showed silicon transfer from the Si_3N_4 counterface balls to the wear track, highlighting the potential for material transfer and surface interaction under sliding conditions. Among all of the coatings studied, the Ti-NbN coating exhibited the best tribological performance with an impressively low COF of ~0.27 ± 0.010. This superior performance is likely due to its excellent hardness, reduced surface roughness, and the formation of a robust oxide layer, providing effective wear resistance. These findings emphasize the critical role of coating composition, processing parameters, and oxide layer formation in enhancing wear resistance and reducing friction, making Ti-NbN a promising candidate for high-performance tribological applications.

Analysis of variance (ANOVA) was performed to statistically validate the differences in hardness and COF among the coatings. The analysis confirmed that the variations in properties across the coatings were statistically significant ($p < 0.05$), emphasizing the effectiveness of the SPS process and the material compositions in enhancing surface performance. These findings underline the critical role of microstructural integrity and composition optimization in achieving reliable and high-performance coatings. The statistical analysis included calculating the mean values and standard deviations (SD) for microhardness across different zones of the Ti-plate, interface, and composite coating and for the COF observed during tribological testing. The microhardness analysis revealed significant differences among the zones and coatings. For instance, the TiC coating exhibited a mean hardness of 230 $\pm$ 25 HV in the Ti-plate region, 466 $\pm$ 25 HV in the interface zone, and 1817 $\pm$ 25 HV in the composite zone. The low SD in the Ti-plate and interface zones indicates uniform hardness in these regions, whereas the higher SD in the composite zone reflects the influence of microstructural imperfections, such as cracks and porosities, which are inherent to ceramic coatings processed via SPS. Similarly, Ti-TaN and Ti-NbN coatings demonstrated enhanced hardness with consistent properties in the composite zone, suggesting their superior microstructural stability. For the tribological performance, the COF values also exhibited notable differences among the coatings. Ti-TaN coatings recorded an average COF of 0.31 $\pm$ 0.020, while Ti-NbN coatings showed the lowest variability with a COF of 0.27 $\pm$ 0.010. The lower SD in these coatings indicates a stable tribological response due to the formation of self-lubricating oxide layers, which minimize wear and friction. Conversely, coatings like Ti-TiN and Ti-TiCN displayed higher COF variability ($\pm$0.025), attributed to localized defects and microstructural inconsistencies during the sintering process.

Porosity significantly influences the COF and overall tribological performance of coatings produced through SPS. It typically acts as a double-edged sword in wear performance. On the one hand, it can serve as a reservoir for lubricating debris, which may reduce friction and enhance wear resistance under specific conditions. On the other hand, excessive porosity weakens the structural integrity of the coating, creating sites for crack initiation and propagation, ultimately accelerating material removal during wear. SEM image analysis using ImageJ Version 1.54 calculated porosity levels ranging from 2.2% to 5.1%, which offers valuable insights into the interplay between microstructural characteristics and tribological behaviour. The Ti-TiN coating, with 2.2% porosity, exhibits a moderate COF of ~0.68, as its low porosity minimizes material removal but allows adhesive wear. Similarly, Ti-TiC, with slightly higher porosity (4.1%), achieves a lower COF (~0.46–0.49) due to its superior hardness and wear resistance. In contrast, coatings with higher porosities, such as Ti-TiCN (5.0%) and Ti-TaN (4.8%), show higher COF values (~0.75 and ~0.28–0.32, respectively) due to increased surface irregularities. However, Ti-TaN compensates for its medium porosity by forming a self-lubricating oxide layer, maintaining a low COF. The Ti-TiNbN coating, with 3.6% porosity, achieves the lowest COF (~0.27) due to its dense microstructure and robust oxide formation.

A similar study was conducted by Ozkan et al. [43], where the substrate was H13 steel, but the coatings applied were AlCrN, ZrN, TioSiN, and TiCrN. TiCN coatings in the current study exhibited a COF of ~0.75, reflecting their high hardness and toughness but also adhesive wear challenges. In contrast, in a study conducted by Ozkan et al. [43], AlCrN coatings achieved a slightly lower COF of ~0.65–0.68 due to the formation of a lubricating oxide layer. This demonstrates the advantage of oxide formation in reducing friction and enhancing stability. The TaN coating outperformed TiCrN with a COF of ~0.32 compared to ~0.45. This difference is attributed to TaN's ability to form a superior self-lubricating tantalum oxide layer during sliding. TiCrN's reliance on Cr-based oxides was less effective in lowering friction, highlighting TaN's tribological advantage. NbN showed an impressively low COF of ~0.27 due to reduced surface roughness and strong nitrogen diffusion. In comparison, the ZrN coating used by Özkan et al. [43] had a higher COF of ~0.40–0.42, likely due to differences in oxide layer characteristics. The robust interfacial bonding in NbN underpins its superior frictional performance. TiC and Ti-TiN coatings demonstrated COF values of ~0.46–0.49 and ~0.68, respectively, influenced by their hardness and toughness. TiSiN coating used by Özkan et al. [43], however, achieved a lower COF of ~0.38, benefiting from a Si-rich lubricating layer that enhanced friction reduction more effectively than TiC and Ti-TiN.

3.3.2. Tribological Behaviour and EDS Mapping

Tribological Behaviour of Ti-Plate

The tribological behaviour of a titanium plate is inherently challenging due to its material properties, which lead to a high COF and susceptibility to various wear mechanisms. Despite its widespread application in aerospace, biomedical, and other industries, pure titanium has limited surface performance characteristics [42]. The high COF of ~0.78 observed for uncoated Ti-plate during pin-on-disk testing highlights its tribological limitations rooted in its mechanical and chemical behaviour. Titanium's high COF is primarily attributed to its strong tendency to form adhesive bonds with contacting surfaces. During sliding contact, the asperities on the titanium surface come into intimate contact with those of the counter body, forming adhesive junctions, as shown in Figure 11a. Titanium's ductile nature further intensifies these adhesive forces, allowing substantial plastic deformation at the contact points. As the sliding process continues, these junctions may fracture, causing material transfer and wear debris generation. This phenomenon, known as adhesive wear, leads to high frictional resistance and contributes significantly to the wear rate of titanium surfaces.

In addition to mechanical wear mechanisms, tribo-chemical interactions significantly influence the tribological behaviour of titanium. The material's high chemical reactivity exacerbates adhesive wear, particularly in environments containing oxygen or moisture. Under such conditions, titanium readily forms thin oxide layers, primarily titanium dioxide (TiO_2), on its surface, as depicted in Figure 11c. Although these oxide layers offer limited protection, their mechanical strength often proves inadequate to resist the shear forces experienced during sliding. Consequently, the oxide layer undergoes breakdown, exposing fresh titanium surfaces that adhere to the counter body. This cycle of oxide formation, degradation, and surface re-exposure results in a fluctuating friction coefficient and material transfer [44].

Figure 11e illustrates this phenomenon, highlighting two distinct regions observed in the EDS map and SEM image. The light grey region corresponds to the TiO_2 layer, which, upon breakdown, is replaced by the formation of a SiO_2 layer. Silicon is introduced from the Si-nitride balls used in this experiment. Titanium's relatively low hardness and inherent softness further exacerbate its tribological challenges, making it susceptible to surface roughening under mechanical loads [42,45]. During sliding, surface asperities experience plastic deformation, forming grooves and wear tracks, which amplify surface roughness, as shown in Figure 11a. This process, known as ploughing, increases frictional forces and contributes to wear. Moreover, the wear debris generated during sliding accumulates within the contact interface, functioning as abrasive particles that accelerate wear and increase the COF. Consequently, titanium's tribological performance presents a significant area for improvement, with research efforts focused on surface modification techniques and advanced coatings to enhance surface properties and mitigate wear.

Tribological Behaviour of Ti-TiCN

Ti-TiCN coatings exhibit tribological properties comparable to those of uncoated Ti-plate due to the synergistic characteristics of carbides and nitrides [46]. Incorporating carbon and nitrogen into the titanium matrix enhances wear resistance and hardness, reducing COF relative to the uncoated Ti-plate [47]. Despite this intent, the measured

COF for Ti-TiCN coatings (~0.75 ± 0.025) represents a modest reduction compared to the Ti-plate (COF = ~0.77). This marginal improvement underscores the complexities and challenges associated with sliding contact, including intricate interfacial interactions. The wear mechanisms differ notably between the two surfaces. Ti-plate primarily undergoes adhesive and ploughing wear, resulting in significant material transfer and surface roughening. Conversely, the Ti-TiCN coating exhibits a combination of adhesive and abrasive wear, with reduced adhesive forces attributed to its hard ceramic composition. However, localized defects within the Ti-TiCN coating introduce inconsistencies in wear behaviour, limiting its overall performance enhancement as shown in Figure 12.

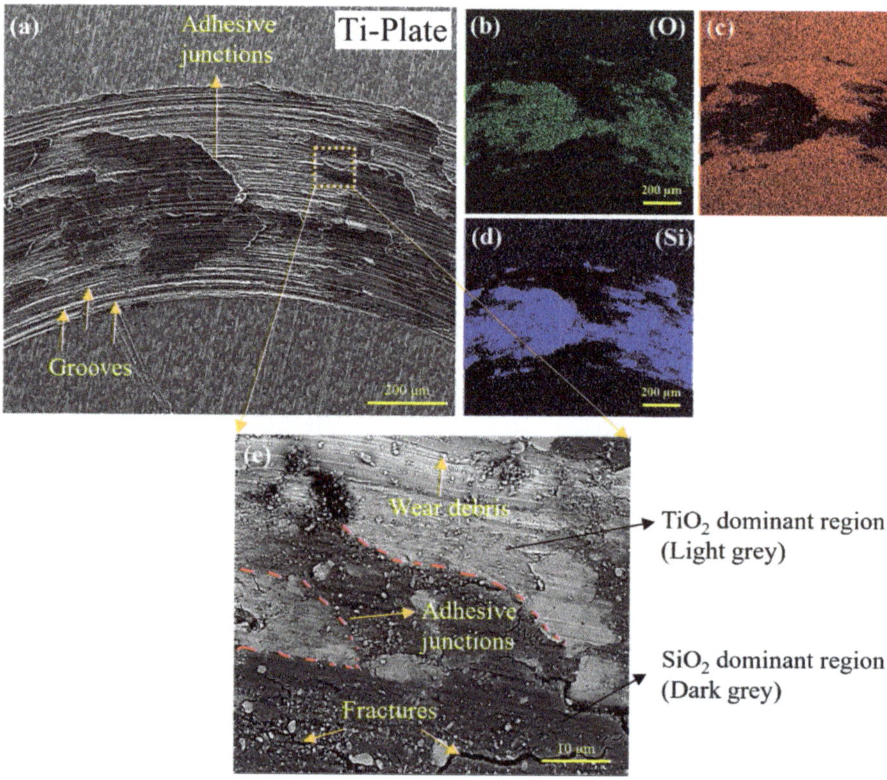

Figure 11. EDS mapping of the wear track on the Ti-plate: (**a**) SEM image of the wear track, showcasing its surface morphology; EDS elemental maps showing the distribution of (**b**) oxygen, (**c**) titanium, and (**d**) silicon. Additionally, (**e**) presents a magnified view of the wear track, highlighting surface details and the corresponding wear mechanism.

Carbonitride phases exhibit high hardness, which effectively reduces surface deformation under applied loads during sliding contact. The increased hardness minimizes the formation of adhesive junctions with the counter body, reducing material transfer and enhancing wear resistance. In contrast to pure titanium, which suffers from extensive plastic deformation and adhesive wear, the Ti-TiCN coating acts as a hard barrier that resists mechanical interlocking and adhesion. As a result, cracks are formed in the wear track, which can be seen in Figure 12. However, the tribological performance of Ti-TiCN coatings is heavily influenced by the quality of the coating microstructure. Achieving full densification of the carbonitride layer during the SPS process is challenging due to high sintering temperatures and limited nitrogen diffusion into the titanium substrate. Incomplete densification results in the presence of porosity or weakly bonded regions within the coating. These microstructural defects only serve as sites for crack initiation [48,49] and propagation during sliding, which compromise the wear resistance of the coating. The tribological properties of Ti-TiCN coatings are also affected by the relative amounts of carbon and nitrogen in the coating [16]. The presence of carbon contributes to the hardness and wear resistance of the coating, while nitrogen improves toughness and ductility. Achieving an optimal balance between carbon and nitrogen content is critical for enhancing the coating's tribological performance. Incomplete diffusion or inhomogeneous distribution of these elements during the SPS process can lead to microstructural variations, which may negatively impact the coating's wear resistance and COF.

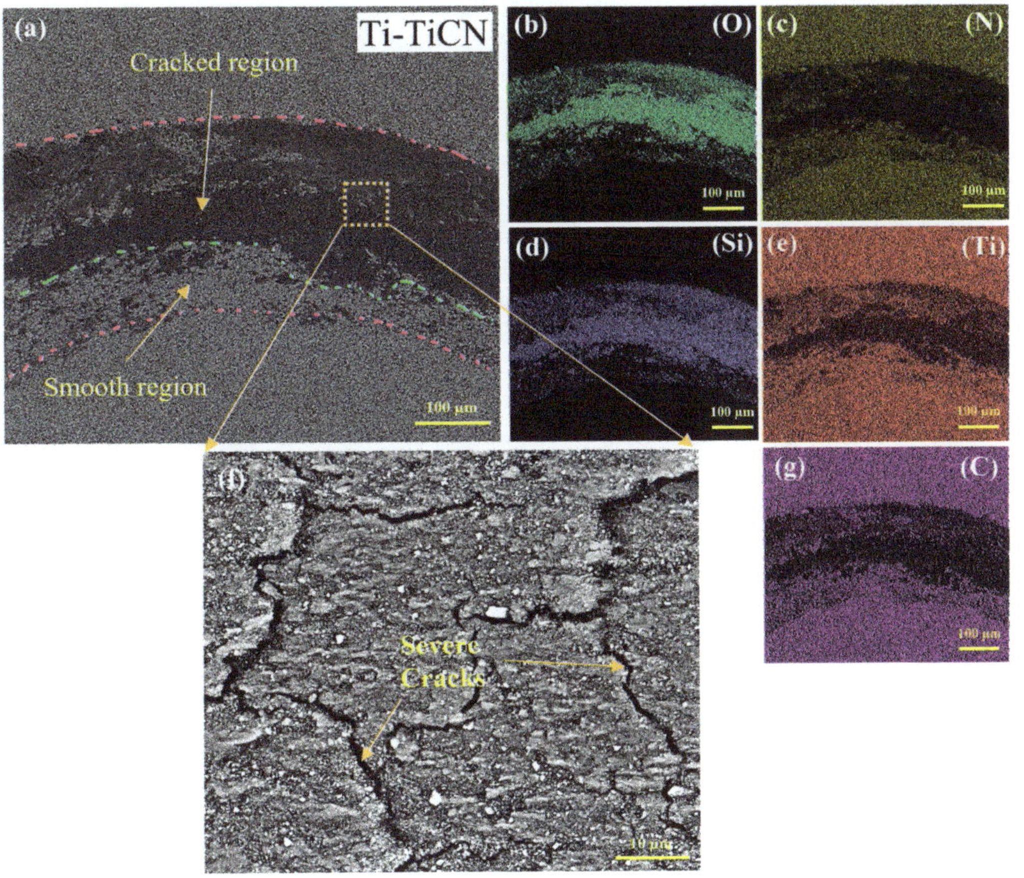

Figure 12. EDS map of wear track on Ti-TiCN. (**a**) SEM image of the wear track showcasing its surface morphology; EDS elemental maps showing the distribution of (**b**) oxygen, (**c**) nitrogen, (**d**) silicon, (**e**) titanium, (**f**) presents a magnified view of the wear track, highlighting severe cracks and (**g**) carbon.

Tribological Behaviour of Ti-TaN

The tribological behaviour of Ti-TaN coatings demonstrates superior friction and wear performance compared to both the uncoated Ti-plate and Ti-TiCN coatings. Tantalum nitride exhibits a significantly lower COF (~0.31) due to its unique properties, including high hardness, excellent chemical stability, and a lubricating oxide layer during sliding contact. This combination of properties makes Ti-TaN coatings particularly effective in reducing wear and frictional forces, resulting in enhanced tribological performance under demanding conditions in applications requiring low friction and high wear resistance [50].

The primary wear mechanism for Ti-TaN coatings is oxidative wear, characterized by the stable formation of a self-lubricating tantalum oxide (Ta_2O_5) layer during sliding, which contributes significantly to the coating's low coefficient of friction (COF), as illustrated in Figure 13. This oxide layer forms due to tribo-chemical reactions at the contact interface, creating a thin, lubricious film that separates the contacting surfaces. The presence of this oxide layer minimizes direct metal-to-metal contact, significantly reducing friction and wear. This wear mechanism contrasts with the adhesive and abrasive wear observed in Ti-plate and the combined adhesive-abrasive wear of Ti-TiCN coatings.

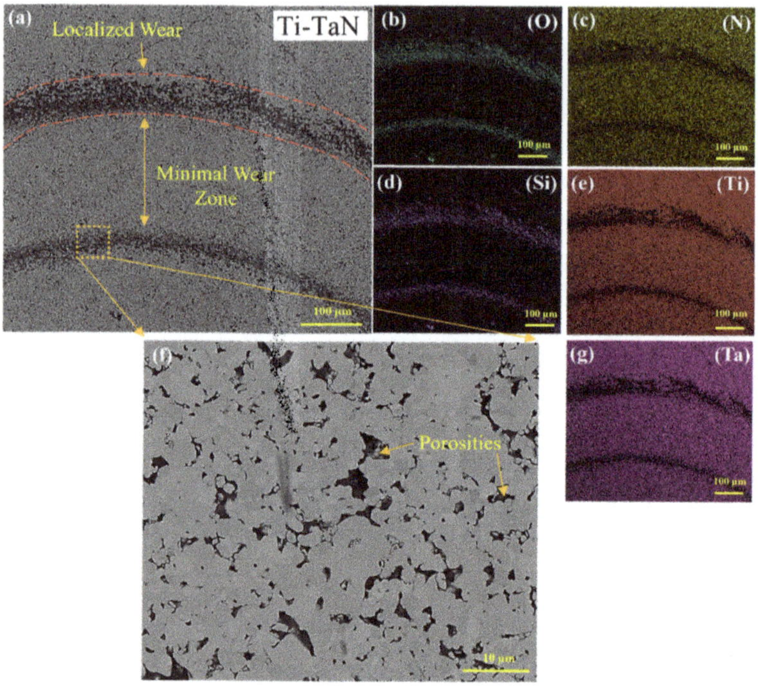

Figure 13. EDS map of wear track on Ti-TaN. (**a**) SEM image of the wear track showcasing its surface morphology; EDS elemental maps showing the distribution of (**b**) oxygen, (**c**) nitrogen, (**d**) silicon, (**e**) titanium, (**f**) presents a magnified view of the wear track, highlighting porosities and (**g**) tantalum.

The high hardness and stability of the tantalum oxide layer reduce the extent of material transfer and wear debris formation, contributing to a smoother and more consistent wear track. The other rationale is that the hard tantalum nitride layer effectively prevents the formation of adhesive solid bonds with the counter body during sliding, reducing material transfer and adhesive wear. This hardness provides a robust barrier against both mechanical interlocking and abrasive wear, offering substantial improvements in wear resistance compared to uncoated Ti-plate and Ti-TiCN coatings.

As shown in Figure 13, EDS mapping revealed that the wear track of Ti-TaN coatings has a high concentration of tantalum and oxygen, indicative of the formation of a tantalum oxide layer at the outer zone of the wear track. The wear track shows a smooth track with minimal wear debris, which is evidence of a continuous protective film. In comparison, Ti-plate wear tracks displayed extensive titanium transfer and oxide formation, while Ti-TiCN coatings exhibited regions of carbon and nitrogen-rich wear debris with signs of localized wear and porosities. The Ti-TaN coating demonstrates superior tribological behaviour compared to both Ti-plate and Ti-TiCN coatings due to its high hardness, stable lubricating oxide formation, and effective resistance to adhesive and abrasive wear.

Tribological Behaviour of Ti-TiN

The tribological behaviour of Ti-TiN coatings significantly enhances hardness and reduces adhesive wear. However, these coatings exhibit a relatively high COF (~0.69) compared to more advanced alternatives like Ti-TaN coatings. While the TiN layer markedly enhances hardness relative to the Ti substrate, its tribological performance is constrained by microstructural deficiencies. Achieving full densification during the coating process proved challenging, mainly due to limited nitrogen diffusion.

Consequently, porosity and microstructural defects developed within the coating, forming localized weak points that compromised wear resistance. These defects are evident in the wear track map and microstructure, as shown in Figure 14, which reveal unaffected zones corresponding to areas of poor wear resistance. Additionally, the high hardness of the TiN coating contributes to its brittleness, leading to elevated contact stresses during sliding. This results in the initiation and propagation of microcracks, as illustrated in Figure 14. These microcracks exacerbate localized wear and increase the frictional force during sliding contact. EDS mapping, as shown in Figure 14, further highlights these limitations. Regions of high wear exhibit reduced nitrogen and titanium concentrations, correlating with porosity zones and microcracks. These defects exacerbate localized wear and contribute to elevated friction. Furthermore, the presence of silicon oxide layers in the wear track, as observed in the EDS analysis, is attributed to material transfer from the silicon nitride ball used in the pin-on-disk tests. These findings underscore the need to further optimize Ti-TiN coating processes to enhance tribological performance, particularly by addressing porosity, brittleness, and material transfer during wear.

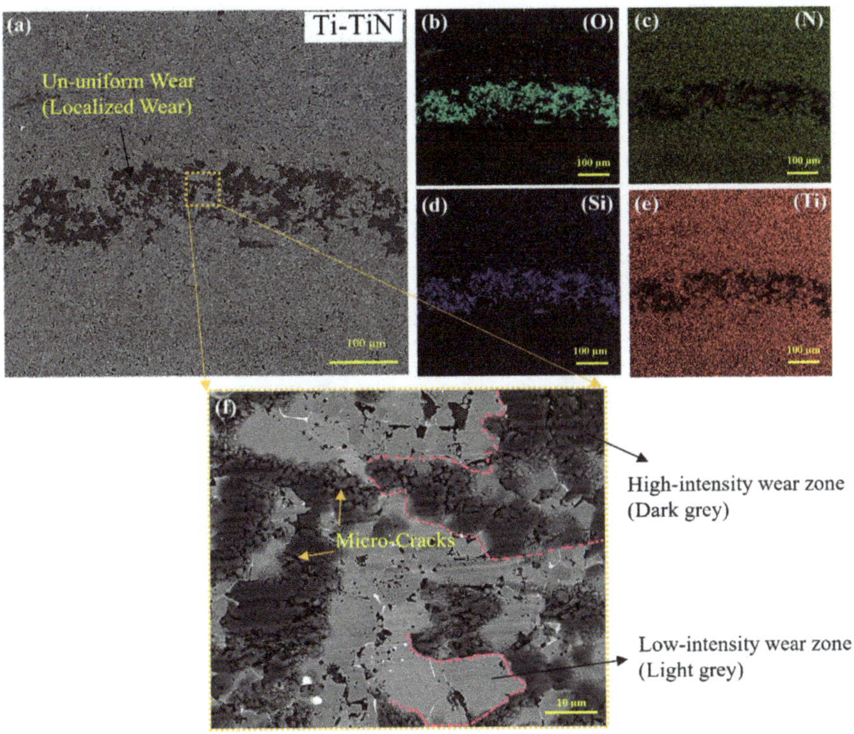

Figure 14. EDS map of wear track on Ti-TiN. (**a**) SEM image of the wear track showcasing its surface morphology; EDS elemental maps showing the distribution of (**b**) oxygen, (**c**) nitrogen, (**d**) silicon, (**e**) titanium, and (**f**) presents a magnified view of the wear track, highlighting micro cracks.

Tribological Behaviour of Ti-TiC

The SEM/EDS images, as shown in Figure 15, reveal a network of cracks within the wear track, which is a critical microstructural characteristic influencing the tribological behaviour of TiC coatings. These cracks primarily result from high contact stresses during sliding, where the brittle nature of the hard TiC layer predisposes it to crack initiation under localized stress concentrations. The initiation of cracks is facilitated by the inherent stiffness and limited ductility of the TiC ceramic layer, combined with surface imperfections such as porosity or microstructural inhomogeneities visible in the wear track. Despite the presence of these cracks, the TiC coating maintains a relatively low COF (~0.47) due to its hard and stable ceramic structure, which limits material transfer and adhesive interactions during sliding. The hard ceramic surface resists severe plastic deformation, and the cracks remain confined to localized regions without compromising the overall wear resistance [51,52]. This behaviour highlights the coating's ability to withstand abrasive wear and minimize frictional forces, even under conditions that promote microcracking.

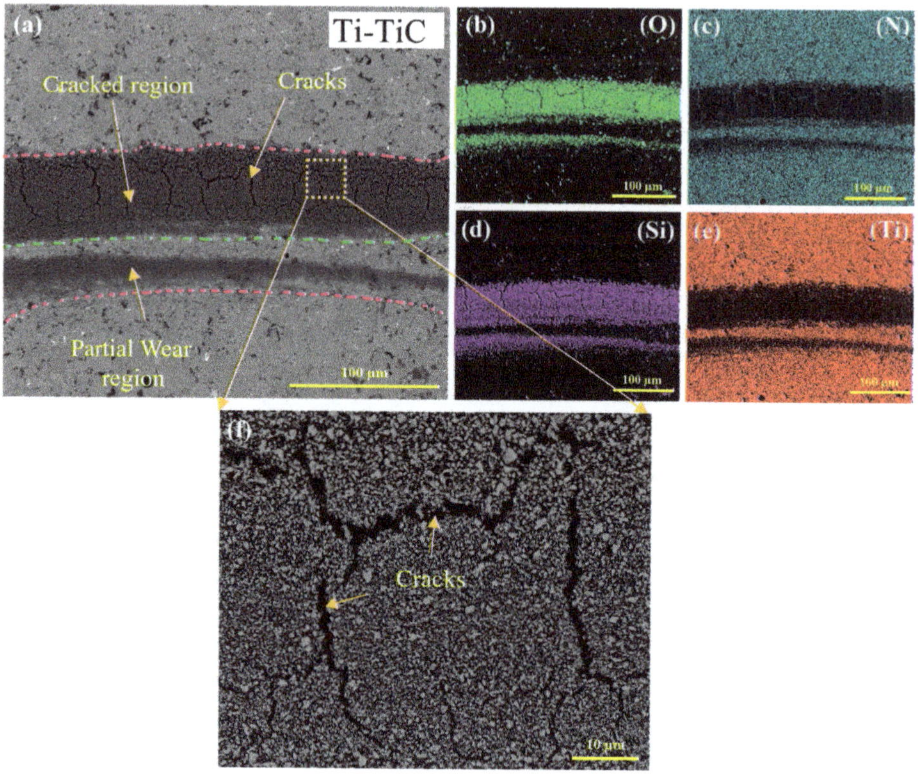

Figure 15. EDS map of wear track on Ti-TiC. (**a**) SEM image of the wear track showcasing its surface morphology; EDS elemental maps showing the distribution of (**b**) oxygen, (**c**) nitrogen, (**d**) silicon, (**e**) titanium, and (**f**) presents a magnified view of the wear track, highlighting cracks.

Moreover, the limited propagation of cracks is attributed to the coating's high hardness and cohesive strength, which prevent catastrophic failure. The cracks serve as stress relief points, reducing the likelihood of larger-scale delamination or spalling [53]. This controlled cracking mechanism contributes to a smoother wear track compared to uncoated titanium or alternative coatings such as Ti-TiN or Ti-TiCN, which often exhibit more pronounced material removal or microstructural defects. The observed cracks within the wear track of TiC coatings highlight a critical interplay between microstructural integrity and tribological properties. The coating's high hardness mitigates the adverse effects of crack initiation, allowing it to maintain superior wear resistance and a low COF. This performance makes TiC coatings suitable for applications where high hardness and resistance to adhesive interactions are prioritized, even in the presence of microcracking.

Tribological Behaviour of Ti-NbN

The Ti-NbN coating analysed using SEM and EDS mapping (Figure 16) exhibits remarkable tribological performance, as evidenced by its low (COF = 0.27) and minimal wear track depth. The SEM images reveal a smooth and homogeneous microstructure within the wear track, indicating minimal material removal or damage. The absence of severe cracking or delamination along the wear path suggests a robust coating structure with exceptional wear resistance. This superior performance is attributed to the synergistic effects of TiN and NbN, which contribute to the coating's high hardness, toughness, and resistance to mechanical degradation under sliding conditions [54]. The elemental distribution observed in the EDS maps confirms the uniform incorporation of titanium (Ti), niobium (Nb), and nitrogen (N) throughout the coating, as shown in Figure 16. This elemental homogeneity eliminates weak points or segregated regions that could serve as initiation sites for wear or crack propagation. Adding niobium into the TiN lattice enhances the overall performance of the coating by forming a strong solid solution, which increases the material's cohesive strength. NbN is known for its ability to improve fracture toughness, providing resistance against crack propagation even under high mechanical stress [20]. This toughened microstructure enables the coating to maintain its integrity under dynamic loading conditions. The combination of TiN and NbN phases in the coating plays a critical role in achieving a balance between hardness and toughness. While TiN provides low hardness and wear resistance, NbN improves the coating's toughness by mitigating the brittleness typically associated with nitride ceramics. The presence of niobium creates microstructural interfaces or ductile phases that absorb mechanical energy, preventing catastrophic failure under stress. This synergy enhances the durability of the coating, making it resistant to crack initiation and propagation even under severe tribological conditions. The Ti-NbN coating's exceptional tribological behaviour is also a result of its optimized microstructure, elemental homogeneity, and the synergistic properties of TiN and NbN. The coating's medium hardness, enhanced toughness, and self-lubricating nature ensure minimal wear and low friction, making it a highly effective material for applications involving extreme contact stresses and sliding conditions. Combining these characteristics highlights the potential of Ti-NbN coatings as a preferred choice in advanced tribological systems.

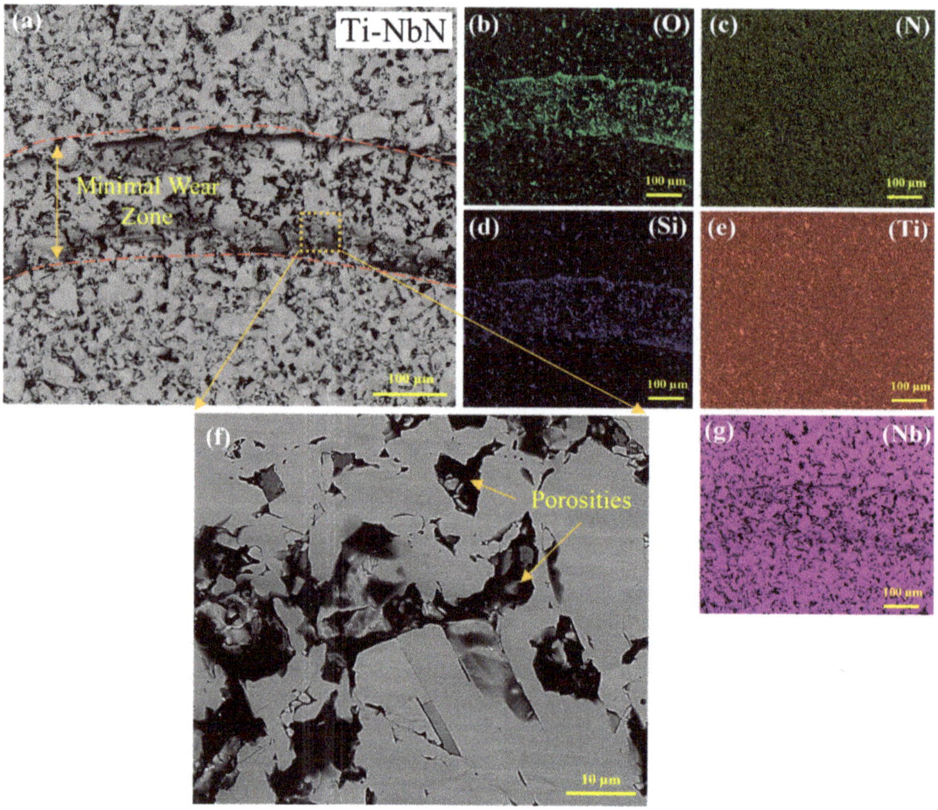

Figure 16. EDS map of wear track on Ti-NbN. (**a**) SEM image of the wear track showcasing its surface morphology; EDS elemental maps showing the distribution of (**b**) oxygen, (**c**) nitrogen, (**d**) silicon, (**e**) titanium, and (**f**) presents a magnified view of the wear track, highlighting porosities and (**g**) niobium.

4. Conclusions

This study successfully demonstrated the potential of spark plasma sintering (SPS) for synthesizing advanced ceramic coatings, such as TiC, TiN, Ti-TaN, Ti-CNT, and Ti-NbN, on titanium substrates to enhance their mechanical and tribological properties. The coatings were characterized by improved surface hardness and wear resistance, addressing the inherent limitations of titanium alloys. The formation of self-lubricating tantalum oxide and niobium oxide layers in Ti-TaN (COF = 0.31 $\pm$ 0.02) and Ti-NbN coatings (COF = 0.27 $\pm$ 0.01), respectively, contributed significantly to their superior wear resistance. Conversely, the performance of Ti-TiN (COF = 0.68 $\pm$ 0.025) and Ti-CNT coatings (COF = 0.75 $\pm$ 0.025) was constrained by porosity and localized microstructural defects, highlighting areas for process improvement. Microstructural analysis showed that SPS produced minimal defect coating–substrate interfaces with well-defined transition zones. TiN had a broader zone due to nitrogen diffusion, while TiC and TiCN had thinner zones from carbon's lower diffusivity.

The scientific novelty of this work lies in leveraging the SPS technique to achieve dense, hard, and wear-resistant coatings while maintaining precise control over the processing parameters. This study also introduced a comparative evaluation of COF and wear mechanisms across multiple ceramic coatings, providing critical insights into the role of the material composition and microstructure in tribological performance. The findings emphasize the advantages of combining nitride and carbide phases to achieve a balanced enhancement of hardness and toughness, particularly for high-performance tribological applications.

Future work should focus on optimizing SPS parameters to reduce porosity, improve coating uniformity, and enhance mechanical and tribological properties. Despite the progress, microstructural issues, such as microcracking in the TiN and TiC layers and voids in the TiCN, impacted their tribological performance. These challenges highlight the importance of refining the sintering parameters, including the temperature and holding time, to improve densification and minimize defects, ultimately boosting the tribological performance of ceramic-coated titanium for demanding applications.

Author Contributions: G.W.: Investigation, Methodology, Data curation, Data analysis; A.C.: Writing—Investigation, Conceptualization, Data curation, Formal analysis; S.D.: Writing—Investigation, Conceptualization, Data curation, Formal analysis; A.B.: Investigation, Data Curation, Methodology, Formal analysis; O.K.: Investigation, Data Curation, Methodology, Formal analysis; P.B.: Investigation, Data Curation, Formal analysis, Writing-review and editing; T.B.: Conceptualization, Review, Editing, Supervision. All authors have read and agreed to the published version of the manuscript.

Funding: This work was supported by Cleveland State University's Undergraduate Summer Research Award (USRA) program. Also, this work was partially funded by the National Science Foundation under Grant No. 1126126. Any opinions, findings, conclusions, or recommendations expressed in this material are those of the author(s) and do not necessarily reflect the views of the National Science Foundation.

Data Availability Statement: The original contributions presented in this study are included in the article. Further inquiries can be directed to the corresponding author.

Conflicts of Interest: Author Praful Bari was employed by ZF Group. The remaining authors declare that the research was conducted in the absence of any commercial or financial relationships that could be construed as a potential conflict of interest. The funders had no role in the design of the study; in the collection, analyses, or interpretation of data; in the writing of the manuscript, or in the decision to publish the results.

References

1. Williams, J.C.; Boyer, R.R. Opportunities and Issues in the Application of Titanium Alloys for Aerospace Components. *Metals* **2020**, *10*, 705. [CrossRef]
2. Chen, L.-Y.; Cui, Y.-W.; Zhang, L.-C. Recent Development in Beta Titanium Alloys for Biomedical Applications. *Metals* **2020**, *10*, 1139. [CrossRef]
3. Zhang, L.C.; Chen, L.Y.; Wang, L. Surface modification of titanium and titanium alloys: Technologies, developments, and future interests. *Adv. Eng. Mater.* **2020**, *22*, 1901258. [CrossRef]
4. Najafizadeh, M.; Yazdi, S.; Bozorg, M.; Ghasempour-Mouziraji, M.; Hosseinzadeh, M.; Zarrabian, M.; Cavaliere, P. Classification and applications of titanium and its alloys: A review. *J. Alloys Compd. Commun.* **2024**, *3*, 100019. [CrossRef]
5. Vishnu, J.; Manivasagam, G. Surface modification and biological approaches for tackling titanium wear-induced aseptic loosening. *J. Bio- Tribo-Corros.* **2021**, *7*, 32. [CrossRef]
6. Bandyopadhyay, A.; Mitra, I.; Goodman, S.B.; Kumar, M.; Bose, S. Improving biocompatibility for next generation of metallic implants. *Prog. Mater. Sci.* **2023**, *133*, 101053. [CrossRef] [PubMed]
7. Chen, Y.; Cheng, T.; Nie, X. Wear failure behaviour of titanium-based oxide coatings on a titanium alloy under impact and sliding forces. *J. Alloys Compd.* **2013**, *578*, 336–344. [CrossRef]
8. Ching, H.A.; Choudhury, D.; Nine, M.J.; Osman, N.A.A. Effects of surface coating on reducing friction and wear of orthopaedic implants. *Sci. Technol. Adv. Mater.* **2014**, *15*, 014402. [CrossRef]
9. Bai, H.; Zhong, L.; Kang, L.; Liu, J.; Zhuang, W.; Lv, Z.; Xu, Y. A review on wear-resistant coating with high hardness and high toughness on the surface of titanium alloy. *J. Alloys Compd.* **2021**, *882*, 160645. [CrossRef]
10. Thakare, J.G.; Pandey, C.; Mahapatra, M.M.; Mulik, R.S. Thermal barrier coatings—A state of the art review. *Met. Mater. Int.* **2021**, *27*, 1947–1968. [CrossRef]
11. Mondal, K.; Nuñez, L., III; Downey, C.M.; Van Rooyen, I.J. Thermal barrier coatings overview: Design, manufacturing, and applications in high-temperature industries. *Ind. Eng. Chem. Res.* **2021**, *60*, 6061–6077. [CrossRef]
12. Vaiani, L.; Boccaccio, A.; Uva, A.E.; Palumbo, G.; Piccininni, A.; Guglielmi, P.; Cantore, S.; Santacroce, L.; Charitos, I.A.; Ballini, A. Ceramic Materials for Biomedical Applications: An Overview on Properties and Fabrication Processes. *J. Funct. Biomater.* **2023**, *14*, 146. [CrossRef] [PubMed]

13. Dutta, B.; Froes, F.S. The additive manufacturing (AM) of titanium alloys. *Met. Powder Rep.* **2017**, *72*, 96–106. [CrossRef]
14. Jang, T.S.; Kim, D.; Han, G.; Yoon, C.B.; Jung, H.D. Powder based additive manufacturing for biomedical application of titanium and its alloys: A review. *Biomed. Eng. Lett.* **2020**, *10*, 505–516. [CrossRef]
15. Zhu, Y.; Wang, W.; Jia, X.; Akasaka, T.; Liao, S.; Watari, F. Deposition of TiC film on titanium for abrasion resistant implant material by ion-enhanced triode plasma CVD. *Appl. Surf. Sci.* **2012**, *262*, 156–158. [CrossRef]
16. Savalani, M.M.; Ng, C.C.; Li, Q.H.; Man, H.C. In situ formation of titanium carbide using titanium and carbon-nanotube powders by laser cladding. *Appl. Surf. Sci.* **2012**, *258*, 3173–3177. [CrossRef]
17. Uddin, G.M.; Jawad, M.; Ghufran, M.; Saleem, M.W.; Raza, M.A.; Rehman, Z.U.; Arafat, S.M.; Irfan, M.; Waseem, B. Experimental investigation of tribo-mechanical and chemical properties of TiN PVD coating on titanium substrate for biomedical implants manufacturing. *Int. J. Adv. Manuf. Technol.* **2019**, *102*, 1391–1404. [CrossRef]
18. Al-Asadi, M.M.; Al-Tameemi, H.A. A review of tribological properties and deposition methods for selected hard protective coatings. *Tribol. Int.* **2022**, *176*, 107919. [CrossRef]
19. Ou, Y.X.; Wang, H.Q.; Hua, Q.S.; Liao, B.; Ouyang, X.P. Tribocorrosion behaviors of superhard yet tough Ti-CN ceramic coatings. *Surf. Coat Tech.* **2022**, *439*, 128448. [CrossRef]
20. Yi, J.; Xiong, J.; Guo, Z.; Yang, T.; Liu, J.; You, Q.; Fang, D.; Yu, Z.; Gou, S.; Chen, S. Growth, mechanical properties, and tribological performance of TiAlN/NbN and NbN/TiAlN bilayer coatings. *Ceram. Int.* **2022**, *48*, 6208–6217. [CrossRef]
21. Digole, S.; Desai, J.; Christopher, C.; Bohara, S.; Witharamage, C.S.; Kothapalli, C.; Gupta, R.K.; Borkar, T. Improved Tribological Performance of Nitride-Reinforced Biocompatible Titanium–Niobium–Zirconium–Tantalum (TNZT) Alloys for Advanced Orthopedic Applications. *Metals* **2024**, *14*, 122. [CrossRef]
22. Yu, H.; Liang, W.; Miao, Q.; Yin, M.; Ma, Y.; Zuo, S. Effect of shot peening pretreatment on the high-temperature tribological behaviours of a TaN coating prepared via double-cathode glow plasma alloying. *Surf. Coat Tech.* **2021**, *427*, 127825. [CrossRef]
23. Li, M.; Huang, K.; Yi, X. Crack Formation Mechanisms and Control Methods of Laser Cladding Coatings: A Review. *Coatings* **2023**, *13*, 1117. [CrossRef]
24. Tokita, M. Progress of Spark Plasma Sintering (SPS) Method, Systems, Ceramics Applications and Industrialization. *Ceramics* **2021**, *4*, 160–198. [CrossRef]
25. Singh, A.; Bakshi, S.R.; Virzi, D.A.; Keshri, A.K.; Agarwal, A.; Harimkar, S.P. In-situ synthesis of TiC/SiC/Ti_3SiC_2 composite coatings by spark plasma sintering. *Surf. Coat. Technol.* **2011**, *205*, 3840–3846. [CrossRef]
26. Digole, S.; Karki, S.; Mugale, M.; Choudhari, A.; Gupta, R.K.; Borkar, T. Spark Plasma Sintering of Pure Titanium: Microstructure and Mechanical Characteristics. *Materials* **2024**, *17*, 3469. [CrossRef] [PubMed]
27. Liparoti, S.; Sofia, D.; Romano, A.; Marra, F.; Pantani, R. Fused filament deposition of PLA: The role of interlayer adhesion in the mechanical performances. *Polymers* **2021**, *13*, 399. [CrossRef]
28. Hamzah, E.; Ali, M.; Toff, M.R.H.M. Effect of substrate bias on friction coefficient, adhesion strength and hardness of tin-coated tool steel. *Surf. Rev. Lett.* **2006**, *13*, 763–771. [CrossRef]
29. Ali, M.; Hamzah, E.; Radzi Toff, M. Friction coefficient and surface roughness of TiN-coated HSS deposited using cathodic arc evaporation PVD technique. *Ind. Lubr. Tribol.* **2008**, *60*, 121–130. [CrossRef]
30. Yumashev, A.; Mikhaylov, A. Development of polymer film coatings with high adhesion to steel alloys and high wear resistance. *Polym. Compos.* **2020**, *41*, 2875–2880. [CrossRef]
31. Vera, E.; Vite, M.; Lewis, R.; Gallardo, E.; Laguna-Camacho, J. A study of the wear performance of TiN, CrN and WC/C coatings on different steel substrates. *Wear* **2011**, *271*, 2116–2124. [CrossRef]
32. Alam, S.; Sasaki, S.; Shimura, H. Friction and wear characteristics of aluminum bronze coatings on steel substrates sprayed by a low pressure plasma technique. *Wear* **2001**, *248*, 75–81. [CrossRef]
33. Panjan, P.; Drnovsek, A.; Gselman, P.; Cekada, M.; Panjan, M. Review of Growth Defects in Thin Films Prepared by PVD Techniques. *Coatings* **2020**, *10*, 447. [CrossRef]
34. Chen, Z.; Zhou, K.; Lu, X.; Lam, Y.C. A review on the mechanical methods for evaluating coating adhesion. *Acta Mech.* **2014**, *225*, 431–452. [CrossRef]
35. Korobenko, A.; Saha, S.; Godfrey, A.T.; Gertsvolf, M.; Naumov, A.Y.; Villeneuve, D.M.; Boltasseva, V.; Shalaev, M.; Corkum, P.B. High-harmonic generation in metallic titanium nitride. *Nat. Commun.* **2021**, *12*, 4981. [CrossRef]
36. Hawker, D.; Lane, C.D.; Cockcroft, S.L.; Maijer, D.M. Numerical modeling of the diffusional transport of nitrogen in multi-phase solid titanium and its application to determine diffusion coefficients. *J. Mater. Res. Technol.* **2021**, *14*, 1089–1098. [CrossRef]
37. Issariyapat, A.; Visuttipitukul, P.; Song, T.; Umeda, J.; Qian, M.; Kondoh, K. Strength-ductility improvement of extruded Ti-(N) materials using pure Ti powder with high nitrogen solution. *Mater. Sci. Eng. A* **2020**, *779*, 139136. [CrossRef]
38. Arvieu, C.; Manaud, J.P.; Quenisset, J.M. Interaction between titanium and carbon at moderate temperatures. *J. Alloys Compd.* **2004**, *368*, 116–122. [CrossRef]
39. Haußmann, L.; Neumeier, S.; Bresler, J.; Keim, S.; Pyczak, F.; Göken, M. Influence of Nb, Ta and Zr on the Interdiffusion Coefficients and Solid Solution Strengthening of γ-TiAl Single Phase Alloys. *Metals* **2022**, *12*, 752. [CrossRef]
40. Kumar, D.; Idapalapati, S.; Wang, W.; Narasimalu, S. Effect of surface mechanical treatments on the microstructure-property-performance of engineering alloys. *Materials* **2019**, *12*, 2503. [CrossRef] [PubMed]
41. Xian, G.; Xiong, J.; Fan, H.; Jiang, F.; Zhao, H.; Xian, L.; Qin, Z. Mechanical and wear properties of TiN films on differently pretreated TiCN-based cermets. *Appl. Surf. Sci.* **2021**, *570*, 151180. [CrossRef]

42. Dong, H. Tribological properties of titanium-based alloys. In *Surface Engineering of Light Alloys*; Woodhead Publishing: Sawston, UK, 2010; pp. 58–80. [CrossRef]
43. Ozkan, D.; Yilmaz, M.A.; Karakurt, D.; Szala, M.; Walczak, M.; Bakdemir, S.A.; Türküz, C.; Sulukan, E. Effect of AISI H13 steel substrate nitriding on AlCrN, ZrN, TiSiN, and TiCrN multilayer PVD coatings wear and friction behaviors at a different temperature level. *Materials* **2023**, *16*, 1594. [CrossRef]
44. Mugale, M.; Choudhari, A.; Karki, S.; Desai, J.; Walunj, G.; Vukkum, V.B.; Schwam, D.; Gupta, R.K.; Borkar, T. Investigation of Protective Coatings for Reducing High-Temperature Oxidation of Steels. *JOM* **2024**, 1–12. [CrossRef]
45. Luo, Y.; Ge, S.; Jin, Z.; Fisher, J. Effect of surface modification on surface properties and tribological behaviours of titanium alloys. *Proc. Inst. Mech. Eng. Part J.* **2009**, *223*, 311–316. [CrossRef]
46. Vasanthakumar, K.; Ghosh, S.; Koundinya, N.T.B.N.; Ramaprabhu, S.; Bakshi, S.R. Synthesis and mechanical properties of TiCx and Ti (C, N) reinforced Titanium matrix in situ composites by reactive spark plasma sintering. *Mater. Sci. Eng. A* **2019**, *759*, 30–39. [CrossRef]
47. Hayat, M.D.; Singh, H.; He, Z.; Cao, P. Titanium metal matrix composites: An overview. *Compos.-A Appl. Sci. Manuf.* **2019**, *121*, 418–438. [CrossRef]
48. Choudhari, A.; Elder, J.; Mugale, M.; Karki, S.; Digole, S.; Omeike, S.; Borkar, T. Enhancing Quality Control: Image-Based Quantification of Carbides and Defect Remediation in Binder Jetting Additive Manufacturing. *Materials* **2024**, *17*, 2174. [CrossRef] [PubMed]
49. Choudhari, A.; Elder, J.; Mugale, M.; Karki, S.; Vukkum, V.B.; Gupta, R.K.; Borkar, T. Additive manufacturing of AISI M2 tool steel by binder jetting (BJ): Investigation of microstructural and mechanical properties. *J. Manuf. Process.* **2024**, *132*, 686–711. [CrossRef]
50. Zhang, Y.; Zheng, Y.; Li, Y.; Wang, L.; Bai, Y.; Zhao, Q.; Xiong, X.; Cheng, Y.; Tang, Z.; Deng, Y.; et al. Tantalum nitride-decorated titanium with enhanced resistance to microbiologically induced corrosion and mechanical property for dental application. *PLoS ONE* **2015**, *10*, e0130774. [CrossRef] [PubMed]
51. Zhevtun, I.G.; Gordienko, P.S.; Mashtalyar, D.V.; Kulchin, Y.N.; Yarusova, S.B.; Nepomnyushchaya, V.A.; Kornakova, Z.E.; Gribanova, S.S.; Gritsuk, D.V.; Nikitin, A.I. Tribological properties of Ti-TiC composite coatings on titanium alloys. *Materials* **2022**, *15*, 8941. [CrossRef] [PubMed]
52. Koricherla, M.V.; Torgerson, T.; Alidokht, S.; Munagala, V.; Chromik, R.; Scharf, T. High temperature sliding wear behavior and mechanisms of cold-sprayed Ti and Ti–TiC composite coatings. *Wear* **2021**, *476*, 203746. [CrossRef]
53. Alidokht, S.A.; Munagala, V.N.V.; Chromik, R.R. Role of third bodies in friction and wear of cold-sprayed Ti and Ti–TiC composite coatings. *Tribol. Lett.* **2017**, *65*, 114. [CrossRef]
54. Chen, J.; Zhang, S.; Zheng, J.; Dong, Y.; Zhang, C.; Li, J.; Chen, Z.; Zhang, J.; Sun, D. Excellent anti-corrosion and conductivity of NbN coated on Ti bipolar plate by controlling N_2 flow rates. *J. Alloys Compd.* **2024**, *976*, 173033. [CrossRef]

metals

MDPI

Article

In Situ Synthesis of $MoSi_2$-SiC Composites by Two-Step Spark Plasma Sintering

Yue-Yao Wang and Guo-Hua Zhang *

State Key Laboratory of Advanced Metallurgy, University of Science and Technology Beijing, Beijing 100083, China; wangyueyao9@163.com
* Correspondence: ghzhang0914@ustb.edu.cn; Tel.: +86-1082377750; Fax: +86-1062332252

Abstract: The effect of different SiC doping content on the properties of $MoSi_2$-based composites was analyzed in this study. The $MoSi_2$-SiC composites were fabricated in situ by the SPS technique, utilizing self-synthesized carbon-containing Mo powder and Si powder as raw materials. A two-step sintering process was employed to ensure the formation of a uniform and dense composite structure. The microstructures and mechanical properties of these composites with various compositions were characterized. The results show that the composites were primarily composed of $MoSi_2$, SiC, and a minor proportion of MoSiC phase. The introduction of SiC as a second phase was found to considerably enhance the mechanical properties of the $MoSi_2$ matrix material. In particular, the $MoSi_2$-26mol.%SiC sample exhibited Vickers hardness, fracture toughness, and flexural strength values of 16.1 GPa, 6.7 $MPa{\cdot}m^{1/2}$, and 496 MPa, respectively, corresponding to increases of 33%, 24%, and 28% compared to the pure $MoSi_2$ material.

Keywords: molybdenum disilicide; $MoSi_2$-SiC composites; microstructure; mechanical properties

1. Introduction

Molybdenum disilicide ($MoSi_2$) is a borderline intermetallic with both metallic and ceramic characteristics [1,2]. It is widely used in aerospace, high-temperature smelting, electronic components, and refractory materials that require high temperature resistance, excellent oxidation resistance, and good thermal stability [3–5]. This can be attributed to several unique properties of $MoSi_2$, including its high melting point, low density, high thermal conductivity, and low coefficient of thermal expansion. Additionally, a protective glass layer is produced on the surface of $MoSi_2$ at high temperatures, which plays the role of hindering oxygen diffusion, so as to maintain a longer service life in high-temperature air environments [6–9]. However, there are a number of problems with its practical applications, which limit the use of $MoSi_2$ in certain areas with high performance requirements. $MoSi_2$ undergoes the "PEST" phenomenon when exposed to oxygen at moderate temperatures, and is rapidly oxidized to powder and volatilizes, leading to the destruction of the structure of the matrix as well as the degradation of its mechanical properties. Although $MoSi_2$ exhibits good strength at high temperatures, its inherent toughness is poor, especially at low temperatures, and the $MoSi_2$ material is susceptible to brittle fracture in environments subjected to shock loads or load variations [10,11]. Therefore, further research is required to enhance the properties of $MoSi_2$ materials.

The properties of $MoSi_2$ can be improved by a variety of methods, such as doping or alloying, process improvements, and surface coating. Numerous enhancement methods

 https://doi.org/10.3390/met15020144

have been proposed by researchers to improve its material properties. These approaches include the incorporation of elements such as metallic W, Al, and Ti, or the addition of reinforcing phases such as carbides, nitrides, and oxides. Furthermore, grain refinement techniques have been employed to control grain size and distribution, thereby enhancing the mechanical properties. Advanced processing techniques have also been utilized to improve the material density and overall mechanical performance [12–15]. Xu et al. successfully prepared $MoSi_2/WSi_2$ solid solution powders with nanostructures by combustion synthesis using a blend of mechanically activated Mo, W, and Si powders [16]. Li et al. prepared $(Mo,Ti)Si_2$ biphasic composite antioxidant coatings on molybdenum alloys by the molten slurry method, and the composite coatings formed during oxidation of the Si-Ti-O composite glass can provide long-term antioxidant protection [17]. Ko et al. synthesized nanoscale $MoSi_2$ and Si_3N_4 powders via high-energy ball milling using Mo_2N and Si powders as raw materials, and then prepared $MoSi_2$-Si_3N_4 compounds by pulsed-current-activated sintering method, with an average hardness and fracture toughness of 12 GPa and 6 $MPa{\cdot}m^{1/2}$, respectively [18]. It is worth noting that SiC, as a strong and high-temperature-resistant material, has a low coefficient of thermal expansion (CTE, 5×10^{-6} K^{-1}) as well as good compatibility with $MoSi_2$ materials [19,20]. Therefore, the addition of SiC to $MoSi_2$ materials can effectively enhance their thermal stability and high-temperature strength, while the composites have a relatively low density, making them advantageous for aerospace and other applications that require lightweighting [21]. Therefore, $MoSi_2$-SiC composites are considered to have broad application prospects and have received extensive attention [22,23]. Chen et al. prepared 10% SiC/$MoSi_2$ composites using in situ pressureless sintering with Mo, Si, and C powders. The flexural strength and fracture toughness of the composites reached 274.5 MPa and 5.5 $MPa{\cdot}m^{1/2}$, respectively, equating to an increase of 40.8% and 30.6%, respectively, over the pure $MoSi_2$ material [24]. Dense SiC/$MoSi_2$ nanocomposites with a relative density exceeding 98% were prepared by Peng et al. using reactive hot pressing. The room temperature flexural strength of the 15 vol% SiC/$MoSi_2$ composite was 610 MPa, which represented a 141% increase compared to that of monolithic $MoSi_2$. Additionally, the yield strength at 1200 °C reached 720 MPa [25]. He et al. fabricated SiC/$MoSi_2$ composite coatings on the surface of C/C composites using the chemical vapor infiltration/reaction technique. The composite coatings effectively protected the samples from oxidation at 1500 °C for 80 h, with only a 1.25% mass loss [26]. Overall, the incorporation of SiC effectively strengthens the comprehensive properties of $MoSi_2$-based composites, positioning them as promising candidates for a wider array of high-temperature environments.

To date, various preparation methods have been developed to synthesize $MoSi_2$ matrix composites, including mechanical alloying (MA) [27], hot pressing (HP) [28], plasma spraying [29], spark plasma sintering (SPS) [30], self-propagating high-temperature synthesis (SHS) [31] and other methods. The results of the current study indicate that powder metallurgy techniques can effectively optimize the microstructure of $MoSi_2$ and improve its mechanical properties by adjusting sintering parameters such as the temperature, holding time, and pressure [32]. Notably, SPS is recognized as one of the most advanced powder metallurgy techniques, capable of achieving nearly full densification while maintaining a fine grain structure during the sintering process [33]. In this study, the $MoSi_2$ materials were strengthened by introducing the SiC second phase, and the $MoSi_2$-based composites were prepared by in situ synthesis using SPS equipment using Si powder and self-synthesized carbon-containing Mo powders as raw materials. $MoSi_2$-SiC composites with different SiC contents were successfully produced and analyzed for their microstructure and various mechanical properties.

2. Experimental Procedures

2.1. Sample Fabrication

The experimental raw materials were ammonium molybdate ($(NH_4)_2Mo_4O_{13}$, Jinduicheng Molybdenum Co., Ltd., Xi'an, China), carbon black (BR193, Tianjin Ebory Chemical Co., Ltd., Tianjin, China), and silicon powder (Si, 99.99% purity, particle size 5 μm, Beijing Xingrongyuan Technology Co., Ltd., Bejing, China). Molybdenum (Mo) powder and carbon-containing Mo powder were first prepared by a two-step reduction method [34]. The detailed preparation process is given as follows.

$(NH_4)_2Mo_4O_{13}$ was calcined at 500 °C for 4 h to obtain molybdenum trioxide, which was mixed with four different molar ratios of carbon black by ball milling (1:1.9, 1:2.2, 1:2.3, and 1:2.4), and then carbothermally reduced to obtain homogeneous and dispersed pre-reduced Mo powders. Finally, a slight portion of residual molybdenum oxide is removed with hydrogen at the deep deoxygenation stage. The C contents of the four powders obtained were 0.01 wt.%, 1.63 wt.%, 3.19 wt.%, and 4.26 wt.%, respectively. The powder mixtures were prepared by adding Si powder according to the powder compositions, which are listed in Table 1. The powder blends were milled by zirconia balls using a high-energy ball mill for 6 h to ensure that the powder mixtures were homogeneous and dispersed. Anhydrous ethanol was used as the grinding medium with a ball to material ratio of 5:1 and a rotational speed of 200 rpm. After the milling treatment, the mixtures were placed into 20 mm diameter cylindrical graphite molds, and a layer of graphite foil was added between the powders and the molds to avoid sticking during the high-temperature sintering procedure. The sintering process was performed using SPS equipment (FIH-452NP, Fuji Electronic Industrial Co., Ltd., Tokyo, Japan) with a two-step sintering process, first at 1200 °C for 5 min, and then at 1600 °C for 10 min. The uniaxial pressure was maintained at 50 MPa during the sintering process, and the temperature ramping rate was 100 °C/min.

Table 1. Designations and raw material compositions for different samples.

Sample Number	Mo/wt.%	C/wt.%	Si/wt.%	Materials
M1	63.07	-	36.93	$MoSi_2$
M2	60.95	1	38.05	$MoSi_2$-12mol.%SiC
M3	58.98	1.94	39.08	$MoSi_2$-21mol.%SiC
M4	57.67	2.57	39.76	$MoSi_2$-26mol.%SiC

2.2. Sample Characterization

The sintered circular flake samples were fully sanded with SiC sandpaper to remove the graphite foil on the surface and polished until the sample surface showed a mirror effect. The theoretical densities of the composites with different compositions were calculated based on the densities of $MoSi_2$ (6.28 g/cm^3) and SiC (3.2 g/cm^3) [35,36]. The actual densities of each group of samples were calculated by Archimedes' method and the relative densities were further obtained by the aspect index of actual density to theoretical density. The Vickers hardness was measured on the surface of the samples by a digital Vickers hardness tester at a load of 1 kg and a holding pressure of 10 s. Every sample was randomly measured at 10 positions and the data were averaged. The Vickers hardness was calculated as $Hv = 1.854 \cdot \frac{f}{d^2}$, where f is the applied load in kg and d is the average of the two diagonal lines of the indentation in mm. Fracture toughness (K_{IC}) data were obtained by measuring the crack length produced on the sample surface under a load of 5 kg and calculated according to the formula of Anstis et al. [37]. The sintered samples of different compositions were all machined into rectangles with 12 mm × 4 mm × 3 mm dimensions by a diamond wire cutting machine. Three-point flexural tests were carried out on a universal testing machine to measure the flexural strength with a loading span of 9 mm and a crosshead speed of 0.5 mm/min, and the samples of each composition were tested three times to calculate the mean value. The electrical resistivity was determined by the four-probe method using a multifunctional digital four-probe tester, and the average value was taken from at least three measurements for each sample.

The phase compositions of the composites were analyzed and determined by an X-ray diffractometer (XRD, TTR III, Rigaku Corporation, Tokyo, Japan) with Cu Kα radiation in the 2θ range of 10–90° at a scanning rate of 0.2 s/step and a step size of 0.02 °/step. The microstructures and chemical compositions of the sintered specimen surfaces and sections were analyzed by field emission scanning electron microscopy (FE-SEM, ZEISS SUPRA 55, Oberkochen, Germany) and energy spectroscopy (EDS).

3. Result and Discussion

3.1. Thermodynamics of $MoSi_2$-SiC System

Based on the composition of the Mo-Si-C mixture powders, the following reactions are assumed to take place during the sintering process:

$$Mo_{(s)} + 2Si_{(s)} = MoSi_{2(s)} \quad (1)$$

$$5Mo_{(s)} + 3Si_{(s)} = Mo_5Si_{3(s)} \quad (2)$$

$$C_{(s)} + Si_{(s)} = SiC_{(s)} \quad (3)$$

In order to further verify the sintered products, the thermodynamic data of the possible reactions were calculated using HSC Chemistry 6.0 software. The Gibbs free energy changes (ΔG^{θ}) for these reactions at various temperatures are depicted in Figure 1, showing that the reactions proceeded spontaneously during both holding stages. Therefore, the composites obtained using two-stage SPS of Mo-Si-C mixed powders are inferred to be composed of $MoSi_2$, Mo_5Si_3, and SiC.

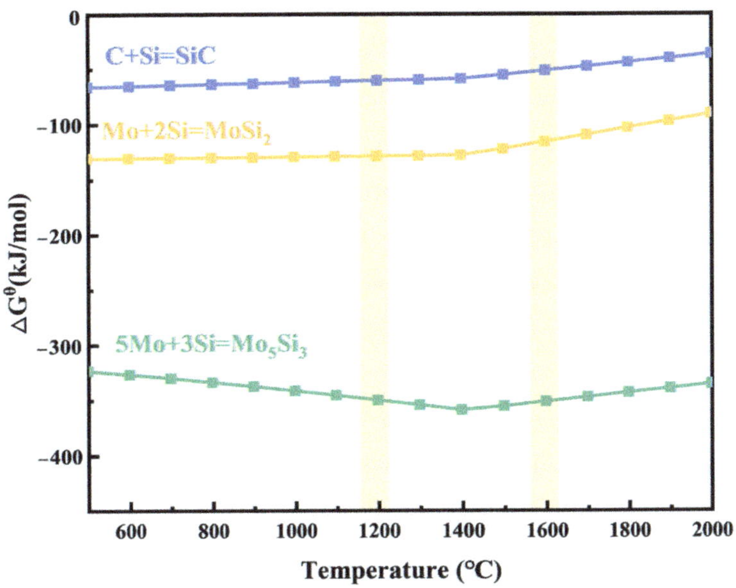

Figure 1. Temperature dependences of the changes in standard Gibbs free energy for Equations (1)–(3).

3.2. Phase Identification

Figure 2 illustrates the XRD patterns of four groups of samples containing different SiC contents after sintering by SPS. The results indicate that the primary phase in the composites is $MoSi_2$, while the SiC phase becomes more obvious as the C content in the powder samples increases. Additionally, weak diffraction peaks corresponding to Mo_5Si_3 were observed. Mo_5Si_3 is a typical by-product in the $MoSi_2$ synthesis process [38,39]. In addition, Hu et al. [40] determined that Mo_5Si_3C is the equilibrium phase within the Mo-Si-C system, as indicated by the following high-temperature reactions:

$$5Mo_{(s)} + 3Si_{(l)} + C = Mo_5Si_3C \tag{4}$$

$$5MoSi_2 + 8C = Mo_5Si_3C + 7SiC \tag{5}$$

The approximate composition of this phase was found to be $Mo_{4.8}Si_3C_{0.6}$. This product phase has been confirmed by other relevant studies [41,42]. It can be observed that $MoSi_2$ and SiC were primarily generated in the reaction process without the residual raw powders.

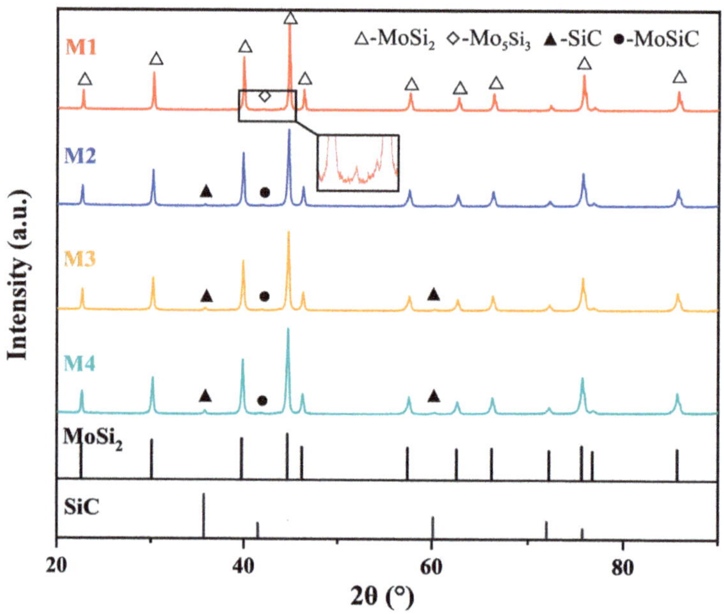

Figure 2. XRD patterns of the as-sintered $MoSi_2$-SiC composites.

3.3. Behavior of Sintering

It is well established that the sintering temperature has an essential role in controlling the microstructure and properties of materials. Therefore, a series of preliminary experiments were conducted by adjusting the sintering temperature (1300 °C–1700 °C) before determining the experimental conditions. The results indicated that at lower sintering temperatures, incomplete reactions of the raw materials and the formation of a porous structure occurred. This is attributed to the slower atomic migration rates and insufficient particle contact, which hindered the formation of sintering necks and reduced the degree of densification. Conversely, higher sintering temperatures promoted excessive grain growth, creating defects and deteriorating the mechanical properties of the material. Additionally, the residual oxides in the samples generate gases during the carbon reduction process at high temperatures, resulting in the formation of numerous pores. When high-temperature sintering was employed directly, rapid densification prevented gas discharge, resulting in the formation of closed pores and consequently increasing the porosity of the sintered samples. In conclusion, the two-step sintering process used in this study offers distinct advantages. In the first step, the material was held at 1200 °C for 5 min to complete the reaction and achieve initial densification, followed by a second step, where the temperature was raised to 1600 °C for 10 min to further enhance the densification and material strength. This approach enabled the gradual optimization of the microstructure, preventing excessive grain growth and sintering inhomogeneity, ultimately producing a more homogeneous and densified material. As shown in Figure 3, the relative densities of the four powders after sintering demonstrate that almost full densification was achieved in all of the samples. The introduction of SiC as a second phase caused a slight reduction in the relative density, attributed to gas evolution during the reduction of SiO_2 by C. However, all of the composites achieved relative densities exceeding 98%.

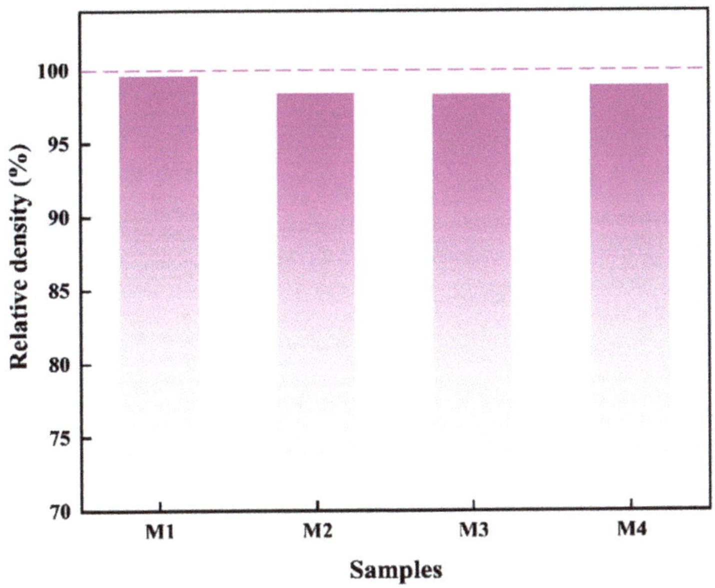

Figure 3. The effect of different SiC contents on the relative densities of the samples.

Figure 4 shows the temperature and shrinkage displacement curves as a function of the sintering time during the two-step sintering process for the Mo-Si-C system. The shrinkage displacement curves represent the changes in the sample thickness, with positive or negative displacements indicating expansion or contraction, respectively. The entire procedure can be separated into five unique stages. In the initial stage, as the temperature and pressure increase, the displacement remains stable after the mixed powder is gradually compacted. In the second stage, as the temperature continues to rise, diffusion and particle rearrangement occur between the powder particles, leading to the formation of preliminary inter-particle connections. Due to the large volume change before and after the reaction, this corresponds to a rapid contraction of the displacement curve in the figure. In the third stage, the rate of displacement contraction slows down during the holding period at 1200 °C. Until the temperature is further increased, in the fourth stage, the diffusion rate between the particles accelerates, and the reaction becomes more intense. The raw material powder is fully reacted and shows an instantaneous shrinkage of the displacement.

During the final holding stage at 1600 °C, the displacement curve remains almost constant and the sintered sample tends to densify completely. During the final densification stage, grain growth occurs. Meanwhile, the particles are more tightly bound and the porosity is reduced. As a result, the sintered samples exhibit high densification degrees, resulting in excellent mechanical properties.

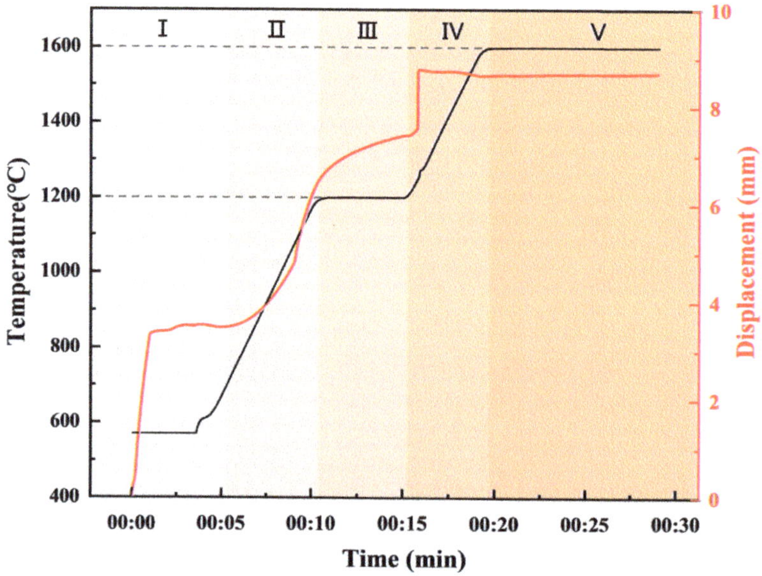

Figure 4. The displacement and temperature variation curves of the M2 sample vs. time during the two-step sintering process.

3.4. Microstructure of Composites

Figure 5 presents the microstructure of the sintered samples of four various compositions, all of which exhibit a uniform and dense microstructure. The pure $MoSi_2$ sample depicted in Figure 5a consists of a grey matrix phase, a white phase, and a nearly round dark phase. Conversely, the Mo-Si-C system samples shown in Figure 5b–d contain, in addition to the grey matrix phase and some white phase, a significant quantity of black phases. These black phases are characterized by irregular shapes and extensive agglomeration, the causes of which will be discussed in detail below.

Figure 6 displays an enlarged view of the sample surface and corresponding EDS results. On the basis of the EDS mapping data, it can be deducted that the grey phase corresponds to the $MoSi_2$ matrix. The white phase, characterized by a high Mo content and low Si content, is likely Mo_5Si_3, while the darker phase, enriched in oxygen, corresponds to SiO_2. Due to the amorphous nature of SiO_2, its diffraction peaks are not detected in XRD. Its generation is related to the inevitable adsorption of oxygen on the powder surface in treatment, and, thus, oxidation of the highly activated Si powder produces the SiO_2 phase. Furthermore, the reduction in the Si content in certain regions causes the creation of the Mo_5Si_3 phase.

In the Mo-Si-C system samples (Figure 6b–d), besides the $MoSi_2$ matrix and a small amount of MoSiC phase, a considerable number of irregular black phases are observed, primarily caused by the agglomeration of SiC particles. Since the melting point of Si is 1410 °C, the elevated sintering temperature and exothermic reactions during sintering cause Si to reach its molten state, increasing its migration rate [41]. As C particles are much lighter, they transfer simultaneously with Si, leading to the agglomeration of SiC. Moreover, as the C content in the sample increases, the surface morphology of the sintered sample remains largely unchanged, but the proportion of SiC phase increases, with numerous aggregates appearing. This process also mitigates the impact of the brittle SiO_2 phase on the mechanical properties of the $MoSi_2$ matrix, as SiO_2 at the grain boundaries is reduced by C to form SiC.

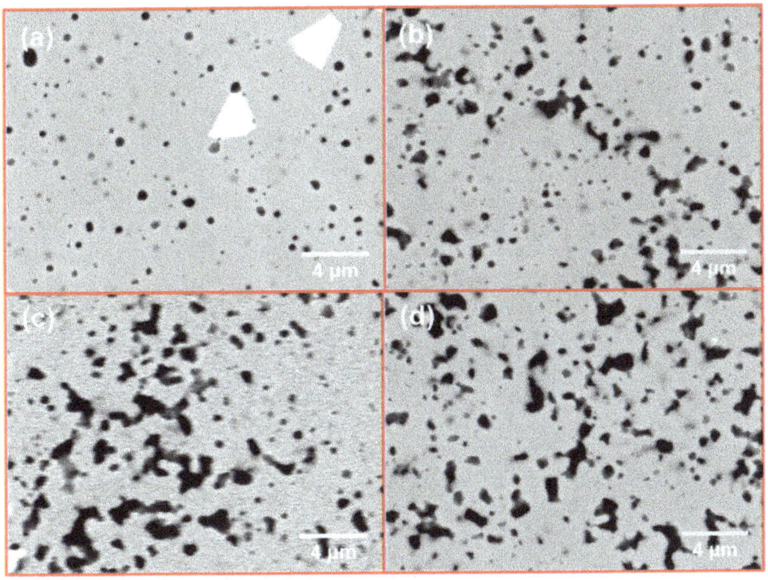

Figure 5. The morphology images of sintered products: (**a**) M1; (**b**) M2; (**c**) M3; (**d**) M4.

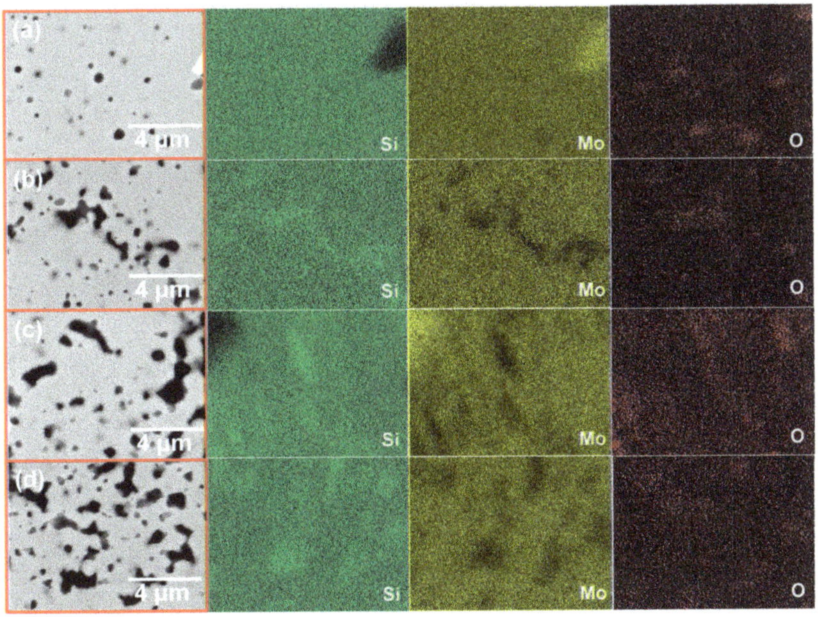

Figure 6. Enlargement of surface topography and EDS mapping analyses of four samples: (**a**) M1; (**b**) M2; (**c**) M3; (**d**) M4.

3.5. Mechanical Properties

Figure 7 illustrates the Vickers hardness and fracture toughness data for four composite sets with varying C contents. It is evident that the presence of carbon in the samples, which results in the generation of SiC as a second phase, significantly enhances both the hardness and toughness of the $MoSi_2$-SiC composites compared to pure $MoSi_2$ materials. As the proportion of SiC increases, the properties of the composites show further improvement. A comparison between the M1 and M4 samples reveals that the hardness increases from 12.1 GPa to 16.1 GPa, representing a 33% enhancement. This demonstrates that the incorporation of SiC effectively raises the hardness of $MoSi_2$-based composites. Although it is generally expected that an increase in hardness would be associated with a decline in the fracture toughness, the data indicate that the fracture toughness of the samples does not follow this trend. In fact, it appears to increase. Specifically, the fracture toughness of the M4 sample reaches 6.7 $MPa \cdot m^{1/2}$, which is 24% higher than the pure $MoSi_2$ sample, which exhibits a toughness of 5.4 $MPa \cdot m^{1/2}$ under similar preparation conditions.

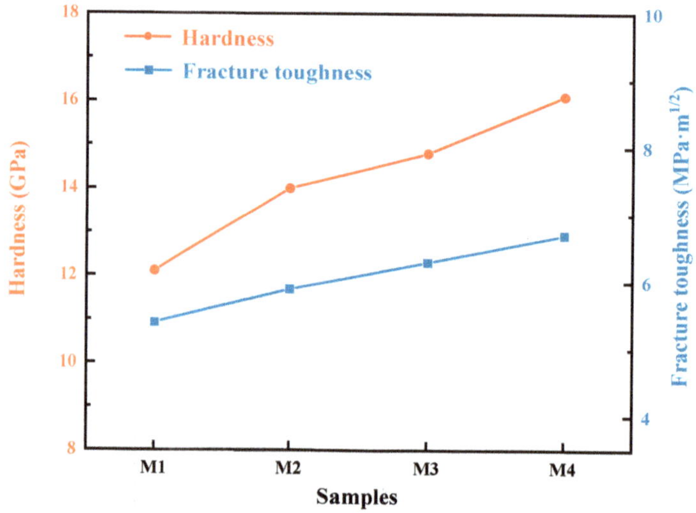

Figure 7. Hardness (Hv) and indentation fracture toughness (K_{IC}) of as-sintered composites versus SiC content.

The trend of increasing fracture toughness in the composites is further analyzed through a comparison of the crack extension paths observed in the four sets of samples. The cracks and their propagation paths resulting from indentations on the surface of the sintered samples are labeled in Figure 8. The crack pattern in the pure $MoSi_2$ sample is characterized by a relatively flat morphology, with crack bridging occurring during crack extension and no significant deflection observed, as shown in Figure 8a. Crack bridging generates resistance to crack propagation, requiring the crack to surmount a higher energy barrier. In composite materials, cracks may exhibit bending or deflection when encountering particles or phase interfaces, thereby increasing the total crack propagation length [13,43]. In the M2 sample (Figure 8b), SiC exists as a second phase, and both crack bridging and deflection are observed during crack extension. With an increase in the SiC phase content, more phase interfaces are formed, causing cracks to propagate primarily through SiC particles while undergoing deflection. This results in a more tortuous crack path, as shown in Figure 8c,d. Both crack bridging and crack deflection mechanisms play a vital part in increasing the toughness of the material, with their combined effects boosting the crack resistance of the composites and consequently decreasing the risk of brittle fracture.

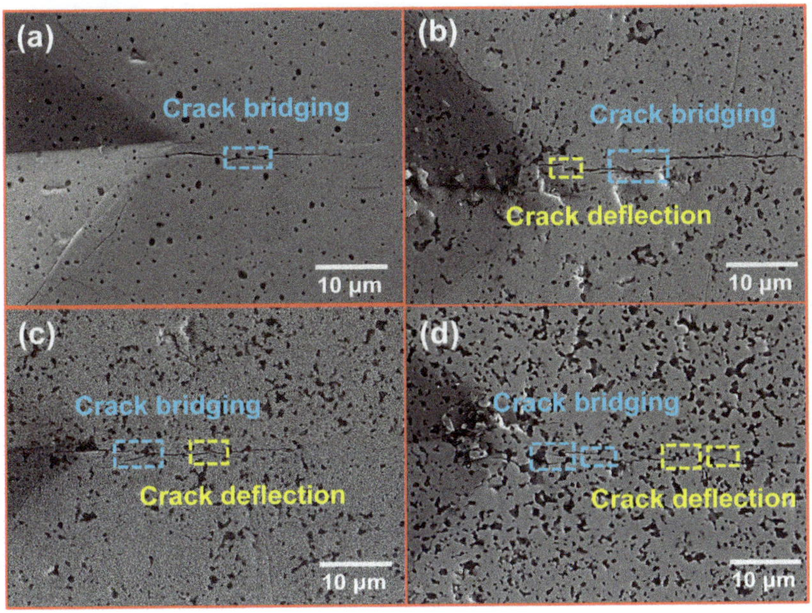

Figure 8. Crack propagation path of $MoSi_2$ composites with various SiC contents: (**a**) M1; (**b**) M2; (**c**) M3; (**d**) M4.

Figure 9 illustrates the fracture surface morphology of the sintered sample. Figure 9a presents a typical brittle fracture characterized by a flat, relatively smooth surface with distinct river patterns and cleavage steps. Based on the uniform crack propagation observed Figure 9a, it can be concluded that the fracture mode of the M1 sample is predominantly transgranular, indicating that the sample has relatively poor toughness, which aligns with the known properties of $MoSi_2$ material. In Figure 9b, the introduction of the second-phase SiC in the M2 sample results in plastic deformation, with the fracture surface appearing uneven. Despite this, the fracture mode remains transgranular overall. With increasing SiC contents, as depicted in Figure 9c,d, the fracture surfaces become noticeably irregular and exhibit an uneven morphology. This can be attributed to plastic deformation, stretching, and slip occurring during the fracture process. The fracture surface reveals distinct grain structures, and at this stage, the fracture mode transitions into a mixed intergranular–transgranular fracture, with intergranular fracture being the dominant mechanism. Additionally, as the SiC content increases, a marked reduction in the grain size is observed. This suggests that the SiC second phase plays a crucial role in refining the $MoSi_2$ matrix, thereby improving the mechanical properties of the composite material.

Figure 10 displays the flexural strength of the pure $MoSi_2$ and composites with various SiC contents. The incorporation of SiC considerably enhances the flexural strength of the samples. The flexural strength increases steadily with the SiC content, with the M4 samples achieving a flexural strength of 496 MPa. This improvement is attributed to the synergistic effects of the high densification degree, fracture behavior, microstructure, and finer grain size of the composites.

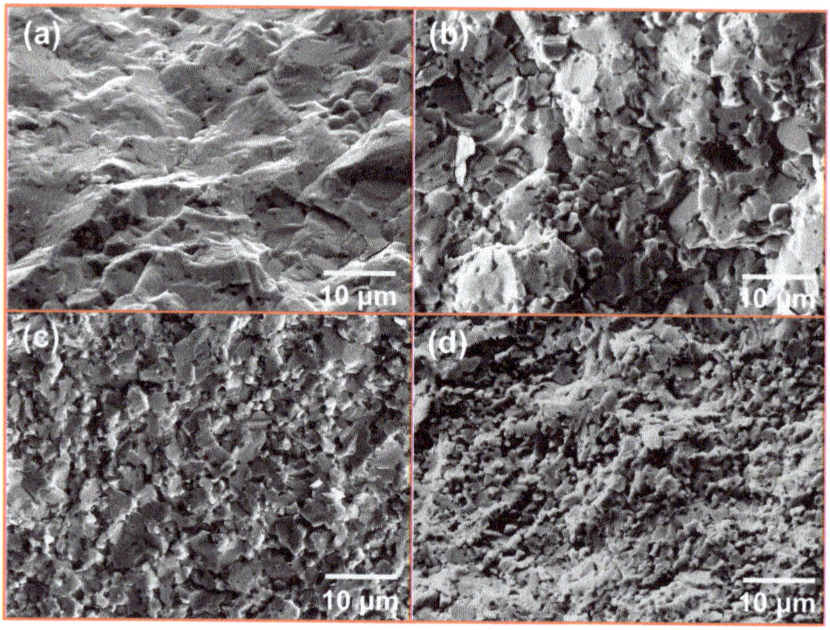

Figure 9. Fracture morphologies of $MoSi_2$ composites with various SiC contents: (**a**) M1; (**b**) M2; (**c**) M3; (**d**) M4.

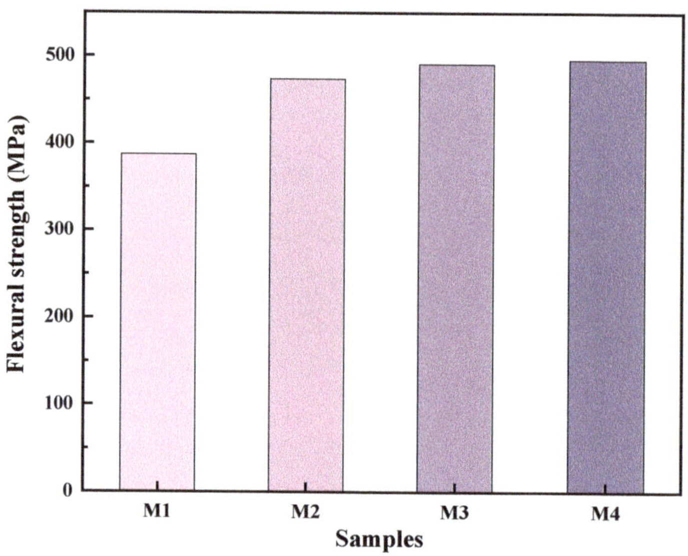

Figure 10. Flexural strength of as-sintered composites versus SiC content.

3.6. Comparison of Mechanical Properties

The relative density, hardness, fracture toughness, and flexural strength of the sintered samples corresponding to the four different compositions in this study are presented in detail in Table 2. The performance data of the $MoSi_2$-based composites produced by different preparation methods and reinforcing phases in the literature are also quoted and further compared with the M4 samples in this study. Compared with the $MoSi_2$-SiC composites obtained by other preparation methods (e.g., microwave reaction sintering, pressureless sintering, and hot press sintering) listed in the table, the M4 samples have higher densities and superior mechanical properties. The fracture toughness or flexural strength values of the individual cited samples are slightly higher than those of the M4 samples, but the data for other properties are much lower than those of the M4 samples. In addition, the $MoSi_2$-SiC composites in this study also exhibit superior mechanical properties compared to $MoSi_2$-based composites that introduce reinforcing phases other than SiC. It can be concluded that the M4 samples synthesized by the two-step SPS method in this study have better overall properties. Notably, the $MoSi_2$-SiC composites demonstrated an enhanced performance, particularly in terms of both the hardness and toughness, which were simultaneously improved.

Table 2. Comparison of the mechanical properties of $MoSi_2$-based composite materials prepared in this study and the literature.

Sample	Preparation Method	Relative Density(%)	Vickers Hardness (GPa)	Fracture Toughness ($MPa{\cdot}m^{1/2}$)	Flexural Strength (MPa)	Reference
M1	SPS	99.6	12.1	5.4	387	-
M2		98.4	14.0	5.9	474	-
M3		98.3	14.8	6.3	491	-
M4		98.9	16.1	6.7	496	-
$MoSi_2$-20vol%SiC	Microwave reaction sintering	91.5	7.86	8.17	251.29	[44]
$MoSi_2$-SiC	High-frequency induction heated combustion	98	13.8	4.8	-	[45]
$MoSi_2$-10vol%SiC	Pressureless sintering	93.1	8.843	5.5	274.5	[24]
$MoSi_2$-SiC	SPS	97.8	13.4	5.1	674	[41]
$MoSi_2$-20vol%SiC	Hot pressing	98	11.5	4.6	240	[46]
$MoSi_2$–5vol%Al_2O_3	Vacuum hot-pressed sintering	96.5	11.14	4.53	435	[47]
$MoSi_2$-40wt%Ti_5Si_3	SPS	99	10.9	5.11	-	[48]
$MoSi_2$-0.4MoB-0.1SiC	Vacuum hot-pressed sintering	97.5	12.5	3.8	537	[49]
$MoSi_2$-Si_3N_4	Pulsed current-activated sintering	97	12	6	-	[18]

3.7. Electrical Properties

The room temperature resistivities of the four groups of sintered samples are detailed in Table 3, where the resistivity of the pure $MoSi_2$ sample is 16.8 $\mu\Omega{\cdot}$cm. The resistivity of the samples shows a gradual increase as the SiC content in the samples increases. The resistivity of the composites is influenced by multiple factors, including the additive composition, porosity, grain size, and grain boundaries [50]. SiC, as a semiconductor, exhibits a room temperature resistivity that can be varied over a wide range, typically between 10^3 and 10^{19} $\mu\Omega{\cdot}$cm for both single-crystal and polycrystalline forms [51]. However, the resistivity of SiC is considerably higher than that of $MoSi_2$ materials. Therefore, the introduction of SiC as a second phase in the $MoSi_2$ matrix leads to an increase in the resistivity of the composite. In addition, as the proportion of the SiC phase increases, more phase interfaces are generated between $MoSi_2$ and SiC, which further increases the resistivity of the composites. In conclusion, the resistivity of the $MoSi_2$-SiC composites is theoretically higher than that of the $MoSi_2$ materials. The room temperature resistivities of the four groups of samples in Table 3 also confirm the effect of the SiC content on the resistivity of the composites. It is observed that the resistivity increases progressively as the molar fraction of SiC increases. Thus, the resistivity of the $MoSi_2$-SiC composites can be effectively tuned by controlling the amount of SiC added.

Table 3. Electrical resistivity versus SiC content.

Sample Number	Electrical Resistivity (μΩ·cm)
M1	16.8
M2	18.9
M3	21.5
M4	23.3

4. Conclusions

$MoSi_2$-SiC composites were successfully prepared from carbon-containing Mo powder and Si powder by in situ synthesis, and the influence of various SiC contents on the microstructure and mechanical properties of the composites were analyzed. The sintering process involved a two-step treatment at 1200 °C and 1600 °C to ensure the formation of uniform, dense, and high-performance composites. The incorporation of SiC as a secondary phase into the $MoSi_2$ matrix effectively enhanced the mechanical properties of the composites. Among the samples, the $MoSi_2$-26mol.%SiC composite demonstrated the highest overall performance, exhibiting a relative density of 98.9%, a Vickers hardness of 16.1 GPa, a fracture toughness of 6.7 $MPa{\cdot}m^{1/2}$, and a flexural strength of 496 MPa. Additionally, the inclusion of SiC was observed to increase the electrical resistivity of the $MoSi_2$ material.

Author Contributions: Y.-Y.W.: formal analysis, investigation, writing—original draft, writing—review and editing; G.-H.Z.: conceptualization, methodology, resources, supervision, funding acquisition, writing—review and editing. All authors have read and agreed to the published version of the manuscript.

Funding: This research received no external funding.

Data Availability Statement: The raw data supporting the conclusions of this article will be made available by the authors on request.

Acknowledgments: The authors gratefully acknowledge the financial support from the State Key Laboratory of Advanced Metallurgy, University of Science and Technology Beijing, China.

Conflicts of Interest: The authors declare that they have no conflicts of interest.

References

1. Ramasesha, S.K.; Tantri, P.S.; Bhattacharya, A.K. $MoSi_2$ and $MoSi_2$-Based Materials as Structural Ceramics. *Met. Mater. Process.* **2000**, *12*, 181–190.
2. Tantri, P.S.; Bhattacharya, A.K.; Ramasesha, S.K. Synthesis and properties of $MoSi_2$ based engineering ceramics. *J. Chem. Sci.* **2001**, *113*, 633–649. [CrossRef]
3. Pan, Y.; Wang, S. Insight into the oxidation mechanism of $MoSi_2$: Ab-initio calculations. *Ceram. Int.* **2018**, *44*, 19583–19589. [CrossRef]
4. Cao, L.; Bai, Z.; Huang, J.; OuYang, H.; Li, C.; Wang, B.; Yao, C. Fabrication of gradient C/C-SiC-$MoSi_2$ composites with enhanced ablation performance. *Ceram. Int.* **2016**, *42*, 12289–12296. [CrossRef]
5. Ding, Z.; Brouwer, J.C.; Zhu, J.-N.; Popovich, V.; Hermans, M.J.; Sloof, W.G. Effects of boron addition on the high temperature oxidation of $MoSi_2$ alloys. *Scr. Mater.* **2023**, *234*, 115580. [CrossRef]
6. Yan, J.-h.; Xu, J.-j.; Wang, Y.; Liu, L.-F. Preparation of agglomerated powders for air plasma spraying $MoSi_2$ coating. *Ceram. Int.* **2015**, *41*, 10547–10556. [CrossRef]
7. Zhang, Y.; Li, Y.; Bai, C. Microstructure and oxidation behavior of Si–$MoSi_2$ functionally graded coating on Mo substrate. *Ceram. Int.* **2017**, *43*, 6250–6256. [CrossRef]
8. Yan, J.; Wang, Y.; Liu, L.; Wang, Y. Oxidation and interdiffusion behavior of Niobium substrate coated $MoSi_2$ coating prepared by spark plasma sintering. *Appl. Surf. Sci.* **2014**, *320*, 791–797. [CrossRef]
9. Erfanmanesh, M.; Bakhshi, S.R.; Khajelakzay, M.; Salekbafghi, M. The effect of argon shielding gas at plasma spray process on the structure and properties of $MoSi_2$ coating. *Ceram. Int.* **2014**, *40*, 4529–4533. [CrossRef]

10. Shan, A.; Fang, W.; Hashimoto, H.; Park, Y.-H. Effect of Mg addition on the microstructure and mechanical properties of $MoSi_2$ alloys. *Scr. Mater.* **2002**, *46*, 645–648. [CrossRef]
11. Wen, S.; Sha, J. Improvement of "Pest" Resistance of $MoSi_2$ Intermetallic Compound at 500 °C and 600 °C via Addition of B Fabricated by Spark Plasma Sintering. *Oxid. Met.* **2019**, *92*, 243–257. [CrossRef]
12. Zhang, H.; Lv, J.; Zhuang, S.; Chen, Y.; Gu, S. Effect of WSi_2 and Si_3N_4 contents on the thermal expansion behaviors of (Mo,W) Si_2-Si_3N_4 composites. *Ceram. Int.* **2017**, *43*, 2847–2852. [CrossRef]
13. Nie, X.; Lu, Q. Fracture toughness of ZrO_2–SiC/$MoSi_2$ composite ceramics prepared by powder metallurgy. *Ceram. Int.* **2021**, *47*, 19700–19708. [CrossRef]
14. Zhu, G.; Wang, X.; Feng, P.; Li, D.; Yang, T.; Akhtar, F. Synthesis, microstructure and mechanical properties of (Mo,Ti) Si_2/Al_2O_3 composites prepared by thermite-reaction-assisted combustion synthesis. *J. Alloys Compd.* **2016**, *688*, 870–877. [CrossRef]
15. Zhu, M.; Zhang, L.; Li, N.; Cheng, D.; Zhang, J.; Yu, S.; Bai, H.; Ma, H. Microstructures and mechanical properties of reactive spark plasma-sintered ZrB_2–SiC–$MoSi_2$ composites. *Ceram. Int.* **2022**, *48*, 27401–27408. [CrossRef]
16. Xu, J.; Wu, H.; Li, B. Synthesis of $MoSi_2$/WSi_2 nanocrystalline powder by mechanical-assistant combustion synthesis method. *Int. J. Refract. Met. Hard Mater.* **2010**, *28*, 217–220. [CrossRef]
17. Li, W.; Fan, J.; Fan, Y.; Xiao, L.; Cheng, H. $MoSi_2$/(Mo,Ti) Si_2 dual-phase composite coating for oxidation protection of molybdenum alloy. *J. Alloys Compd.* **2018**, *740*, 711–718. [CrossRef]
18. Ko, I.-Y.; Kang, H.-S.; Doh, J.-M.; Yoon, J.-K.; Shon, I.-J. Properties and densification of nanocrystalline $MoSi_2$–Si_3N_4 composite from mechanically alloyed powders by pulsed current-activated sintering. *J. Alloys Compd.* **2010**, *502*, L10–L13. [CrossRef]
19. Peng, K.; Han, G.; Li, J.; Wang, L.; Zhang, W.; Wang, W.; Zhang, J. Anti-oxidation properties and phase composition evolution of the $MoSi_2$-HfO_2 duplex bond coat in air and steam/oxygen atmosphere at 1450 °C. *Surf. Coat. Technol.* **2024**, *492*, 131235. [CrossRef]
20. Chen, P.; Zhu, L.; Ren, X.; Kang, X.; Wang, X.; Feng, P. Preparation of oxidation protective $MoSi_2$–SiC coating on graphite using recycled waste $MoSi_2$ by one-step spark plasma sintering method. *Ceram. Int.* **2019**, *45*, 22040–22046. [CrossRef]
21. Wang, X.; Gao, X.; Zhang, Z.; Cheng, L.; Ma, H.; Yang, W. Advances in modifications and high-temperature applications of silicon carbide ceramic matrix composites in aerospace: A focused review. *J. Eur. Ceram. Soc.* **2021**, *41*, 4671–4688. [CrossRef]
22. Zhang, X.; Lu, Z.; Jin, Z. Electrical resistivity and microstructure of pressureless reactive sintered $MoSi_2$–SiC composite. *Mater. Chem. Phys.* **2004**, *86*, 16–20. [CrossRef]
23. Esmaeily, S.; Kermani, M.; Razavi, M.; Rahimipour, M.R.; Zakeri, M. An investigation on the in situ synthesis–sintering and mechanical properties of $MoSi_2$–xSiC composites prepared by spark plasma sintering. *Int. J. Refract. Met. Hard Mater.* **2015**, *48*, 263–271. [CrossRef]
24. Chen, F.; Xu, J.; Hou, Z. In situ pressureless sintering of SiC/$MoSi_2$ composites. *Ceram. Int.* **2012**, *38*, 2767–2772. [CrossRef]
25. Peng, K.; Yi, M.; Ran, L.; Ge, Y. Reactive hot pressing of SiC/$MoSi_2$ nanocomposites. *J. Am. Ceram. Soc.* **2007**, *90*, 3708–3711. [CrossRef]
26. He, Z.-b.; Li, H.-j.; Shi, X.-h.; Fu, Q.-g.; Heng, W. Formation mechanism and oxidation behavior of $MoSi_2$–SiC protective coating prepared by chemical vapor infiltration/reaction. *Trans. Nonferrous Met. Soc. China* **2013**, *23*, 2100–2106. [CrossRef]
27. Zakeri, M.; Yazdani-Rad, R.; Enayati, M.; Rahimipour, M. Synthesis of nanocrystalline $MoSi_2$ by mechanical alloying. *J. Alloys Compd.* **2005**, *403*, 258–261. [CrossRef]
28. Guo, S.-Q.; Nishimura, T.; Mizuguchi, T.; Kagawa, Y. Mechanical properties of hot-pressed ZrB2–$MoSi_2$–SiC composites. *J. Eur. Ceram. Soc.* **2008**, *28*, 1891–1898. [CrossRef]
29. Wang, C.-C.; Li, K.-Z.; He, D.-Y.; Shi, X.-H. Oxidation behavior of plasma-sprayed $MoSi_2$-Yb_2O_3 composite coating at 1700 °C. *Ceram. Int.* **2020**, *46*, 9538–9547. [CrossRef]
30. Zhu, L.; Zhang, S.; Ye, F.; Ren, X.; Feng, P. Recycling of $MoSi_2$-based industrial solid wastes for the fabrication and high-temperature oxidation behavior of $MoSi_2$–$ZrSi_2$–SiC composite coating. *Compos. Part B Eng.* **2024**, *274*, 111281. [CrossRef]
31. Potanin, A.Y.; Vorotilo, S.; Pogozhev, Y.S.; Rupasov, S.; Lobova, T.; Levashov, E. Influence of mechanical activation of reactive mixtures on the microstructure and properties of SHS-ceramics $MoSi_2$–HfB_2–MoB. *Ceram. Int.* **2019**, *45*, 20354–20361. [CrossRef]
32. Cabouro, G.; Chevalier, S.; Gaffet, E.; Grin, Y.; Bernard, F. Reactive sintering of molybdenum disilicide by spark plasma sintering from mechanically activated powder mixtures: Processing parameters and properties. *J. Alloys Compd.* **2008**, *465*, 344–355. [CrossRef]
33. Kermani, M.; Razavi, M.; Rahimipour, M.R.; Zakeri, M. The effect of temperature on the in situ synthesis–sintering and mechanical properties of $MoSi_2$ prepared by spark plasma sintering. *J. Alloys Compd.* **2014**, *585*, 229–233. [CrossRef]
34. Sun, G.-D.; Zhang, G.-H.; Ji, X.-P.; Liu, J.-K.; Zhang, H.; Chou, K.-C. Size-controlled synthesis of nano Mo powders via reduction of commercial MoO3 with carbon black and hydrogen. *Int. J. Refract. Met. Hard Mater.* **2019**, *80*, 11–22. [CrossRef]
35. Xu, J.; Jiang, G.; Li, W.; Zhuang, H.; Zhang, B.; Chen, L. In situ synthesis of SiCW/$MoSi_2$ composite through SPS process. *J. Alloys Compd.* **2008**, *462*, 170–174. [CrossRef]

36. Singh, S.; Bhaskar, R.; Narayanan, K.B.; Kumar, A.; Debnath, K. Development of silicon carbide (SiC)-based composites as microwave-absorbing materials (MAMs): A review. *J. Eur. Ceram. Soc.* **2024**, *44*, 7411–7431. [CrossRef]
37. Anstis, G.; Chantikul, P.; Lawn, B.R.; Marshall, D. A critical evaluation of indentation techniques for measuring fracture toughness: I, direct crack measurements. *J. Am. Ceram. Soc.* **1981**, *64*, 533–538. [CrossRef]
38. Kermani, M.; Razavi, M.; Rahimipour, M.R.; Zakeri, M. The effect of mechanical alloying on microstructure and mechanical properties of $MoSi_2$ prepared by spark plasma sintering. *J. Alloys Compd.* **2014**, *593*, 242–249. [CrossRef]
39. Zhang, J.; Wang, Z.; Luo, J.; Wang, S.; Liang, B.; Chen, W. Microstructure, properties and toughening mechanisms of $MoSi_2$@ ZrO_2 core shell composites prepared by spark plasma sintering. *Mater. Charact.* **2023**, *195*, 112510. [CrossRef]
40. Hu, Q.; Luo, P.; Yan, Y. Microstructures and densification of $MoSi_2$–SiC composite by field-activated and pressure-assisted combustion synthesis. *J. Alloys Compd.* **2009**, *468*, 136–142. [CrossRef]
41. Han, X.-x.; Wang, Y.-l.; Xiong, X.; Heng, L.; Chen, Z.-k.; Wei, S. Microstructure, sintering behavior and mechanical properties of SiC/$MoSi_2$ composites by spark plasma sintering. *Trans. Nonferrous Met. Soc. China* **2018**, *28*, 957–965. [CrossRef]
42. Guan, S.; Liang, H.; Liu, Y.; Lin, W.; He, D.; Peng, F. Production of silicon carbide reinforced molybdenum disilicide composites using high-pressure sintering. *Ceram. Int.* **2020**, *46*, 23643–23650. [CrossRef]
43. Nazari, M.; Shakeri, E.; Mohammadi, M. Fabrication and optimization process of mechanical properties of $MoSi_2$ composites reinforced by carbon nanotubes (CNT) using Taguchi method. *Ceram. Int.* **2024**, *50*, 25568–25577. [CrossRef]
44. Xu, J.; Chen, F.; Tan, F. In-situ preparation of SiC–$MoSi_2$ composite by microwave reaction sintering. *Ceram. Int.* **2012**, *38*, 6895–6898. [CrossRef]
45. Oh, D.Y.; Kim, H.C.; Yoon, J.K.; Shon, I.J. One step synthesis of dense $MoSi_2$–SiC composite by high-frequency induction heated combustion and its mechanical properties. *J. Alloys Compd.* **2005**, *395*, 174–180. [CrossRef]
46. Mitra, R.; Mahajan, Y.; Prasad, N.E.; Chiou, W. Processing—Microstructure—Property relationships in reaction hot-pressed $MoSi_2$ and $MoSi_2$/SiCp composites. *Mater. Sci. Eng. A* **1997**, *225*, 105–117. [CrossRef]
47. Wang, J.; Feng, P.; Niu, J.; Guo, R.; Liu, Z.; Akhtar, F. Synthesis, microstructure and properties of $MoSi_2$–5 vol% Al_2O_3 composites. *Ceram. Int.* **2014**, *40*, 16381–16387. [CrossRef]
48. Kasraee, K.; Tayebifard, S.A. Mechanical and Oxidation Behavior of $MoSi_2$-40wt.% Ti_5Si_3 Composites Produced by MASHS and SPS. *Ceram. Int.* **2024**, *51*, 1888–1899. [CrossRef]
49. Zhu, G.; Wang, X.; Feng, P.; Liu, Z.; Niu, J.; Akhtar, F. Synthesis and properties of $MoSi_2$–MoB–SiC ceramics. *J. Am. Ceram. Soc.* **2016**, *99*, 1147–1150. [CrossRef]
50. Kim, Y.-W.; Kim, Y.-H.; Kim, K.J. Electrical properties of liquid-phase sintered silicon carbide ceramics: A review. *Crit. Rev. Solid State Mater. Sci.* **2020**, *45*, 66–84. [CrossRef]
51. Kim, Y.; Lee, J.Y.; Cho, M.K.; Ahn, J.-P.; Kim, Y.-W. Electrical resistivity at the micron scale in a polycrystalline SiC ceramic. *Ceram. Int.* **2021**, *47*, 27100–27106. [CrossRef]

metals

MDPI

Article

Mechanical Properties and Corrosion Resistance of La_2O_3/A356 Composites Fabricated by Ultrasonic-Assisted Casting

Hao Wan, Luming Shuai, Lishibao Ling, Zhi Hu and Hong Yan *

School of Advanced Manufacturing, Nanchang University, Nanchang 330031, China;
412400230054@email.ncu.edu.cn (H.W.); ncu1371922663@163.com (L.S.); nealllm@163.com (L.L.);
huzhi@ncu.edu.cn (Z.H.)
* Correspondence: hyan@ncu.edu.cn; Tel.: +86-0791-8396-9622

Abstract: Mechanical properties and corrosion resistance of La_2O_3/A356 composites with different contents of La_2O_3 were investigated by optical microscopy, X-ray diffractometry, scanning electron microscopy, electrochemical tests, and immersion corrosion tests. The results show that the addition of La_2O_3 refined the α-Al phase of the A356 matrix, and the long stripe-like Si phase and β-Al_5FeSi phase were transformed into short rod-like forms. The La_2O_3/A356 composites with 1.0 wt.% La_2O_3 exhibited the most optimal mechanical properties and corrosion resistance. The yield strength, ultimate tensile strength, and elongation of La_2O_3/A356 composites with 1.0 wt.% La_2O_3 were higher than those of the matrix. The results of electrochemical experiments and the immersion corrosion test show that the corrosion potential of La_2O_3/A356 composites with 1.0 wt.% La_2O_3 was 72 mV higher than that of the matrix, the corrosion current density was 84.8% lower than that of the matrix, and the impedance Z was improved by 59.1% compared to the matrix. The addition of La_2O_3 improved the mechanical properties of the A356 matrix by refining the grains, inhibiting the nucleation of eutectic Si, and promoting the twinning growth mechanism. Moreover, the effect of La_2O_3 on the micro-galvanic corrosion behavior of A356 was discussed.

Keywords: La_2O_3/A356 composites; mechanical properties; corrosion resistance; strengthening mechanism

1. Introduction

Because of their high specific strength, good castability, and good corrosion resistance, Al–Si alloys are often used in industrial manufacturing, electronics, aerospace, and automotive industries [1–4]. Although the cost of using industrially produced, low-cost cast Al–Si-based alloys to manufacture complex, thin-walled, and highly airtight parts is relatively low, the resulting cast products often contain casting defects, such as porosity and inclusions, which can reduce the mechanical properties of the alloy [5]. In addition, Al–Si alloys can contain large α-Al grains and eutectic Si with long strips as well as lath-like Fe phases during the production process, which are detrimental factors that reduce the mechanical properties of the alloys, such as tensile resistance and elongation, as well as corrosion resistance [6–8].

In recent years, in order to improve the mechanical properties and corrosion resistance of Al–Si alloys, many researchers have improved the microstructure of alloys by introducing rare earth modifiers and ceramic nanoparticles, such as Ce [9], Yb [10], TiB_2 [11], TiC [12], etc., to the alloys to achieve the improvement of the mechanical properties and corrosion resistance of aluminum alloys. Arrabal et al. [13] found that the addition of Nd to A356 alloy refined the eutectic Si and formed intermetallic compounds containing Nd, which greatly improved the pitting corrosion resistance of the alloy, and the A356 alloy containing Nd had a more refined microstructure. Jiang et al. [14] found that the addition of 0.2 wt.% of rare earth elements greatly reduced the size of α-Al and eutectic Si of the as-cast A357 alloy and lowered the secondary dendrite arm spacing (SDAS) value, while its tensile strength and elongation increased by 15.4% and 6.3%, respectively, compared to the matrix. Wang [15] modified the A356 alloy by adding Sr. The results showed that the addition of Sr reduced the size and aspect ratio of eutectic Si particles, and the addition of Sr exhibited higher ductility compared to unmodified A356 alloy. Colombo et al. [16] showed that the addition of 0.3 wt.% Er improved the morphology of eutectic Si and the volume fraction of intermetallic compounds was low, which is favorable for improving the corrosion resistance of A356 alloy. In addition to the incorporation of rare earth densifiers to improve the mechanical and corrosion properties of alloys, many researchers have incorporated ceramic nanoparticles to improve the properties of composites. Shin et al. [17] investigated the refining effect of oxide nanoparticles on the organization of A356 alloy by adding ZnO nanoparticles to the alloy, and the coarse primary aluminum dendritic organization was transformed into a finer equiaxed organization, which improved the mechanical properties of the A356 alloy. Shayan et al. [18] prepared AA 2024-xTiO_2 (np) (x = 0, 0.5, and 1 vol.%) nanocomposites by the stir casting method, and the experimental results showed that the grain size and dendrite arm spacing of TiO_2 nanoparticles with 1 vol.% addition were reduced by about 66% and 31%, respectively. In addition, the hardness, ultimate tensile strength, yield strength, and elongation were increased by 25%, 28%, 4%, and 163%, respectively, over the matrix material. Abdizadeh et al. [19] prepared A356/ZrO_2 composites with ZrO_2 volume fractions of 5%, 10%, and 15%, respectively, and the results showed that the fracture of all the specimens was due to inter-dendritic cracking of the matrix alloy. In addition, the A356/ZrO_2 composites prepared at 750 °C with a ZrO_2 content of 15% had the highest mechanical properties. Sajjadi et al. [20] produced A356/Al_2O_3 nanometer and micrometer composites with different particle mass fractions by both stir casting and composite casting processes. The experimental results showed that the particles in the micro- and nano-composites are uniformly distributed with grain refinement and low porosity. The incorporation of Al_2O_3 (micro and nano) increased the yield strength, ultimate tensile strength, compressive strength, and hardness of the material.

However, researchers have mainly focused on the study of rare earth modifiers and ceramic nanoparticles on the mechanical properties and corrosion resistance of aluminum alloys, while the effect of rare earth oxides on the mechanical properties and corrosion resistance of aluminum alloys has been studied very little, and the cost of rare earth oxides is relatively inexpensive. Therefore, the development of high-performance rare earth oxides-reinforced aluminum alloy to reduce cost is of great significance for industrial production. In this paper, La_2O_3/A356 composites with different contents of La_2O_3 were prepared by the high-energy ultrasonic stirring method, and the effects of La_2O_3 on the mechanical properties and corrosion resistance of the A356 matrix were investigated.

2. Materials and Methods

2.1. Experimental Materials

A356 aluminum alloy has good strength and is easy to cast and process, and further enhancement of its performance is an effective means to extend its application range. In this paper, A356 aluminum alloy is selected as the matrix material, and its chemical composition is shown in Table 1. The main chemical composition in A356 aluminum alloy is Al, Si, and Mg, with small amounts of Zn, Fe, Cu, Mn, and other elements. Aluminum powder used in the preparation of composites was ≥99% pure, with a particle size of 75–150 μm. The main reinforcing phase added to the composites was La_2O_3, with purity $\geq$ 99% and a particle size of 500 nm^{-1} μm.

Table 1. Main chemical composition of A356 aluminum alloy (%).

Elements	Si	Mg	Zn	Fe	Cu	Mn	Ti	Al
(wt.%)	7.1749	0.4162	0.0716	0.0934	0.0036	0.0016	0.1010	Bal.

2.2. Preparation of A356-xLa_2O_3 Composites

Before the experiment started, the ball milling jar, aluminum powder, and La_2O_3 powder were dried. Then, the aluminum powder and La_2O_3 powder were weighed and put into the ball milling jars at a ratio of 9 to 1. The ball material ratio was 9 to 1, and 1.0 wt.% stearic acid was added as a process control agent for ball milling to prevent the occurrence of cold welding. The planetary ball mill was turned on in an argon atmosphere at a rotational speed of 200 r/min for a running time of 2 h. Finally, the ball-milled powder was cold pressed and shaped under a pressure of 500 MPa in a hydraulic press to obtain the powder mixtures (Al-10 wt.% La_2O_3). In order to ensure the addition efficiency of the Al-10 wt.% La_2O_3 powder mixtures, they were cut into small pieces, weighed to the expected addition amount, and wrapped in aluminum foil. Subsequently, the graphite crucible was preheated to about 400 °C in a pit-type resistance furnace; after that, the weighed block of A356 aluminum alloy was put into the graphite crucible, and the temperature was raised to 750 °C so that it was completely melted; then, the weighed Al-10 wt.% La_2O_3 prefabricated blocks were added to the inside of the alloy solution (the additions were set as 0.5 wt.%, 1.0 wt.%, 1.5 wt.%), stirred thoroughly, and held for 60 min to ensure its complete melting. In order to improve the dispersion of its La_2O_3 in A356 alloy, it was ultrasonically stirred using an ultrasonic process. After slagging, the alloy melt was ultrasonicated for 15 min using an ultrasonic frequency of 21 kHz and an ultrasonic power of 2.8 kW. After the ultrasonic treatment, it was held to 750 °C for slag and gas removal, and finally, the prepared mixed melt was poured into a metal mold that had been preheated to 200 °C. The experimental flow chart is shown in Figure 1.

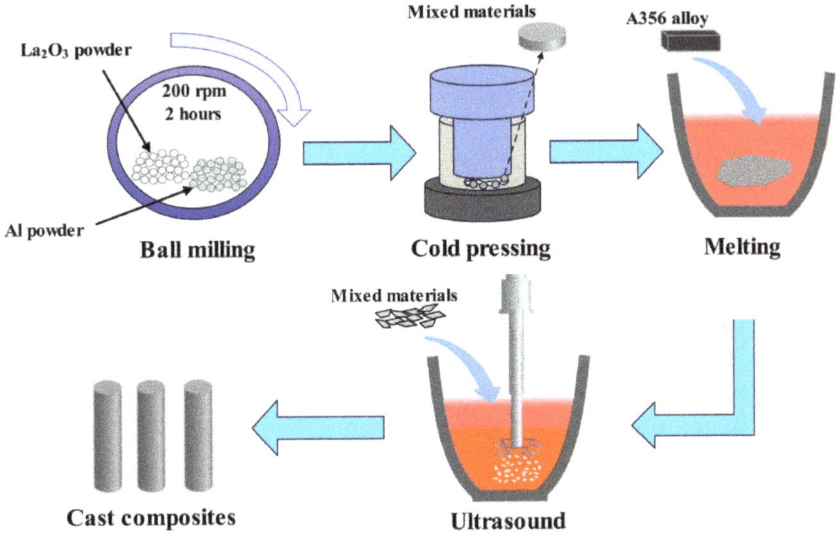

Figure 1. Experimental flow chart of A356-xLa_2O_3 composites.

2.3. Microstructure Characterization

The specimens were corroded with 0.5 vol.% HF on the polished surface before observation, and then, the microstructures and fracture surfaces of the specimens were observed by optical microscopy (Nikon Eclipse MA200, Nikon Co., Ltd., Tokyo, Japan), scanning electron microscopy (Quanta 200, FEI Metrology Co., Ltd., Hillsboro, OR, USA), and EDS. Metallographic specimens that were prepared for optical microscopy were taken for XRD analysis. The model of X-ray diffraction analyzer used in this paper was the Bruker D-8 X-ray diffraction analyzer (Bruker Co., Ltd., Karlsruhe, Germany). The source of radiation used for the test was Cu-Kα, the scanning range was 5–90°, the measurement step was set to 0.03°, the scanning speed was 2°/min, and the accelerating voltage was 36 kV.

In the specimen preparation process, a standardized tensile specimen bar was produced, and then the model SUNS UTM5105 electronic universal tensile testing machine (Shenzhen Suns Technology Stock Co., Ltd., Shenzhen, China) was used for the different tensile tests. The measured specimen at the beginning of the scale distance was 50 mm, and the tensile rate was set to 1 mm/min; different tensile specimen bars were employed for the selection of five specimens to reduce the error of the experiment. After the specimen bars were pulled off the scale, the distance was measured using vernier calipers and recorded, followed by the calculation of elongation and other parameters. The corresponding tensile strength and elongation values were recorded, taking the corresponding average of each group of data. The fracture areas of the specimen bars were cut into short bars of the same length within 12 h, and the morphology was observed and analyzed by SEM to explore the fracture mechanism of different specimens.

2.4. Electrochemical Experiments and Immersion Corrosion Tests

In this experiment, the potentiodynamic polarization (Tafel) curves were measured in an electrolyte of 3.5 wt.% NaCl solution at a room temperature of 25 °C using Princeton P4000 electrochemical workstation (Ametek Co., Ltd., Montvale, NJ, USA) test equipment equipped with a flat-plate three-electrode electrolytic cell. The working electrode material was a specimen with an exposed area of 1 cm^2, the counter electrode material was a platinum sheet, and the reference electrode was a saturated calomel electrode. To ensure that a stable open-circuit potential was obtained, all the specimens were immersed in 3.5 wt.% NaCl solution for 10 min before the experiments. The scanning rate of the polarization curve parameter was set to 1 mV/s, the scanning range was ±300 mV, and the frequency scanning range of the impedance spectroscopy parameter was set to 0.01 Hz to 100 kHz, with an amplitude of 10 mV. The electrochemical experiments were performed five times at room temperature for each group of samples.

Immersion corrosion experiments were conducted using inlay samples with only the square bottom surface exposed, and all specimens were ground and polished against the square bottom surface. All specimens were ultrasonically cleaned for 15 min using an ethanol solution (95 wt.%) prior to the immersion experiments. After air-drying by a chiller, the samples were placed in a beaker filled with 3.5 wt.% NaCl solution and immersed at a room temperature of 25 °C for 15 days. After the immersion test, the corrosion products were removed by using a mixed solution of 20 g/L CrO_3 and 1.69 g/mL H_3PO_4 and finally cleaned ultrasonically for 15 min using anhydrous ethanol, and the corrosion morphology was observed and analyzed by using a scanning electron microscope.

3. Results

3.1. Microstructure

Figure 2 shows the optical microscopic microstructure of La_2O_3/A356 composites with different La_2O_3 additions. As shown in Figure 2a, the main phases of the A356 matrix alloy include large-sized α-Al grains, long stripe-like eutectic Si phases, and lath-like β-Al_5FeSi phases. As shown in Figure 2b–d, when different contents of La_2O_3 are added, most of the α-Al grain shapes are transformed into ellipsoids, the grain boundaries are gradually sharpened, and most of the long stripe-like Si phases are transformed into short rod-like Si phases and granular Si phases. It can be found that the metamorphic modification effect is the best in La_2O_3/A356 composites with 1.0 wt.% La_2O_3. When the content of La_2O_3 in La_2O_3/A356 composites is increased to 1.5 wt.%, a slight increase in the size of α-Al grains is observed, while the short rod-like Si particles tend to aggregate and increase in size.

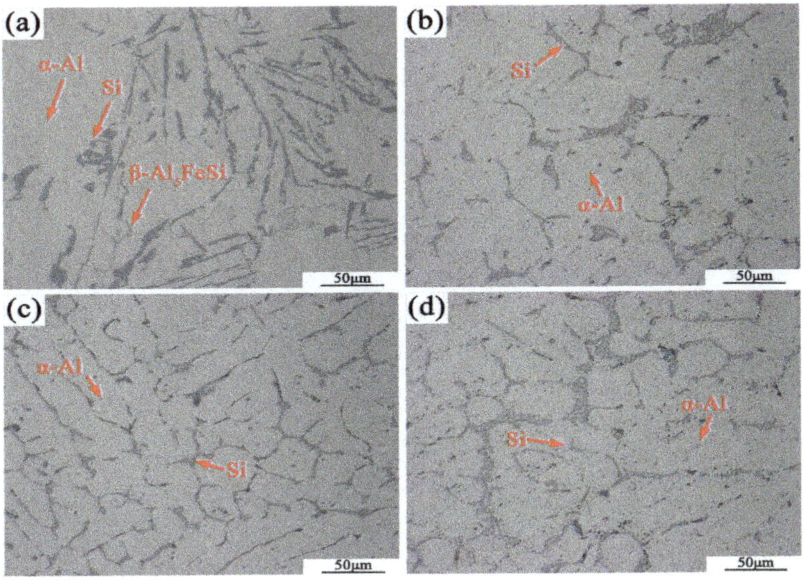

Figure 2. Microstructures of La_2O_3/A356 composites with different contents of La_2O_3: (**a**) 0.0 wt.%; (**b**) 0.5 wt.%; (**c**) 1.0 wt.%; (**d**) 1.5 wt.%.

Figure 3 shows the backscattered SEM microstructure of La_2O_3/A356 composites with different contents of La_2O_3 added. As shown in Figure 3a, the SEM image of A356 aluminum alloy mainly shows the white Chinese character-like phase pointed by the arrow and a small amount of black phase inside the circle, and the corresponding Fe and Mg elemental distribution maps are also given in Figure 3a, which can be observed in the labeled places of the corresponding Fe and Mg elemental aggregates. Combined with the XRD results in Figure 4 and the morphological analysis of the phases, it is hypothesized that the white Chinese character-like phase is the α-Al_8Fe_2Si phase and the black phase is the Mg_2Si phase. From Figure 3b–d, a new bright white massive phase is observed in the composites, and its aggregation is observed in the circle labeled corresponding to the elemental distribution diagram of La. This La-rich rare earth phase exists around the α-Al_8Fe_2Si phase, according to a related report [21], and may consist of phases such as $Al_{11}La_3$, Al_4La, and others. And, two bright white patches are observed in Figure 3c, and a clear aggregation of La can also be found in the La elemental distribution diagram. Combined with the Fe element distribution diagram, it can be found that the Fe phase is uniformly distributed in the composites and covered by a bright white rare earth phase so that the morphology is refined and improved. As shown in Figure 3d, it can be observed that the size of the bright white phase becomes larger, with a predominance of massive morphology, which produces a certain amount of agglomeration, and at the same time, the Fe-rich phase also agglomerates into a large-size phase. Compared to Figure 3c, there is an excess of the second phase in the composites, which generates a greater number of rare earth phases with larger sizes, which can lead to a decrease in the mechanical properties of the composite.

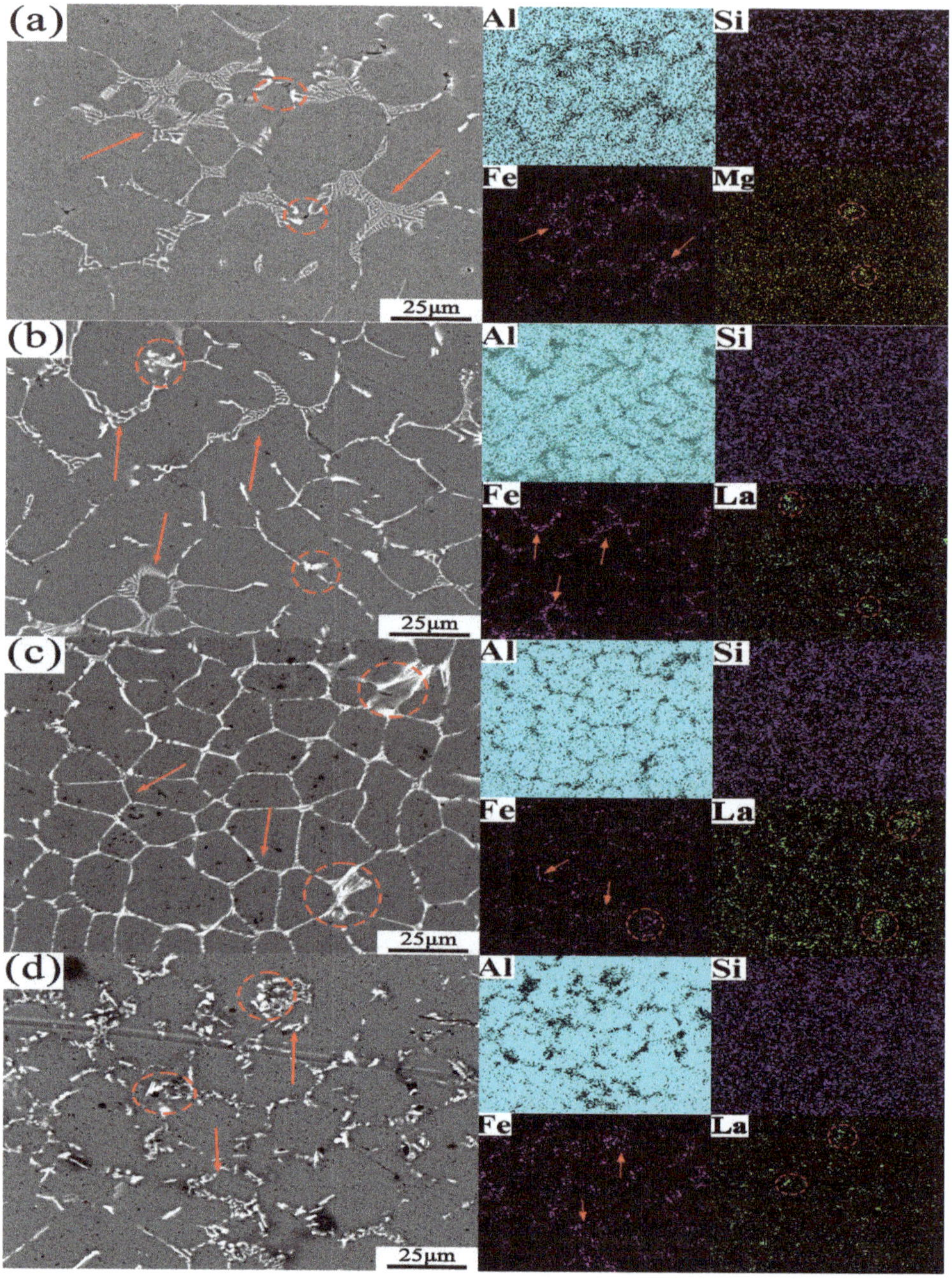

Figure 3. Backscatter SEM image and EDS diagram of La_2O_3/A356 composites with different La_2O_3 contents: (**a**) 0.0 wt.%; (**b**) 0.5 wt.%; (**c**) 1.0 wt.%; (**d**) 1.5 wt.%.

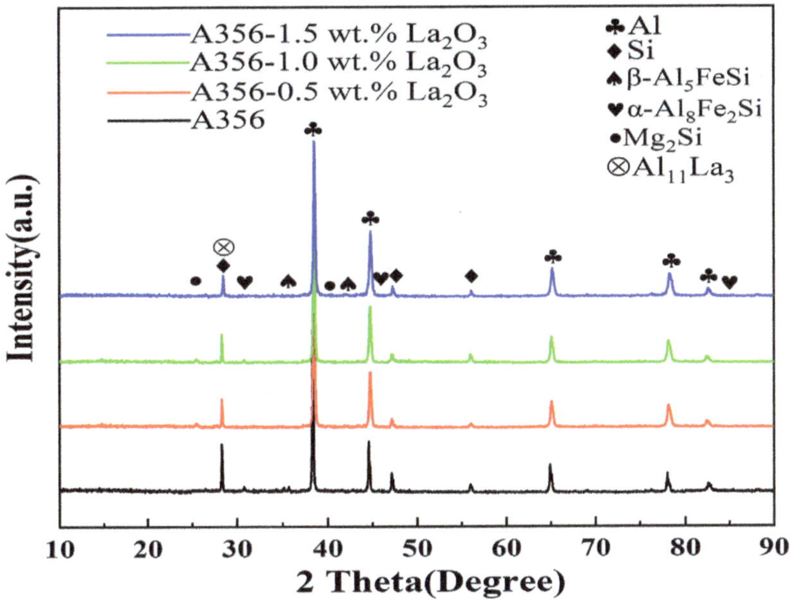

Figure 4. XRD analysis of La_2O_3/A356 composites with different La_2O_3 contents.

Figure 5a shows a high-magnification SEM image of La_2O_3/A356 composites with the addition of 1.0 wt.% La_2O_3. Referring to the point-scan elemental analysis in Figure 5b,c, it can be concluded that the short rod-like bright white phase is the La-rich rare earth phase and the fishbone-like grayish white phase is the Fe-rich phase, and combined with the atomic ratios, the Fe-rich phase can be inferred to be the α-Al_8Fe2Si phase based on the morphology. The figure illustrates that the La-rich rare earth phase has a good adsorption property compared to the Fe-rich phase, which can better hinder the growth of the Fe phase, while the small size of the La-rich phase can improve the mechanical properties of the composite.

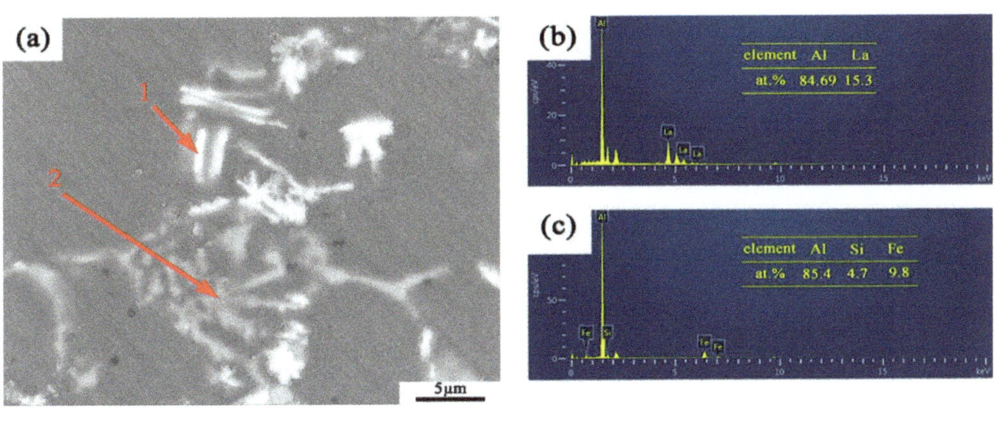

Figure 5. (**a**) High-rate SEM image of La_2O_3/A356 composites with 1.0 wt.% La_2O_3 added; (**b**) point-scan image of point 1 in (**a**); (**c**) point-scan image of point 2 in (**a**).

3.2. Mechanical Properties

It has been shown [22] that the ultimate tensile strength and elongation of Al-Si alloys are mainly affected by the amount, morphology, and distribution of the second phase in the alloy. The addition of rare earth elements can effectively refine the size of eutectic Si in the alloy and improve its morphology, thus enhancing the mechanical properties of the alloy. The engineering stress–strain curves and mechanical properties of La_2O_3/A356 composites with different La_2O_3 contents are shown in Figure 6a,b. The engineering yield strength (YS), ultimate tensile strength (UTS), and elongation (EL) are summarized in Table 2. The results show that the addition of La_2O_3 increases the tensile strength, yield strength, and elongation of La_2O_3/A356 composites. The reason for this enhancement effect is that the morphology of eutectic Si in the composite changes from long stripe-like to short rod-like, and the size of eutectic Si decreases. In addition, the size of the α-Al_8Fe2Si phase is refined, and the formation of the rare earth phase further strengthens the ultimate tensile strength of the composite. The composite has the best mechanical properties when the La_2O_3 content is 1.0 wt.% and the yield strength, ultimate tensile strength, and elongation of the composite further increase to 141.3 MPa, 192.4 MPa, and 5.3%, respectively. However, after increasing the content of La_2O_3 to 1.5 wt.%, it is the excessive addition of La_2O_3 that produces the agglomeration phenomenon, leading to the reduction of mechanical properties.

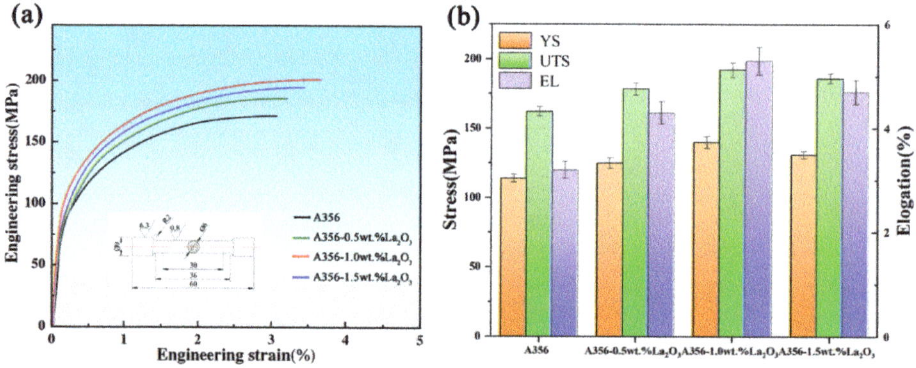

Figure 6. (**a**) Engineering stress–strain curves of A356-xLa_2O_3 composites; (**b**) mechanical properties of A356-xLa_2O_3 composites.

Table 2. Mechanical properties of A356-xLa_2O_3 composites.

Specimen	YS (MPa)	UTS (MPa)	EL (%)
A356	115.8 ± 3	162.1 ± 4	3.2 ± 0.2
A356-0.5wt.% La_2O_3	127.6 ± 4	178.4 ± 5	4.3 ± 0.3
A356-1.0wt.% La_2O_3	141.3 ± 4	192.4 ± 4	5.3 ± 0.3
A356-1.5wt.% La_2O_3	132.4 ± 3	186.2 ± 5	4.7 ± 0.3

3.3. Fracture Analysis

The fracture characteristics of La_2O_3/A356 composites with different La_2O_3 contents are shown in Figure 7. From Figure 7a, it can be found that on the fracture of A356 aluminum alloy, there are large cleavage planes and fewer dimples, as well as some cracks, and the whole shows obvious brittle fracture characteristics. When 0.5 wt.% La_2O_3 is added to A356 alloy, the fracture surface is still distributed with many cleavage planes, but some small dimples have appeared where the blue ellipse is marked, and the whole fracture surface is still characterized by brittle fracture, but there is a tendency to transform into ductile fracture. When the addition of La_2O_3 is increased to 1.0 wt.%, the fracture surface is distributed with a small size and a large number of dimples in the vast majority of the area. There are still some cleavage planes distributed, but no obvious cracks are observed, and the tendency of the fracture surface of La_2O_3/A356 composites to change from brittle fracture to ductile fracture is further strengthened. Further addition of 1.5 wt.% La_2O_3 result in a rise in the number of cleavage planes on the fracture surface and a significant expansion of the area, and obvious cracks are observed, making the brittle fracture characteristics of the composites more apparent. The SEM results of the fracture of the tensile specimens show that the addition of 1.0 wt.% La_2O_3 to the matrix material is the most favorable for the mechanical properties of the composite in terms of the fracture mode, whereas the excessive addition of La_2O_3 leads to a tendency for the composite to fracture brittlely. These experimental results show that La_2O_3/A356 composites with 1.0 wt.% La_2O_3 possesses the best tensile properties.

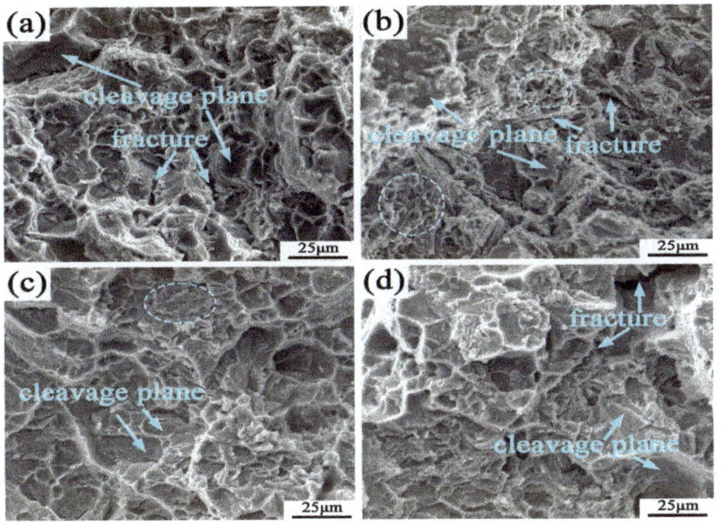

Figure 7. Tensile fracture diagram of La_2O_3/A356 composites with different La_2O_3 contents: (**a**) 0.0 wt.%; (**b**) 0.5 wt.%; (**c**) 1.0 wt.%; (**d**) 1.5 wt.%.

3.4. Corrosion Resistance Analysis

The Tafel polarization curves of La_2O_3/A356 composites with different contents of La_2O_3 are shown in Figure 8, and at higher values of potential, it can be observed that the anodic current is higher than the cathodic current, which suggests that the corrosion by cathodic reaction plays a major role. It is also observed that the anodic and cathodic curves of the polarization curves are gradually shifted to the left, and it can be concluded that the addition of La_2O_3 reduces the corrosion rate. The corrosion data fitted by extrapolation using Tafel polarization curves are shown in Table 3. It can be observed that the corrosion potential of the composite increases and then decreases as the La_2O_3 content increases, while the corrosion current density decreases and then increases. When 1.0 wt.% La_2O_3 is added, the corrosion potential of the composite reaches the maximum value (−719.391 mV), the corrosion current density reaches the minimum value (1.62 $\mu A \cdot cm^{-2}$), the corrosion potential is 72mV higher than that of the matrix, and the corrosion current density is 84.8% lower than that of the matrix, which has the best corrosion resistance.

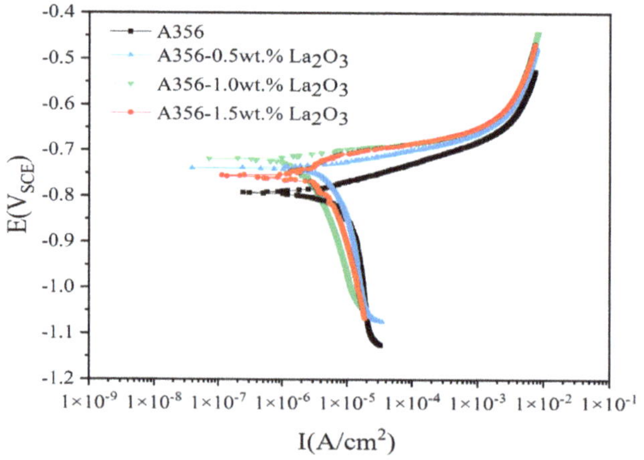

Figure 8. Polarization curves of La_2O_3/A356 composites with different La_2O_3 contents.

Table 3. Corrosion potential and corrosion current density of La_2O_3/A356 composites with different La_2O_3 contents.

Specimen	E_{corr} (mV_{SCE})	I_{corr} ($\mu A \cdot cm^{-2}$)
A356	−791.756	10.67
A356-0.5 wt.% La_2O_3	−738.535	5.69
A356-1.0 wt.% La_2O_3	−719.391	1.62
A356-1.5 wt.% La_2O_3	−755.786	7.69

Electrochemical impedance spectroscopy (EIS) was applied to study and analyze La_2O_3/A356 composites with different contents of La_2O_3 added in 3.5 wt.% NaCl solution, and the test results are shown in Figure 9. Similar capacitance loops are shown in the Nyquist plots of the samples in Figure 9a, based on the facts that the different radii of the capacitance loops between the samples are strongly correlated with the dissolution of the metal during corrosion and the radii are correlated with the charge transfer resistance, suggesting that the corrosion process is mainly governed by charge transfer [23]. The difference in impedance between samples can be determined by comparing the size of the capacitive loop radius [24]. Comparing the size of the loop radius in Figure 9a, it can be concluded that in terms of impedance, A356 < A356-1.5 wt.% La_2O_3 < A356-0.5 wt.% La_2O_3 < A356-1.0 wt.% La_2O_3, which indicates that La_2O_3/A356 composites with 1.0 wt.% La_2O_3 have the best corrosion resistance.

The Bode phase plots and Bode magnitude of the tested composites are shown in Figure 9b,c. From Figure 9b, it can be found that the impedance modulus $|Z|$ shows a trend of increasing and then decreasing with the increase of La_2O_3 content, and the impedance modulus $|Z|$ reaches the maximum value at the addition of 1.0 wt.% of La_2O_3. At the same time, the larger the impedance modulus $|Z|$ of the composite is in the low-frequency region, the better the corrosion resistance is [25]. From the Bode amplitude plot in Figure 9c, it can be seen that the mid-frequency phase angle becomes wider with increasing La_2O_3 content, and it is widest at 1.0 wt.% La_2O_3. The equivalent circuit used to fit the impedance spectral curve is shown in Figure 9d, where Rs denotes the electrolyte resistance between the working and reference electrodes, the value of which depends only on the conductivity of the test medium (a 3.5 wt.% NaCl solution). Rp denotes the charge transfer resistance, and CPE is the capacitance of the double electric layer at the interface between the metal surface and the corrosive medium because of the inhomogeneity of the surface reaction of the composite. The relationship between the CPE impedance and frequency can be expressed by the following equation [26]:

$$Z_{CPE} = \frac{1}{T(i2\pi f)^{n}} \quad (1)$$

where T is the CPE constant, f is the frequency, and the value of n ranges from 0 to 1. When the CPE acts as a pure resistor, the value of n is 0, and when CPE acts as a pure capacitor, n is 1. Furthermore, the value of impedance Z can be calculated by the following equation [27]:

$$Z = R_S + R_P \quad (2)$$

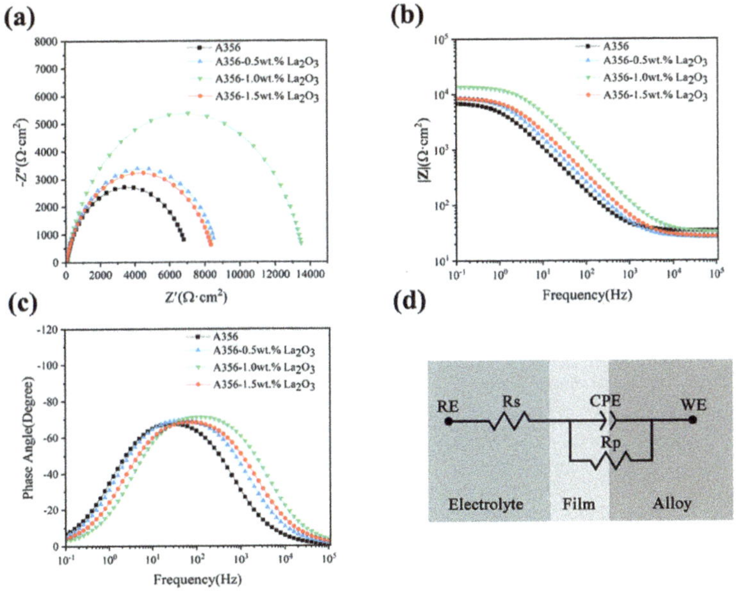

Figure 9. (**a**) Nyquist plots; (**b**,**c**) Bode plots; (**d**) equivalent circuit.

Impedance Z is a very important physical parameter; it is a physical quantity that acts as a barrier to the current in a circuit. Because impedance Z is inversely proportional to the corrosion rate and directly proportional to the corrosion resistance, impedance Z can be used to measure the corrosion rate and the corrosion resistance of samples; the higher the impedance value, the slower the corrosion rate, and the better the corrosion resistance. The parameters of each device in the fitted equivalent circuit are listed in Table 4. From Table 4, we know that the impedance Z value of La_2O_3/A356 composites is 8584.61 $\Omega \cdot cm^2$, and the Z value of A356 alloy after the addition of La_2O_3 is improved compared to the matrix. Impedance Z is the largest (13,655.33 $\Omega \cdot cm^2$) when adding 1.0 wt.% La_2O_3, which is 59.1% higher compared to the matrix A356, which indicates that the addition of La_2O_3 can increase the impedance Z value of A356 alloy and improve the corrosion resistance of the alloy, and La_2O_3/A356 composites with 1.0 wt.% La_2O_3 has the best corrosion resistance. This indicates that the addition of La_2O_3 can increase the impedance Z value of A356 alloy and improve the corrosion resistance of the alloy, and the corrosion resistance of La_2O_3/A356 composites with 1.0 wt.% La_2O_3 is the best.

Table 4. Fitting parameter data obtained from the EIS data of A356-x La_2O_3 composites.

Specimen	R_s ($\Omega\cdot cm^2$)	R_p ($\Omega\cdot cm^2$)	CPE-T ($\mu F\cdot s^p\cdot cm^{-2}$)	CPE-P	Z ($\Omega\cdot cm^2$)
A356	32.61	8552	3.0812×10^{-5}	0.80882	8584.61
A356-0.5wt.%La_2O_3	26.03	8773	1.7974×10^{-5}	0.83716	8799.03
A356-1.0wt.%La_2O_3	32.33	13,623	5.5691×10^{-6}	0.84987	13,655.33
A356-1.5wt.%La_2O_3	26.71	8628	1.3624×10^{-5}	0.82711	8654.71

Figure 10 shows the corrosion morphology of the A356 matrix and La_2O_3/A356 composites with 1.0 wt.% La_2O_3 after 15 days immersion in 3.5 wt.% NaCl solution. As shown in Figure 10a, large corrosion pits are formed on the surface of the A356 matrix without the addition of La_2O_3, and the surface area of the corrosion pits is large. However, after the addition of 1.0 wt.% La_2O_3, there are only a few relatively small corrosion pits on the surface of the composite, and the corrosion depth is shallow. The results show that the addition of La_2O_3 can effectively improve the corrosion resistance of A356 aluminum alloy.

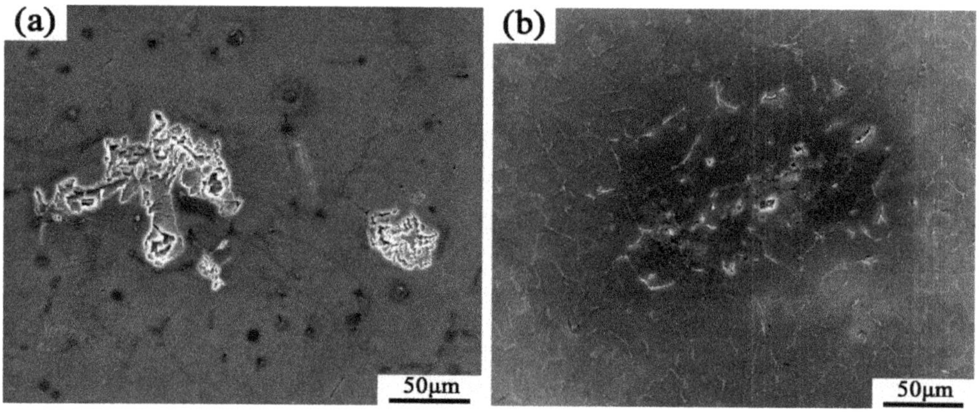

Figure 10. Corrosion morphology of A356 matrix and La_2O_3/A356 composites with 1.0 wt.% La_2O_3 addition in 3.5 wt.% NaCl solution for 15 days: (**a**) A356; (**b**) A356-1.0 wt.% La_2O_3.

4. Discussion

Since oxides usually have good stability, reference is made to the rules for the Al and oxide reaction system [28–31], as well as for the Al and rare earth oxide system ($4Al + 3CeO_2 \rightarrow 2Al_2O_3 + 3Ce$, $2Al + Y_2O_3 = Al_2O_3 + 2Y$) [32,33], which generally require excitation at high temperatures for this reaction to occur. In addition, in Zhou et al.'s study [34], with high energy input, Al and La_2O_3 occur to form Al_2O_3 and $Al_{11}La_3$. In this experiment, ball milling and ultrasonic vibration are used to prepare composite materials, and ultrasonic vibration introduced into the melt produces a transient high temperature that can reach 10,000 K [35]. Based on the above reports as well as the experimental results, under high energy and high temperature generated by ball milling and ultrasonic vibration, Al and La_2O_3 react to form Al_2O_3 and $Al_{11}La_3$ ($28Al + 3La_2O_3 \rightarrow 3Al_2O_3 + 2Al_{11}La_3$) [34].

The addition of La_2O_3 can improve the mechanical properties of A356 aluminum alloy; its main strengthening mechanisms are grain refinement and the second phase strengthening. The presence of the $Al_{11}La_3$ phase can significantly refine the α-Al grains, and grain refinement produces more grain boundaries, and the greater the number of grain boundaries, the greater the hindrance to the movement of dislocations, which makes the composite's resistance to deformation improved. However, the addition of excess La_2O_3

leads to the aggregation of rare earth phases into a coarser second phase, which reduces the strength of the composite.

Although the addition of rare earth elements can inhibit the nucleation of eutectic Si, it still cannot change the eutectic Si from long stripe-like to fibers or short rod-like forms [36]. Therefore, other metamorphic mechanisms exist to refine eutectic Si, among which the impurity-induced twin growth mechanism (IIT mechanism) is one of the mechanisms often used to explain the formation of twins in eutectic Si. According to the IIT mechanism [37], the ratio of the atomic radii of the metastable atoms to those of the Si atoms is 1.646, which induces the production of twins, whereas the ratio of the atomic radii of La to those of Si is 1.59, which is very close to this value, suggesting that the incorporation of La_2O_3 promotes the formation of twins. By adsorbing on the {111} surface of Si, the La atoms make the eutectic Si produce twin branches in different directions, and the La atoms continue to adsorb on the interface of the twins and finally make the eutectic Si change from long stripe-like to fibers and short rod-like forms; the eutectic Si is refined, and the mechanical properties of the composites are improved.

The microstructure of the composites is an important factor that affects their corrosion resistance [38]. A356 aluminum alloy matrix organization yields a coarse eutectic Si phase and β-Al_5FeSi phase, and its surface potential is higher than α-Al. So, in the corrosion process, it acts as the cathodic phase to form a more serious micro-galvanic corrosion. As shown in Figure 11a, the anodic reaction is the dissolution of Al, with the loss of three electrons to form Al^{3+}, which reacts as follows [39,40]:

$$Al \rightarrow Al^{3+} + 3e^- \tag{3}$$

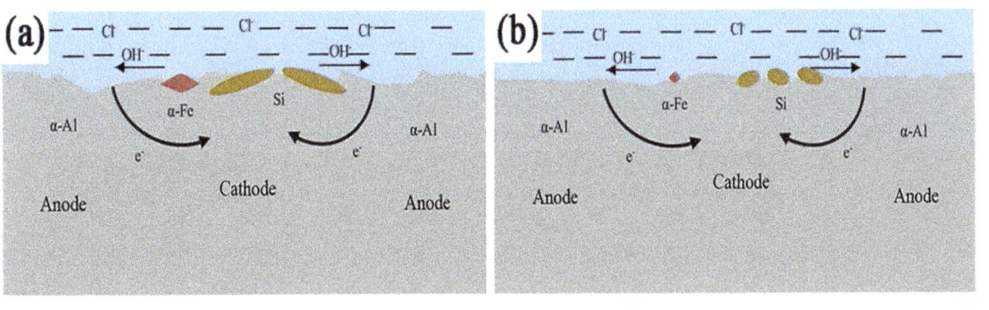

Figure 11. Corrosion mechanism diagram of A356 matrix and La_2O_3/A356 composites with 1.0 wt.% La_2O_3.

Oxygen reduction reaction occurs at the cathode because of NaCl in the corrosive environment, where Cl^- and H_2O are mainly present in the corrosive medium:

$$O_2 + 2H_2O + 4e^- \rightarrow 4OH^- \tag{4}$$

In environments where there is less oxygen in the corrosion pits, a water reduction reaction occurs:

$$2H_2O + 2e^- \rightarrow H_2\uparrow + 2OH^- \tag{5}$$

The overall response is as follows:

$$2Al + 6H_2O \rightarrow 2Al(OH)_3\downarrow + 3H_2\uparrow \tag{6}$$

The second phase in the composite acts as the cathodic phase to gain electrons and reacts with the α-Al phase in a corrosive reaction to form an $Al(OH)_3$ precipitate, which is converted to Al_2O_3 in a subsequent reaction:

$$2Al(OH)_3 \rightarrow Al_2O_3 + 3H_2O \quad (7)$$

The main second phase in A356 aluminum alloy is the Si phase, which is large in number and coarse in size and acts as the main cathodic phase to form micro-galvanic corrosion with the nearby α-Al phase, while the Fe-rich phase, which has a higher surface potential, is also used as the cathodic phase during the corrosion process to promote the corrosion reaction, although its content is less. The anode α-Al gradually dissolves with the corrosion reaction, and after a certain time of corrosion occurrence, the exposed cathodic phase falls off to form corrosion holes. While adding La_2O_3 to A356 aluminum alloy, the rare earth phase itself is also higher than the surface potential of α-Al [41], which can be used as the cathodic phase, which increases the number of cathodic phases and promotes the corrosion of A356 matrix. The optimal addition of 1.0 wt.% La_2O_3 effectively refines the α-Al grains as well as the size of eutectic Si and improves the morphology. The mechanistic diagram of the corrosion reaction is shown in Figure 11b. The eutectic Si phase and the Fe-rich phase are still the main cathodic phases, but the corrosion primary cell that can be formed by the refined cathodic phase and the anodic phase becomes smaller, the microcurrent that passes through it is smaller, and the ratio of the cathodic area to the anodic area decreases, which results in a decrease in the corrosion current density. As a result, it has improved corrosion resistance.

5. Conclusions

In this paper, the mechanical properties and corrosion resistance of La_2O_3/A356 composites with different contents of La_2O_3 are investigated, and the main conclusions are as follows:

(1) The addition of La_2O_3 refines the α-Al phase of the A356 matrix, and the long stripe-like Si and β-Al_5FeSi phases are transformed into short rod-like particles. Concurrently, the rare earth-reinforced phase $Al_{11}La_3$ was generated. The metamorphic effects of La_2O_3/A356 composites with 1.0 wt.% La_2O_3 were the best at this time.

(2) The addition of different contents of La_2O_3 improved the mechanical properties of the A356 matrix to some extent. The yield strength, ultimate tensile strength, and elongation of La_2O_3/A356 composites with 1.0 wt.% La_2O_3 were best. The addition of La_2O_3 made the A356 matrix have a tendency to shift from brittle fracture to ductile fracture, and the shift to ductile fracture was most obvious when the addition amount was 1.0 wt.%.

(3) With the addition of La_2O_3 to the A356 matrix, the cathodic phases (eutectic Si phase and β-Al5FeSi phase) in the microcells were refined, reducing the corrosion rate in the cathodic region. The corrosion potential (−719.391 mV) of La_2O_3/A356 composites with 1.0 wt.% La_2O_3 was 72 mV higher than that of the matrix (−791.756 mV), the corrosion current density (1.62 $\mu A \cdot cm^{-2}$) was 84.8% lower than that of the matrix (10.67 $\mu A \cdot cm^{-2}$), and the impedance Z (13,655.33 $\Omega \cdot cm^2$) was 59.1% higher than that of the matrix (8584.61 $\Omega \cdot cm^2$) and had the best corrosion resistance.

Author Contributions: Conceptualization, H.W., L.S. and H.Y.; methodology, H.W. and L.S.; software, H.W. and L.L.; validation, H.W. and Z.H.; investigation, H.W. and L.S.; writing—original draft preparation, H.W.; writing—review and editing, L.L., Z.H. and H.Y.; supervision, H.Y.; funding acquisition, H.Y. All authors have read and agreed to the published version of the manuscript.

Funding: This research was funded by the National Natural Science Foundation of China (No. 51965040).

Data Availability Statement: The original contributions presented in this study are included in the article. Further inquiries can be directed to the corresponding author.

Conflicts of Interest: The authors declare no conflicts of interest.

References

1. Wu, W.Q.; Gong, M.Y.; Wei, B.Q.; Misra, A.; Wang, J. Atomistic modeling of interface strengthening in Al-Si eutectic alloys. *Acta Mater.* **2022**, *225*, 117586. [CrossRef]
2. Xi, H.L.; Xiao, W.L.; Li, H.; Fu, Y.; Yi, G.; Qie, J.; Ma, X.; Ma, C. Effects of submicron-sized TiC particles on the microstructure modification and mechanical properties of Al-Si-Mg alloy. *J. Alloys Compd.* **2023**, *968*, 171963. [CrossRef]
3. Pramod, S.L.; Prasada Rao, A.K.; Murty, B.S.; Bakshi, S.R. Effect of Sc addition on the microstructure and wear properties of A356 alloy and A356-TiB2 in situ composite. *Mater. Des.* **2015**, *78*, 85–94. [CrossRef]
4. Garza Elizondo, G.H.; Elsharkawi, E.A.; Samuel, A.M.; Doty, H.W.; Samuel, F.H. Effects of Alloying Elements Additions on Ambient Temperature Performance of Al–Si–Cu–Mg Base Alloys. *Int. J. Met.* **2021**, *15*, 1385–1401. [CrossRef]
5. Liao, H.C.; Lu, L.Z.; Li, G.J.; Huang, Y.L.; Yang, L.L.; Guo, H.T.; Wu, F. Influence of minor addition of La and Ce on the ageing precipitation behavior of Sr-modified Al-7Si-0.6Mg alloy. *Mater. Today Commun.* **2024**, *39*, 108825. [CrossRef]
6. Liang, Y.G.; Cui, C.X.; Geng, H.T.; Liu, L.; Cui, S.; Yang, S.C. Effects of In-Situ BaB6/Al Composite Inoculant on the Mechanical and Corrosion Behavior of Al-7Si-0.3Mg Alloy. *Met. Mater. Int.* **2022**, *29*, 968–982. [CrossRef]
7. Heusler, L.; Schneider, W. Influence of alloying elements on the thermal analysis results of Al-Si cast alloys. *J. Light Met.* **2002**, *2*, 17–26. [CrossRef]
8. Yang, Y.; Zhou, W.F.; Tong, Z.P.; Chen, L.; Yang, X.Q.; Larson, E.A.; Ren, X.D. Electrochemical Corrosion Behavior of 5083 Aluminum Alloy Subjected to Laser Shock Peening. *J. Mater. Eng. Perform.* **2019**, *28*, 6081–6091. [CrossRef]
9. Kang, J.; Su, R.; Wu, D.Y.; Liu, C.H.; Li, T.; Wang, L.S.; Narayanaswamy, B. Synergistic effects of Ce and Mg on the microstructure and tensile properties of Al-7Si-0.3Mg-0.2Fe alloy. *J. Alloys Compd.* **2019**, *796*, 267–278. [CrossRef]
10. Amer, S.M.; Glavatskikh, M.V.; Barkov, R.Y.; Khomutov, M.G.; Pozdniakov, A.V. Phase composition and mechanical properties of Al-Si based alloys with Yb or Gd addition. *Mater. Lett.* **2022**, *320*, 132320. [CrossRef]
11. Liu, N.; Jiang, B.; Wang, Y.; Ji, Z.; Hu, M.; Xu, H. Influence of trace amount chromium on microstructure and corrosion behavior of A356-5vol.%TiB2 alloy. *Mater. Lett.* **2022**, *314*, 131798. [CrossRef]
12. Yang, R.; Wang, L.; Zhu, Q.; Zhang, Z.; Liu, M. TiC nanoparticles enhanced ultrasonic treatment for microstructure refinement of A356 alloy. *Mater. Lett.* **2020**, *277*, 128358. [CrossRef]
13. Arrabal, R.; Mingo, B.; Pardo, A.; Mohedano, M.; Matykina, E.; Merino, M.C.; Rivas, A. Microstructure and corrosion behaviour of A356 aluminium alloy modified with Nd. *Mater. Corros.* **2015**, *66*, 534–541. [CrossRef]
14. Jiang, W.M.; Fan, Z.T.; Dai, Y.C.; Li, C. Effects of rare earth elements addition on microstructures, tensile properties and fractography of A357 alloy. *Mater. Sci. Eng. A* **2014**, *597*, 237–244. [CrossRef]
15. Wang, Q.G. Microstructural effects on the tensile and fracture behavior of aluminum casting alloys A356/357. *Metall. Mater. Trans. A* **2003**, *34*, 2887–2899. [CrossRef]
16. Colombo, M.; Buzolin, R.H.; Gariboldi, E.; Vallant, R.; Sommitsch, C. Characterization of localized corrosion of heat treated Er- and Zr-containing A356 alloys in 3.5 wt% NaCl aqueous solution. *Mater. Corros.* **2019**, *70*, 246–258. [CrossRef]
17. Shin, J.H.; Jeon, J.H.; Bae, D.H. Microstructure refining of aluminum alloys using aluminothermic reaction with ZnO nanoparticles. *Mater. Lett.* **2015**, *151*, 96–99. [CrossRef]
18. Shayan, M.; Eghbali, B.; Niroumand, B. Fabrication of AA2024−TiO_2 nanocomposites through stir casting process. *Trans. Nonferrous Met. Soc. China* **2020**, *30*, 2891–2903. [CrossRef]
19. Abdizadeh, H.; Baghchesara, M.A. Investigation on mechanical properties and fracture behavior of A356 aluminum alloy based ZrO_2 particle reinforced metal-matrix composites. *Ceram. Int.* **2013**, *39*, 2045–2050. [CrossRef]
20. Sajjadi, S.A.; Ezatpour, H.R.; Torabi Parizi, M. Comparison of microstructure and mechanical properties of A356 aluminum alloy/Al_2O_3 composites fabricated by stir and compo-casting processes. *Mater. Des.* **2012**, *34*, 106–111. [CrossRef]
21. Zhou, S.H.; Napolitano, R.E. Phase equilibria and thermodynamic limits for partitionless crystallization in the Al–La binary system. *Acta Mater.* **2006**, *54*, 831–840. [CrossRef]
22. Yan, H.; Chen, F.H.; Li, Z.H. Microstructure and mechanical properties of $AlSi_{10}Cu_3$ alloy with (La+Yb) addition processed by heat treatment. *J. Rare Earths* **2016**, *34*, 938–944. [CrossRef]
23. Silva, T.C.; Barros, A.S.; Filho, J.C.; Moreira, A.L.; Barbosa, C.R.; Rocha, O.L. Correction: Study of Electrochemical Corrosion in Samples of a Horizontally Solidified AlCuSi Alloy. *Int. J. Metalcast.* **2023**, *17*, 2349. [CrossRef]

24. Bertin, F.; Joshi, G.R.; Kittel, J.; Sagnard, C.; Ropital, F.; Martinez, M.; Bosch, C.; Wolski, K. Electrochemical impedance response of a thick and porous calcium carbonate layer deposited by thermal growth on a carbon steel electrode. *Corros. Sci.* **2024**, *227*, 111778. [CrossRef]
25. Yang, J.; Peng, J.; Nyberg, E.A.; Pan, F.-S. Effect of Ca addition on the corrosion behavior of Mg–Al–Mn alloy. *Appl. Surf. Sci.* **2016**, *369*, 92–100. [CrossRef]
26. Peng, Y.C.; Huang, B.; Zhong, Y.F.; Su, C.C.; Tao, Z.S.; Rong, X.C.; Li, Z.Y.; Tang, H.Q. Electrochemical corrosion behavior of 6061 Al alloy under high rotating speed submerged friction stir processing. *Corros. Sci.* **2023**, *215*, 111029. [CrossRef]
27. Wang, Z.; Dong, L.; Hu, B.; Chen, B. The Effect of Cu Addition on Corrosion Resistance of Al-Si-Mg-Cr Alloy. *Metals* **2023**, *13*, 795. [CrossRef]
28. Zhu, H.; Dong, K.; Huang, J.; Li, J.; Wang, G.; Xie, Z. Reaction mechanism and mechanical properties of an aluminum-based composite fabricated in-situ from Al–SiO_2 system. *Mater. Chem. Phys.* **2014**, *145*, 334–341. [CrossRef]
29. Dikici, B.; Gavgali, M. The effect of sintering time on synthesis of in situ submicron α-Al_2O_3 particles by the exothermic reactions of CuO particles in molten pure Al. *J. Alloys Compd.* **2013**, *551*, 101–107. [CrossRef]
30. Zhang, Q.; Xiao, B.L.; Wang, W.G.; Ma, Z.Y. Reactive mechanism and mechanical properties of in situ composites fabricated from an Al–TiO_2 system by friction stir processing. *Acta Mater.* **2012**, *60*, 7090–7103. [CrossRef]
31. Zhu, H.; Min, J.; Li, J.; Ai, Y.; Ge, L.; Wang, H. In situ fabrication of (α-Al_2O_3+Al_3Zr)/Al composites in an Al–ZrO_2 system. *Compos. Sci. Technol.* **2010**, *70*, 2183–2189. [CrossRef]
32. Zou, X.; Cheng, Z.; Wang, J.; Yan, H. Enhanced interfacial bonding and mechanical properties of CeO_2-coated graphene oxide reinforced 7075 composites. *Mater. Sci. Eng. A* **2024**, *910*, 146892. [CrossRef]
33. Moussa, M.E.; El-Hadad, S.; Khalifa, W. Influence of chemical modification by Y_2O_3 on eutectic Si characteristics and tensile properties of A356 alloy. *Trans. Nonferrous Met. Soc. China* **2019**, *29*, 1365–1374. [CrossRef]
34. Zhou, C.; Lv, M.; Zan, Y.N.; Liu, Y.; Shao, X.H.; Wang, Q.Z.; Wang, D.; Xiao, B.L.; Ma, Z.Y. Microstructure and mechanical properties of aluminum matrix composites produced by Al-La_2O_3 in-situ reaction. *Mater. Charact.* **2022**, *188*, 111887. [CrossRef]
35. Xiong, J.J.; Yan, H. Microstructure and mechanical properties of ADC12 composites reinforced with graphene nanoplates prepared by ultrasonic assisted casting. *Trans. Nonferrous Met. Soc. China* **2020**, *30*, 3210–3225. [CrossRef]
36. McDonald, S.D.; Nogita, K.; Dahle, A.K. Eutectic nucleation in Al–Si alloys. *Acta Mater.* **2004**, *52*, 4273–4280. [CrossRef]
37. Lu, S.Z.; Hellawell, A. The mechanism of silicon modiffcation in aluminum-silicon alloys-impurity induced twinning. *Metall. Mater. Trans. A* **1987**, *18*, 1721–1733. [CrossRef]
38. Cui, L.; Guo, Y.; Wang, X.; Du, Z.; Cheng, F. Dissolution kinetics of aluminum and iron from coal mining waste by hydrochloric acid. *Chin. J. Chem. Eng.* **2015**, *23*, 590–596. [CrossRef]
39. Sivaiah, M.V.; Petit, S.; Brendlé, J.; Patrier, P. Rapid synthesis of aluminium polycations by microwave assisted hydrolysis of aluminium via decomposition of urea and preparation of Al-pillared montmorillonite. *Appl. Clay Sci.* **2010**, *48*, 138–145. [CrossRef]
40. Abouei, V.; Saghafian, H.; Shabestari, S.G.; Zarghami, M. Effect of Fe-rich intermetallics on the wear behavior of eutectic Al–Si piston alloy (LM13). *Mater. Des.* **2010**, *31*, 3518–3524. [CrossRef]
41. Arthanari, S.; Jang, J.C.; Shin, K.S. Corrosion performance of high pressure die-cast Al-Si-Mg-Zn alloys in 3.5 wt% NaCl solution. *J. Alloys Compd.* **2019**, *783*, 494–502. [CrossRef]

metals

MDPI

Article

Microstructure and Performance of Body-Centered Cubic-Based Dual-Phase Composite Eutectic High-Entropy Alloys Prepared by Si Doping

Saike Liu, Aoxiang Li, Kaiwen Kang, Jinshan Zhang, Di Huang, Chunning Che, Yiteng Jiang, Mingkun Xu, Borui Zhang, Yaqing Li and Gong Li *

State Key Laboratory of Metastable Materials Science and Technology, Yanshan University, Qinhuangdao 066004, China; liusaike@stumail.ysu.edu.cn (S.L.); aoxiangli@stumail.ysu.edu.cn (A.L.); kaiwenkang@stumail.ysu.edu.cn (K.K.); zhangjs@stumail.ysu.edu.cn (J.Z.); huangdi@stumail.ysu.edu.cn (D.H.); ccning@stumail.ysu.edu.cn (C.C.); jiangyiteng@stumail.ysu.edu.cn (Y.J.); xumingkun@stumail.ysu.edu.cn (M.X.); boruizhang@stumail.ysu.edu.cn (B.Z.); liyaqing202321020125@stumail.ysu.edu.cn (Y.L.)

* Correspondence: gongli@ysu.edu.cn

Abstract: AlCrFeNi-based high-entropy alloys (HEAs) have emerged as a prominent research system, attracting significant interest due to their compositional diversity and the tunability of their phase structures. However, in practical applications, single-phase AlCrFeNi-based HEAs often face a trade-off between toughness and strength. Therefore, designing multi-phase composite eutectic high-entropy alloys (EHEAs) to optimize their mechanical properties and microstructure has become a key research focus. Si, a common non-metallic element, plays a significant role in strengthening metal materials. In this paper, AlCrFeNi with Si doping strengthening (AlCrFeNi)100-xSix composite EHEAs were successfully fabricated. A systematic analysis was conducted to investigate the impacts of Si doping on the microstructure and mechanical properties of AlCrFeNi-based composite EHEAs. This study shows that with increasing Si content, the biphasic lamellar composite structure at the grain boundaries gradually expands, forming flower petals. The precipitate structure within the grains evolves into flower disks, which form a sunflower-like composite structure in the alloy. The volume fraction of lamellar structures increases in the petals, accompanied by grain refinement. Furthermore, the yield strength of the alloy increases from 1131 MPa to 1360 MPa with increasing Si content. This provides guidance for the design of high-performance composite EHEAs.

Keywords: composite eutectic high-entropy alloys; Si doping; mechanical properties; composite structure

1. Introduction

High-entropy alloys (HEAs) contain multiple principal elements in significant proportions, with their concentrations evenly distributed, resulting in a significantly higher configurational entropy [1]. This high entropy helps suppress grain boundary formation and enhances the stability of the alloy [2–4]. Moreover, due to the unique multi-principal-element composition, HEAs exhibit more complex phase stability and microstructural characteristics [4–8]. For instance, traditional alloys are generally composed of a single element (iron, aluminum, or copper), and properties are enhanced by adding small amounts of alloying elements [9]. In contrast, HEAs form simple solid-solution phases (such as body-centered cubic (BCC) or face-centered cubic (FCC) phases) under the influence of high-entropy effects. This solid-solution structure is responsible for the superior mechanical properties of HEAs, high strength [10–13], excellent high-temperature property [14–19], and outstanding corrosion resistance [20,21]. However, single-phase HEAs often struggle to achieve a balance between strength and toughness. Therefore, designing high-performance multi-phase composite HEAs has become both a key focus and a challenge in materials science.

AlCrFeNi-based HEAs, a prevalent research system, have attracted significant attention due to their compositional diversity and the tunability of their phase structures. This system typically exhibits simple solid-solution phases such as BCC or FCC, which are closely related to the high-entropy effect and sluggish diffusion [22]. However, single-phase AlCrFeNi-based HEAs encounter difficulties in balancing strength and toughness. Therefore, the design of AlCrFeNi-based HEAs with multi-phase composite structures has become a key area of research. EHEAs, due to the formation of a regular eutectic structure during solidification, allow for a uniform distribution of different phases at the microscale, thereby significantly enhancing the material's strength, toughness, and stability [23]. The ordered BCC structure, known as the B2 phase, effectively enhances the alloy hardness through solid-solution strengthening [24]. In contrast, the disordered BCC structure, known as the A2 phase, contributes to the alloy's excellent plasticity [25], improving its ductility. Among multi-phase composite structures, the B2/A2 dual-phase system has attracted considerable interest due to its favorable balance of strength and toughness [26]. The AlCrFeNi-based HEA system, with its complex interactions among the constituent elements, provides a strong foundation for designing B2/A2 dual-phase eutectic structures. Si, a commonly used non-metallic element, plays a crucial role in strengthening metal materials. The addition of Si not only improves the stability of the B2 phase but also refines the eutectic microstructure and enhances the interface bonding strength, thus improving both the strength and ductility of the material [27]. The atomic radius of Si significantly affects the crystal structure, phase composition, and microstructure of the alloy. In Al0.3CoCrFeNi HEAs, increasing Si content causes a transformation of the alloy microstructure from a single FCC phase to a mixed FCC + BCC phase [28]. Our previous studies have shown that Si doping in various HEA systems can significantly alter their phase structures and properties [29]. In CoCrFeNi-based HEAs, an optimal Si content enhances the formation of the BCC phase, thereby improving both the hardness and corrosion resistance of the material [29,30]. However, the specific effects of Si doping at different concentrations on AlCrFeNi-based dual-phase composite eutectic high-entropy alloys (EHEAs) are not yet fully understood, especially concerning systematic studies on crystal structure, microstructure, and mechanical properties, which are still relatively limited.

In the (AlCrFeNi)100-xSix dual-phase composite EHEA system, varying Si contents may induce complex microstructural evolutions [31], including changes in the stability of the solid-solution phase, the formation and distribution of precipitates, and grain size variation. These microstructural changes directly affect the strength, hardness, plasticity, and fracture toughness. Therefore, a systematic study of the effects of Si doping on the microstructure and mechanical properties of AlCrFeNi-based dual-phase composite EHEAs is essential for elucidating the inherent relationship between composition and microstructure in multi-principal–element systems.

2. Experimental Materials and Methods

2.1. Alloy Preparation

The (AlCrFeNi)100-xSix (x = 0, 1, 2, 3, denoted as AlSi0, AlSi1, AlSi2, AlSi3) composite EHEAs were prepared using vacuum induction melting. The raw materials included Al, Cr, Fe, Ni, and Si (purity > 99.95 wt.%). The alloy ingots were subjected to melting in a vacuum induction furnace under an argon atmosphere. Each ingot was remelted at least six times to ensure a uniform component in the resulting alloy ingots.

2.2. Material Characterization and Testing

X-ray diffraction (XRD) was performed using a D/max-2500/PCX X-ray diffractometer with a Cu target. The operating conditions were set to a voltage of 40 KV and a current of 200 mA. The scan rate was 4°/min, and the X-ray wavelength was 1.54056 Å. The microstructure of the alloy samples was observed using the Hitachi S-3400 tungsten filament SEM. The energy dispersive X-ray spectroscopy (EDS) attached to the SEM was used for phase composition analysis. The microstructural analysis at the nanoscale was conducted using a Talos F200X transmission electron microscope (TEM). Mechanical properties were tested using an INSTRON-5982 universal testing machine to evaluate the room-temperature compression behavior of the alloys. The strain rate during the compression tests was set to 5×10^{-4} s^{-1}, and the elongation of the sample during the compression was measured using an extensometer. At least three tests were conducted for each alloy to ensure data accuracy. Vickers hardness was measured using a FM-ARS 9000 automatic Vickers hardness tester, with a loading force of 1.96133 N and a hold time of 10 s. Each sample had 10 points measured, and the average was taken as the hardness value. Nano-hardness testing of the dendritic region in the alloy microstructure was conducted using a TriboIndenter Ti-900 nanoindentation system (Hysitron, Minneapolis, MN, US). The instrument has a load range of 1 μN to 10 mN, with a maximum indentation depth of 20 μm, a displacement resolution of 0.0002 nm, and a thermal drift of less than 0.05 nm/s. The drive-loading mode used in this study was the constant loading rate mode, with both the loading and unloading rates set at 1000 μN/s, and the loading time was 8 s. To minimize experimental errors, 20 points were tested for each phase of the sample, and the average value was taken as the nano-hardness for that phase.

3. Results and Discussion

3.1. Phase and Microstructure

Figure 1 shows the XRD patterns of (AlCrFeNi)100-xSix (x = 0, 1, 2, and 3) composite EHEAs with varying Si contents. The results reveal diffraction peaks corresponding to both the B2 and A2 crystal structures in all the alloys. Overlapping peaks are observed in the XRD patterns, suggesting that the ordered Ni-Al (B2) phase and the disordered Fe-Cr (A2) phase have similar crystal structures. According to previous studies, this phenomenon can be attributed to spinodal decomposition [32,33]. The content of B2 phase in the alloy can be characterized by the characteristic peak at (100). As demonstrated in Figure 1b, the intensity of the (100) diffraction peak increases with the addition of Si, while shifting toward smaller angles. This suggests that the large atomic radius mismatch between Si and other elements in the AlCrFeNi HEAs cause lattice distortion, leading to changes in the diffraction angle. Additionally, the increase in the lattice constant of the B2 phase in AlCrFeNi HEAs due to Si addition is also evident. Therefore, the doping of Si increases the B2 phase content in the AlCrFeNi HEAs, indicating that as the Si content increases, the B2 unit cell volume also expands. The lattice parameters and microstrain of the A2 and B2 phases of the AlSi alloy were estimated based on the position of the diffraction peaks, and the results are shown in Table 1.

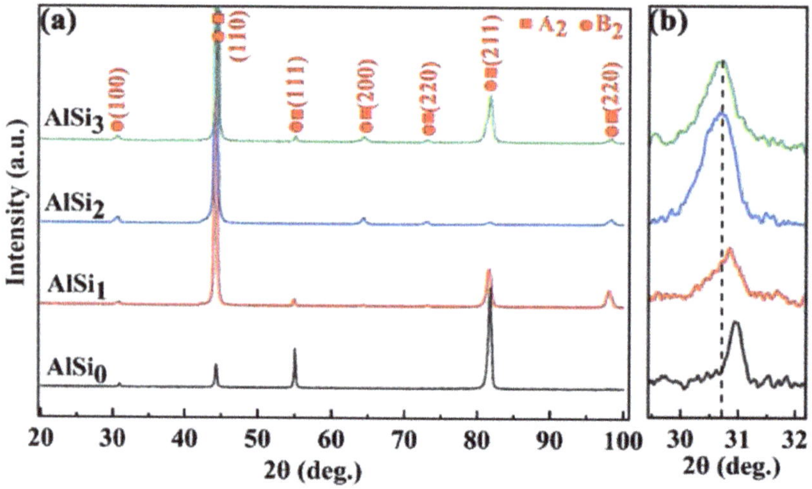

Figure 1. The as-cast (AlCrFeNi)100-xSix composite EHEAs. (**a**) XRD patterns, (**b**) magnified view of the (100) diffraction peak.

Table 1. The lattice parameters and micro-strain of the A2 and B2 phases of the AlSi alloy.

Alloys	Lattice Constant (Å)		Micro-Strain (%)	
	A2	B2	A2	B2
AlSi0	2.888 ± 0.005	2.9990 ± 0.005	0.197 ± 0.0367	0.186 ± 0.0353
AlSi1	2.889 ± 0.005	2.989 ± 0.005	0.196 ± 0.0382	0.181 ± 0.0365
AlSi2	2.887 ± 0.005	2.998 ± 0.005	0.192 ± 0.0325	0.152 ± 0.0320
AlSi3	2.887 ± 0.005	2.998 ± 0.005	0.220 ± 0.0302	0.214 ± 0.0302

Figure 2 presents the SEM images of the as-cast (AlCrFeNi)100-xSix (x = 0, 1, 2, 3) composite EHEAs. In Figure 2a, the AlSi0 alloy exhibits a nearly identical lamellar eutectic structure. The high-magnification SEM image of the alloy, shown in Figure 2b, reveals dense lamellae near the grain boundaries. A substantial quantity of rod-shaped lamellar precipitates, distributed uniformly throughout the alloy, is evident. An XRD analysis indicates that these rod-shaped precipitates correspond to the B2 structure, while the matrix consists of BCC phases. Furthermore, it is evident that with an increase in Si content, the lamellar structure at the grain boundaries expands. As shown in Figure 2a,e,g, the grain size of the AlSi2 and AlSi3 alloys is smaller than that of the AlSi0 alloy.

To analyze the distribution of the B2 phase and the Cr-Fe-rich A2 phase in the grains, SEM-EDS surface analyses were conducted on the AlSi1 and AlSi3 alloys. As shown in Figures 3 and 4, the lamellar composite structure at the grain boundaries consists of alternating layers of the B2 phase enriched in Al-Ni and the A2 phase enriched in Cr-Fe. In the flower disk structure of Figure 4, there are particle-like Cr-Ni-rich phases surrounded by Al-Ni-rich phases. Comparing the grains in Figures 3 and 4, it can be observed that with increasing Si content, the B2 lamellar structure enlarges as the grain boundary regions expand, which is consistent with the shift in the (100) diffraction peak of the B2 phase to smaller angles. In the AlSi1 alloy, the distribution of elements at the grain boundaries is uniform, with no distinct B2 or A2 phases. In the AlSi3 alloy, a well-defined alternation of the A2 and B2 phases is observed in both the grains and the grain boundaries. The central particles and the petal-shaped lamellae radiating outward from the center are Cr-Fe-rich A2 phases, while the other parts of the center and the non-radiating lamellae are B2 phases.

Moreover, the volume fraction of lamellar structures at the grain boundaries increases, leading to an overall increase in the grain boundary regions with higher Si content. This corresponds to an increase in the B2 phase content, which aligns with the XRD phase analysis results.

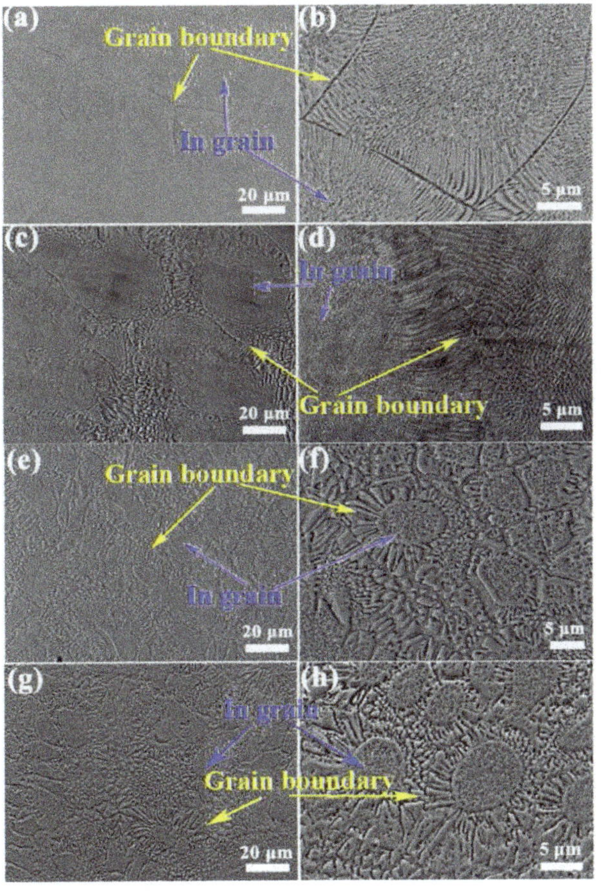

Figure 2. SEM images of the as-cast (AlCrFeNi)100-xSix composite EHEAs. (**a**,**b**) AlSi0, (**c**,**d**) AlSi1, (**e**,**f**) AlSi2, (**g**,**h**) AlSi3.

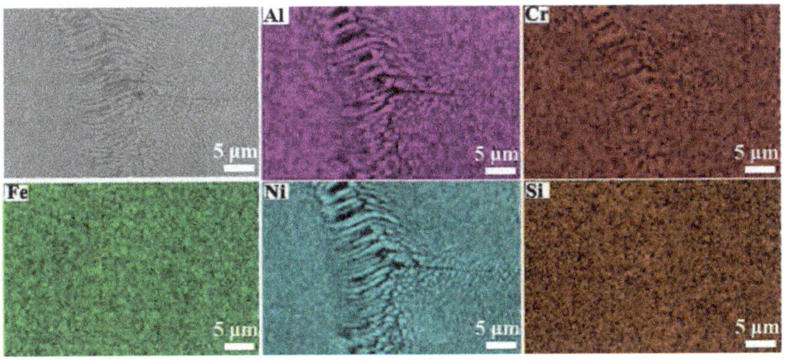

Figure 3. SEM images and EDS elemental maps of the as-cast AlSi1 alloy.

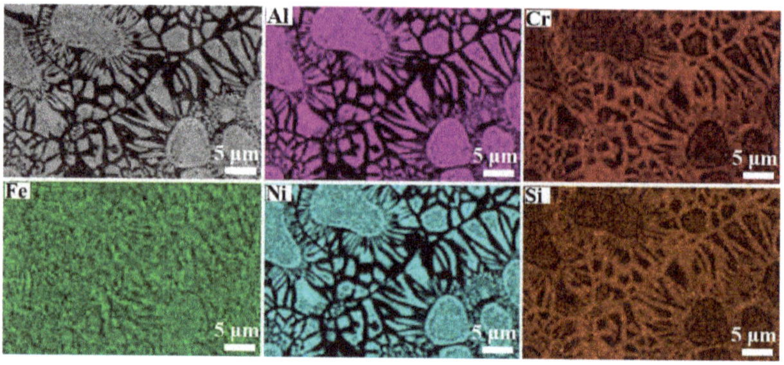

Figure 4. SEM images and EDS elemental maps of the as-cast AlSi3 alloy.

The Si element is primarily distributed around the B2 particles at the grain boundaries. Therefore, the addition of Si in AlCrFeNi HEAs promotes precipitation of the B2 phase at these boundaries, which is consistent with the increase in the intensity of the (100) diffraction peak. The expansion of the lamellar composite structure at the grain boundaries, along with the reduction in grain size and its tendency to become more rounded, ultimately leads to the formation of a sunflower-like composite structure (as shown in Figure 2f,h). The petal-like lamellar composite structure radiates outward from the central disc region, while sunflower-seed-like precipitates are uniformly distributed within the central disc. According to the literature, the formation of such structures can be attributed to the spinodal decomposition process [32,33], where the A2 and B2 phases grow in a lamellar form from the boundaries perpendicular to the previously precipitated phase after grain formation.

To further analyze and identify the microstructure, phase composition, and element distribution of the alloy, a TEM analysis was performed on the AlSi3 alloy as an example. Figure 5 shows the TEM images of the sunflower structure in the AlSi3 alloy and the corresponding selected area electron diffraction (SAED) patterns. Figure 5a presents a low-magnification TEM bright-field image, revealing that the alloy structure consists of two regions: the sunflower-like disk and petal regions. Figure 5b shows a high-angle annular dark-field (HAADF) TEM image at the boundary between the sunflower disk and petal regions. The alternating light and dark bands in the petal region indicate that adjacent lamellae have different phase structures. A considerable quantity of submicron-sized sunflower seed-like particles are precipitated in the sunflower disk region. Figure 5c shows a high-magnification image of rod-shaped precipitates in the petal region, showing a significant number of nanoscale precipitates of the B2 phase on the A2 phase lamellae. Figure 5d is a high-magnification TEM image of the petal region, and Figure 5e–h present the SAED patterns along the crystal zone axis. In Figure 5a, the light gray region in the sunflower disk separates all the elliptical grains. The enlarged view in Figure 5b shows that the disk region contains spherical or elliptical-shaped precipitates, while the rod-shaped precipitates in the petal region alternate. Figure 5d provides an enlarged view of the rod-shaped precipitates in the petal region. A diffraction analysis confirms that the grains in the sunflower disk region correspond to the A2 phases (Figure 5e), while the rest are characterized by the B2 phases (Figure 5f). The compositions of adjacent rod-shaped precipitates in the petal region differ, with the diffraction spots in Figure 5g,h showing that these rod-shaped precipitates are the B2 and A2 phases, respectively.

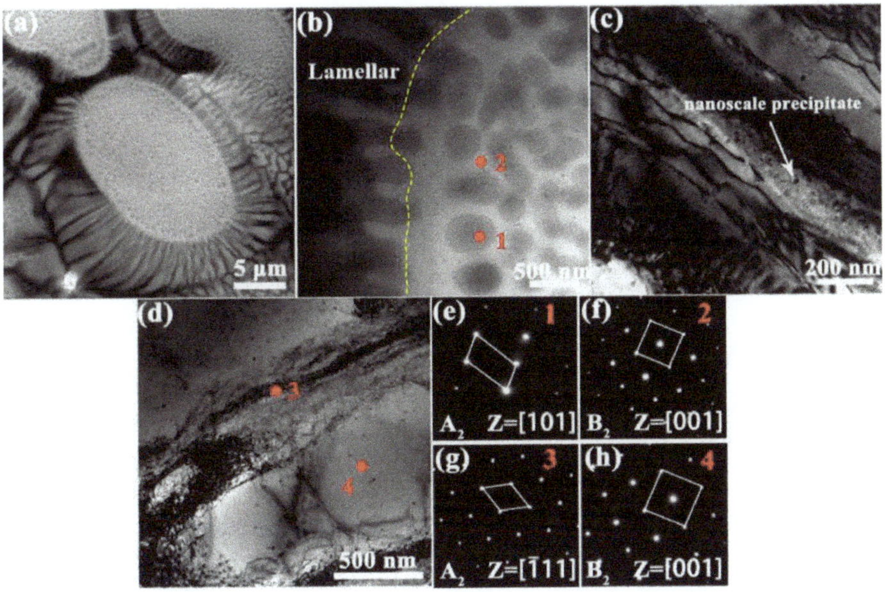

Figure 5. TEM images of the as-cast AlSi3 alloy: (**a**) low-magnification image, (**b**) high-magnification image of the boundary between the flower disk and petal regions, (**c**) high-magnification image of the nanometer-sized precipitates within the rod-shaped precipitates in the petal region, (**d**) high-magnification image of the boundary of the rod-shaped precipitates in the petal region, (**e**–**h**) SAED patterns from the corresponding regions.

The TEM-EDS elemental analysis was conducted on the region shown in Figure 5b, and the resulting elemental distribution map is presented in Figure 6. The analysis reveals that the two phases are enriched in different elements: the bright A2 phase is rich in Fe and Cr, while the darker B2 phase is primarily composed of Ni and Al. Both phases contain Si, but the Si content is notably higher in the B2 phase than in the A2 phase.

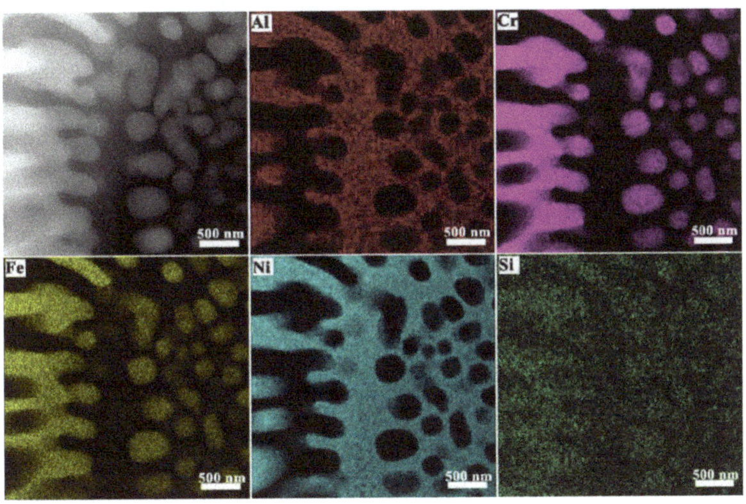

Figure 6. HAADF image and EDS elemental maps of the as-cast AlSi3 alloy.

3.2. Mechanical Properties

3.2.1. Compression Properties and Vickers Hardness of (AlCrFeNi)100-xSix Composite EHEAs

Figure 7a shows the compression stress–strain curves for the as-cast (AlCrFeNi)100-xSix composite EHEAs with varying Si content. As illustrated, compared to the AlSi0 alloy, the other three alloys exhibit minimal changes in plasticity and toughness but show a significant increase in yield strength (YS). Notably, the YS of the AlSi3 alloy reaches 1360 MPa.

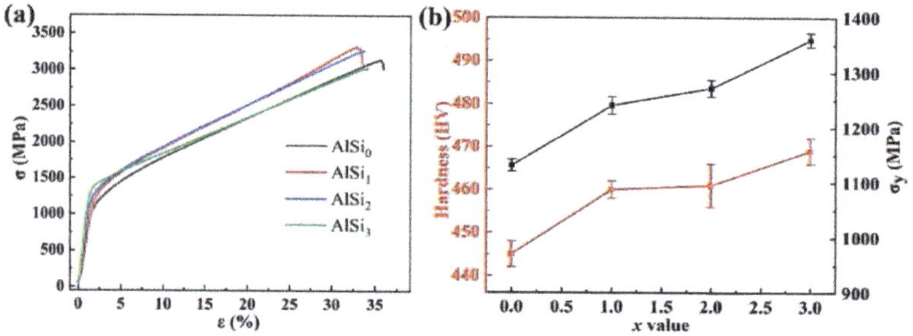

Figure 7. (**a**) Compression curves of the as-cast (AlCrFeNi)100-xSix composite EHEAs, (**b**) variation trends of hardness (HV) and YS (σy) as a function of the Si content (x).

Figure 7b illustrates the variation in Vickers hardness (HV) and YS (σy) with respect to the Si content (x) in the as-cast (AlCrFeNi)100-xSix composite EHEAs. It is evident that both HV and YS follow a similar increasing trend with increasing Si content. From the AlSi0 to AlSi3 alloys, HV and YS show a steady rise. Specifically, there is a sharp increase from AlSi0 to AlSi1, a moderate increase from AlSi1 to AlSi2, and a more pronounced increase from AlSi2 to AlSi3. These four alloys can be divided into two groups: one group includes AlSi0 and AlSi1, while the other consists of AlSi2 and AlSi3. The most salient structural characteristic of the latter group is the presence of a substantial quantity of sunflower composite eutectic structures. This suggests that the addition of an appropriate amount of Si in the (AlCrFeNi)100-xSix composite EHEAs system promotes the formation of sunflower composite structures, which significantly enhance the mechanical properties. Previous studies on sunflower composite structures have also indicated their positive effect on improving the compressive properties of alloys [34].

Table 2 summarizes the mechanical properties of the four alloys. The compression test results show that the YS of the AlSi0, AlSi1, AlSi2, and AlSi3 alloys are 1131 MPa, 1241 MPa, 1272 MPa, and 1360 MPa, respectively. The alloys fracture strengths are 3135 MPa, 3259 MPa, 3261 MPa, and 3019 MPa, with corresponding fracture strain values of 35.6%, 33.3%, 33.8%, and 34.3%, respectively. The HV values of these four alloys range from 445 HV to 469 HV (445 HV, 460 HV, 461 HV, and 469 HV, respectively). Overall, the AlSi3 alloy exhibits superior mechanical properties.

Table 2. Mechanical properties of the as-cast (AlCrFeNi)100-xSix composite EHEAs.

Alloys	Vickers Hardness (HV)	Yield Strength σ_y (MPa)	Fracture Strength σ_f (MPa)	Fracture Strain ε_f (%)
AlSi$_0$	445 ± 3	1131 ± 11	3135 ± 21	35.6 ± 0.3
AlSi$_1$	460 ± 2	1241 ± 16	3259 ± 25	33.3 ± 0.7
AlSi$_2$	461 ± 5	1272 ± 15	3261 ± 24	42.3 ± 0.5
AlSi$_3$	469 ± 3	1360 ± 13	3019 ± 27	46.5 ± 0.2

3.2.2. Nanoindentation Testing of (AlCrFeNi)100-xSix Composite EHEAs

To further investigate the micro-mechanical properties of different regions in the as-cast (AlCrFeNi)100-xSix composite EHEAs, nanoindentation tests were performed on both the grain boundary (petal) and the grain interior (flower disk) of each alloy. To ensure accuracy, 20 indents were made in each phase, and the average value was taken as the nano-hardness for that phase. The results of these tests are shown in Figure 8 and Table 3. Figure 8a,b show typical nanoindentation load–displacement curves for the grain interior and grain boundary regions of the as-cast (AlCrFeNi)100-xSix (x = 0, 1, 2, 3) composite EHEAs. It is clear that the hardness values of the grain interior and grain boundary regions of the AlSi3 alloy are the highest, measuring 7.06 GPa and 8.10 GPa, respectively. The hardness values of the AlSi2 alloy are close to those of AlSi3, with hardness values of 8.02 GPa and 6.92 GPa at the grain boundary and grain interior, respectively. The hardness values at the grain boundary of AlSi0 and AlSi1 alloys are nearly identical, at 7.26 GPa and 7.25 GPa, whereas their hardness values in the grain interior are 6.73 GPa and 6.45 GPa, respectively. Figure 8c shows the atomic force microscope (AFM) image used to identify suitable locations for indentation on the alloy surface, in conjunction with the nano-indenter. The positions of the sunflower disk and petal regions were determined from the AFM image for indentation testing.

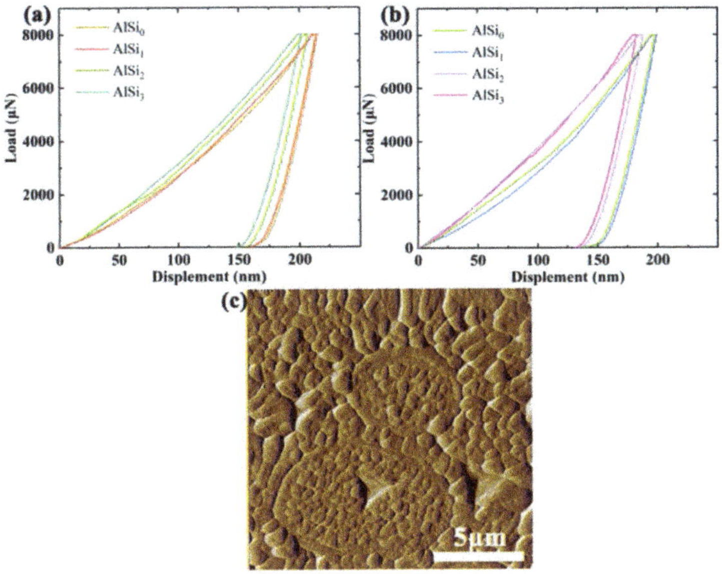

Figure 8. Nanoindentation load–displacement curves of the as-cast (AlCrFeNi)100-xSix (x = 0, 1, 2, 3) composite EHEAs for (**a**) the DR region and (**b**) the ID region. (**c**) Atomic force microscopy image showing the selected indentation points in the flower disk and petal regions.

Table 3. Nanoindentation test results of the as-cast (AlCrFeNi)100-xSix (x = 0, 1, 2, 3) composite EHEAs for the crystal interior and grain boundary regions.

Alloys	Hc (GPa)	
	DR Region	ID Region
$AlSi_0$	6.73 ± 0.12	7.26 ± 0.15
$AlSi_1$	6.45 ± 0.13	7.25 ± 0.11
$AlSi_2$	6.92 ± 0.11	8.02 ± 0.13
$AlSi_3$	7.06 ± 0.09	8.10 ± 0.12

Figure 9 shows the trend of hardness values in the grain interior and grain boundary regions of the experimental alloys as a function of Si content. The results indicate that the hardness in the grain boundary regions is higher than in the grain interior regions for all four alloys. This can be attributed to the addition of Si, which causes a significant amount of lamellar B2 phase precipitation at the grain boundaries, thereby enhancing hardness through a precipitation strengthening effect. Additionally, Figure 9 indicates a significant increase in hardness for both regions when x = 2. This substantial hardness increase occurs just after the formation of the sunflower composite structure. The sunflower composite structure refines the microstructure of the alloy, leading to higher hardness values for the AlSi2 and AlSi3 alloys compared to AlSi0 and AlSi1.

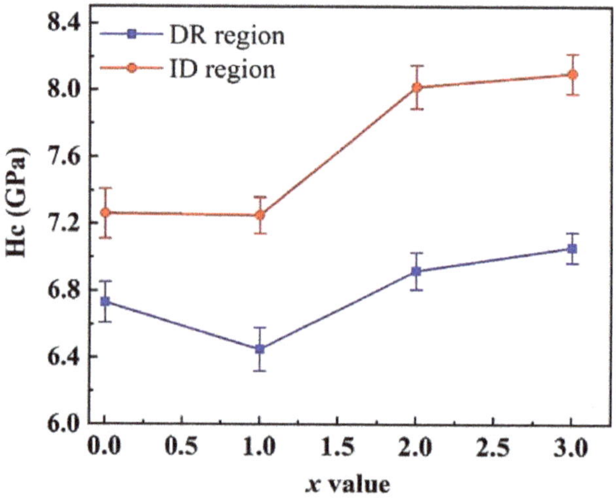

Figure 9. Line graph showing the variation trends of hardness values in the crystal interior and grain boundary regions of the as-cast (AlCrFeNi)100-xSix (x = 0, 1, 2, 3) composite EHEAs as a function of x.

3.3. *Strengthening Mechanisms of (AlCrFeNi)100-xSix Composite EHEAs*

It is assumed here that precipitation strengthening induced by uniformly distributed, ordered B2 precipitate phases enhances the YS of as-cast (AlCrFeNi)100-xSix composite EHEAs. Previous studies have shown [35,36] that precipitation strengthening of the solid-solution matrix can occur through either the Orowan dispersion mechanism or the dislocation shear mechanism, with the effectiveness of the Orowan mechanism relying on the interaction between moving dislocations and the precipitate phase. Typically, the dominance of the shear mechanism and the Orowan mechanism is governed by the size of the precipitates and their compatibility with the matrix, respectively [37]. In AlSi2 and AlSi3 alloys with graded B2 precipitates embedded in an A2 matrix, the shear mechanism prevails in the nanoscale B2 precipitates, while the Orowan dispersion mechanism dominates in the submicron B2 precipitates.

The enhancement in YS from the precipitation shear mechanism in nanoscale precipitates is attributed to three factors: coherency strengthening (ΔσCS), modulus strengthening (ΔσMS), and ordering strengthening (ΔσOS). The classical equations for calculating their contributions are given in [36,38]:

$$\Delta\sigma_{CS} = M\alpha_{\varepsilon}(G\varepsilon_c)^{\frac{3}{2}}\left(\frac{rf}{0.5Gb}\right)^{\frac{1}{2}} \quad (1)$$

$$\Delta\sigma_{MS} = 0.0055M(\Delta G)^{\frac{3}{2}}\left(\frac{r}{b}\right)^{\frac{3m}{2}-1} \quad (2)$$

$$\Delta\sigma_{OS} = 0.81M\left(\frac{\gamma_{apb}}{2b}\right)\left(\frac{3\pi f}{8}\right)^{\frac{1}{2}} \quad (3)$$

In the equation, M = 2.73 (M is the Taylor factor [36]). $\alpha\varepsilon$ = 2.6 [39]. εc is the lattice misfit under constraint, defined as $\varepsilon c = 2\varepsilon/3$. The expression $\varepsilon = \Delta\alpha/\alpha = 0.0038$ is utilized to represent the misfit between the matrix and the precipitate phase [40]. b is the magnitude of the Burgers vector, which is expressed as $3\alpha/2$. R denotes the average radius of the nanoscale B2 precipitates, measured at 11 nm. F represents the volume fraction of the nanoscale precipitates [41], which is 15.8%. The shear modulus G (using α-Fe) is 83 GPa, while ΔG denotes the shear modulus mismatch between the A2 matrix and the B2 precipitates, valued at 3 GPa (for the B2-NiAl phase, G = 80 GPa) [42]. It has been established that the constant for edge dislocations is m = 0.85 [43]. The anti-phase boundary energy for the B2 phase is represented by γapb, which is equivalent to 0.25 J/m^{-2} [44]. It is important to note that these strengthening processes are sequential, with the first two factors ($\Delta\sigma$CS and $\Delta\sigma$MS) acting before dislocation shear of the precipitates, while the last factor ($\Delta\sigma$OS) contributes during the shear process. In this case, the overall strength increase from the shear mechanism is determined by the sum of ($\Delta\sigma$CS + $\Delta\sigma$MS) or $\Delta\sigma$OS. The strength increases from ($\Delta\sigma$CS + $\Delta\sigma$MS) in the A2 matrix due to the 22 nm B2 nano-dispersions were found to be $\Delta\sigma$CS = 487 MPa, $\Delta\sigma$MS = 100.9 MPa, and $\Delta\sigma$OS = 276.3 MPa. The total shear mechanism strength increase is thus 587.9 MPa.

When the particles are large or incoherent with the matrix, the Orowan dispersion mechanism becomes dominant. The Orowan-type mechanism ($\Delta\sigma$Orowan) can be calculated using the following equation [45]:

$$\Delta\sigma_{Orowan} = 0.4MGb\frac{\ln(2\sqrt{\frac{2r}{3}}/b)}{\lambda_P\pi\sqrt{1-\nu}} \quad (4)$$

In this equation, ν is the Poisson's ratio of the BCC metal, with a value of 0.3, where r = 208 n, f = 26.5%, and λP denotes the precipitation spacing of the submicron-sized precipitates, defined as follows:

$$\lambda_P = 2\sqrt{\frac{2}{3}}r\left(\sqrt{\frac{\pi}{4f}}-1\right) \quad (5)$$

By incorporating these values into the above equation, the YS increase induced by the submicron B2 precipitates was obtained, with $\Delta\sigma$Orowan = 263.5 MPa.

On this basis, the strengthening due to precipitate hardening is estimated to be approximately 851.4 MPa, and the true YS of the as-cast AlSi2 alloy (~1272 MPa) is primarily attributed to approximately 66.93%. With precipitation strengthening, the remaining strength is mainly attributed to mechanisms such as the lattice friction strength of the A2 matrix, solid-solution strengthening, and grain boundary strengthening. The analysis reveals that shear mechanism strengthening contributes a stress increment of about 440.8 MPa, while Orowan dispersion strengthening contributes a stress increment of 518.1 MPa.

4. Conclusions

This paper prepared cast ingots of (AlCrFeNi)100-xSix (x = 0, 1, 2, 3) composite EHEAs, with varying Si contents and systematically investigated the effect of Si doping on the microstructure and mechanical properties of AlCrFeNi-based composite EHEAs. The following conclusions were drawn:

1. The microstructure of (AlCrFeNi)100-xSix (x = 0, 1, 2, 3) composite EHEAs consists of two regions: the grain interior (flower disk) and grain boundaries (flower petals). The grain interior exhibits an alternating A2/B2 lamellar composite structure with B2 phase nano-dispersed precipitates. As the Si content increases, the lamellar structure at the grain boundaries expands, forming flower petals, while the precipitates within the alloys evolve into flower disks and the alloy exhibits a sunflower-like composite structure. The volume fraction of lamellar structures increases in the petals, accompanied by grain refinement.
2. With increasing Si content, the yield strength of the alloy increases from 1131 MPa for the AlSi0 alloy to 1360 MPa for the AlSi3 alloy. The shear strengthening mechanism caused by nano-precipitates and the Orowan dispersion strengthening caused by submicron precipitates contribute approximately 587.9 MPa and 267.5 MPa, respectively.
3. A significant number of nano/sub-micron precipitates on lamellar structures at grain boundaries enhance precipitation strengthening, resulting in greater hardness at the grain boundaries compared to the grain interiors in (AlCrFeNi)100-xSix (x = 0, 1, 2, 3) composite EHEAs. The formation of the sunflower-like composite structure refines the alloy's microstructure, leading to higher hardness in AlSi2 and AlSi3 alloys compared to the AlSi0 and AlSi1 alloys.

Author Contributions: Conceptualization, Y.J., B.Z. and Y.L.; Methodology, S.L., A.L., K.K., J.Z., D.H., C.C., Y.J. and M.X.; Software, D.H., C.C., Y.J., M.X., B.Z. and Y.L.; Formal analysis, S.L., A.L. and J.Z.; Investigation, S.L. and A.L.; Resources, G.L.; Writing—original draft, S.L.; Writing—review and editing, G.L.; Supervision, K.K.; Funding acquisition, G.L. All authors have read and agreed to the published version of the manuscript.

Funding: This research was funded by National Key Research and Development Program of China (grant number 2022YFA1603800), National Natural Science-Foundation of China (grant number 12274362), and Songshan Lake Materials Laboratory (grant number 2022SLABFK01).

Data Availability Statement: The original contributions presented in this study are included in this article. Further inquiries can be directed to the corresponding author.

Conflicts of Interest: The authors declare no conflicts of interest.

References

1. Wu, Y.; Liaw, P.K.; Li, R.; Zhang, W.; Geng, G.; Yan, X.; Liu, G.; Zhang, Y. Relationship between the unique microstructures and behaviors of high-entropy alloys. *Int. J. Miner. Metall. Mater.* **2024**, *31*, 1350–1363. [CrossRef]
2. Yadav, Y.K.; Shaz, M.A.; Mukhopadhyay, N.K.; Yadav, T.P. Formation of B2 phase and its stability in equiatomic Al-Cu-Fe-Ni-Ti high entropy alloy. *J. Alloys Metall. Syst.* **2024**, *8*, 100137. [CrossRef]
3. Rao, K.R.; Mohan, M.; Dewangan, S.K.; Nagarjuna, C.; Lee, K.; Ahn, B. Microstructural characterization and thermal stability of AlCrFeNiTi + Y_2O_3 high-entropy alloy nanocomposites prepared by mechanical alloying. *Mater. Lett.* **2024**, *372*, 137018.
4. Kotan, H.; Koç, R.C.; Batıbay, A.B. Remarkable thermal stability of nanocrystalline CoCrFeNi high entropy alloy achieved through the incorporation of rare-earth element samarium. *Intermetallics* **2025**, *178*, 108608. [CrossRef]
5. Zhang, J.; Yoon, K.N.; Kim, M.S.; Ahn, H.S.; Kim, J.Y.; Ryu, W.H.; Park, E.S. Manipulation of Microstructure and Mechanical Properties in N-Doped CoCrFeMnNi High-Entropy Alloys. *Metals* **2021**, *11*, 1487. [CrossRef]
6. Kovalenko, E.; Krasanov, I.; Valdaytseva, E.; Stankevich, S.; Klimova-Korsmik, O.; Gushchina, M. Microstructure and Mechanical Properties of High-Entropy Alloy FeCoNiCr(X) Produced by Laser Directed Energy Deposition Process: Effect of Compositional Changes. *Metals* **2024**, *15*, 26. [CrossRef]
7. Pourmohammadi, S.; Mohammadnejad, A.; Bahrami, A.; Anijdan, S.H.M.; Park, N.; Ghosh, M.; Stability, P. Microstructure, and Mechanical Properties of Spark Plasma Sintered Nanocrystalline Boron-Doped AlCoFeMnNi High-Entropy Alloy. *Metals* **2023**, *13*, 1025. [CrossRef]

8. Jiang, B.; Huang, Z.; Liu, C.; Wang, H.; Shu, F.; Zhao, Y.; Lei, H. Optimization of Process Parameters and Microstructure of CoCrFeNiTiAl High-Performance High-Entropy Alloy Coating. *Metals* **2024**, *14*, 1384. [CrossRef]
9. Zhang, H.; Zhang, G.-H. Preparation and performances of molybdenum—Copper alloy with additions of tungsten, nickel, iron and cerium dioxide. *Int. J. Refract. Met. Hard Mater.* **2024**, *121*, 106691. [CrossRef]
10. Pan, Y.; Li, B.; Xu, Y.; Wu, D.; Hou, X.; Gao, Y.; Bai, P.; Liang, C. Effect of Sigma and $L2_1$ phases co-precipitation on the mechanical properties of $CoCr_{0.6}NiV_{0.6}Al_{0.35}Tix$ high entropy alloy. *Mater. Sci. Eng. A* **2025**, *923*, 147689. [CrossRef]
11. Tan, M.; Meng, L.; Lin, C.; Ke, L.; Liu, Y.; Qu, J.; Qi, T. Variation of microstructures and properties of $Co_{0.2}CrAlNi$ high entropy alloy doped Si. *J. Alloys Compd.* **2022**, *927*, 167081. [CrossRef]
12. Huang, L.; Wang, X.; Jia, F.; Zhao, X.; Huang, B.; Ma, J.; Wang, C. Effect of Si element on phase transformation and mechanical properties for FeCoCrNiSix high entropy alloys. *Mater. Lett.* **2021**, *282*, 128809. [CrossRef]
13. Liao, L.; Cheng, Y.; Dai, S.; Khan, M.A.; Zhang, H.; Li, F. Effect of cold-rolling and annealing temperature on microstructure, texture evolution and mechanical properties of FeCoCrNiMn high-entropy alloy. *J. Mater. Res. Technol.* **2024**, *33*, 683–697. [CrossRef]
14. Odabas, O.; Ozgurluk, Y.; Karaoglanli, A.C. Microstructural evolution and high temperature hot corrosion behaviour of AlCoCrFeNiTi high-entropy alloy coatings. *Mater. Today Commun.* **2024**, *41*, 110910. [CrossRef]
15. Yi, H.; Zhang, Y.; Xie, R.; Bi, M.; Wei, D. High-Temperature Deformation Behaviors of the C-Doped and N-Doped High Entropy Alloys. *Metals* **2021**, *11*, 1517. [CrossRef]
16. Liu, D.; Jin, X.; Yang, H.; Qiao, J.; Zhang, Y. High-Temperature Mechanical Behavior of Cobalt-Free FeMnCrNi(Al) High-Entropy Alloys. *Metals* **2023**, *13*, 1885. [CrossRef]
17. Wang, Y.; Li, D.; Wang, S.; Zhang, M.; Gong, P.; Hu, Z.; Li, B. Effect of Cr content on the high temperature oxidation behavior of FeCoNiMnCr porous high-entropy alloys. *J. Mater. Res. Technol.* **2024**, *33*, 3324–3333. [CrossRef]
18. Wang, M.; Lu, Y.; Lan, J.; Wang, T.; Zhang, C.; Cao, Z.; Li, T.; Liaw, P.K. Lightweight, ultrastrong and high thermal-stable eutectic high-entropy alloys for elevated-temperature applications. *Acta Mater.* **2023**, *248*, 118806. [CrossRef]
19. Wang, M.; Lu, Y.; Wang, T.; Zhang, C.; Cao, Z.; Li, T.; Liaw, P.K. A novel bulk eutectic high-entropy alloy with outstanding as-cast specific yield strengths at elevated temperatures. *Scr. Mater.* **2021**, *204*, 114132. [CrossRef]
20. Samoilova, O.; Pratskova, S.; Suleymanova, I.; Shaburova, N.; Moghaddam, A.O.; Trofimov, E. Effect of Pt Addition on the Oxidation and Corrosion Resistance of $Al_{0.25}CoCrFeNi$ High-Entropy Alloy. *Metals* **2023**, *13*, 1709. [CrossRef]
21. Qi, W.; Wang, W.; Yang, X.; Xie, L.; Zhang, J.; Li, D.; Zhang, Y. Effect of Zr on phase separation, mechanical and corrosion behavior of heterogeneous CoCrFeNiZrx high-entropy alloy. *J. Mater. Sci. Technol.* **2022**, *109*, 76–85. [CrossRef]
22. Zhou, Y.; Liu, X.; Chen, J.; Wang, A.; Fu, H.; Zhang, H.; Zhu, Z. Strategies for optimizing mechanical properties of refractory high entropy alloys induced by solid solution strengthening mechanism. *Mater. Sci. Eng. A* **2025**, *923*, 147696. [CrossRef]
23. Ye, X.; Diao, Z.; Lei, H.; Wang, L.; Li, Z.; Li, B.; Feng, J.; Chen, J.; Liu, X.; Fang, D. Multi-phase FCC-based composite eutectic high entropy alloy with multi-scale microstructure. *Mater. Sci. Eng. A* **2024**, *889*, 145815. [CrossRef]
24. Chen, Y.; Tang, C.; Jiang, J.-Z. Bulk metallic glass composites containing B2 phase. *Prog. Mater. Sci.* **2021**, *121*, 100799. [CrossRef]
25. Aizenshtein, M.; Strumza, E.; Brosh, E.; Hayun, S. Precipitation kinetics, microstructure, and equilibrium state of A2 and B2 phases in multicomponent $Al_{2.75}CoCrFeNi$ alloy. *J. Mater. Sci.* **2020**, *55*, 7016–7028. [CrossRef]
26. Aizenshtein, M.; Hayun, S. Synthesis of A2 and B2 phases in $Al_xCoCrFeNi$ multi-component system. *Metallogr. Microstruct. Anal.* **2020**, *9*, 305–311. [CrossRef]
27. Callegari, B.; Lima, T.N.; Coelho, R.S. The influence of alloying elements on the microstructure and properties of Al-Si-based casting alloys: A review. *Metals* **2023**, *13*, 1174. [CrossRef]
28. Gu, X.; Zhuang, Y.-X.; Jia, P. Evolution of the microstructure and mechanical properties of as-cast $Al_{0.3}CoCrFeNi$ high entropy alloys by adding Si content. *Mater. Sci. Eng. A* **2022**, *840*, 142983. [CrossRef]
29. Yang, H.; Liu, X.; Li, A.; Li, R.; Xu, S.; Zhang, M.; Yu, P.; Yu, S.; Jiang, M.; Huo, C.; et al. Effect of silicon addition on the corrosion resistance of $Al_{0.2}CoCrFe_{1.5}Ni$ high-entropy alloy in saline solution. *J. Alloys Compd.* **2023**, *964*, 171226. [CrossRef]
30. Luo, W.; Yuan, X.; Zhang, Z.; Cheng, C.; Liu, H.; Qiu, H.; Cheng, X. Effect of volumetric energy density on the mechanical properties and corrosion resistance of laser-additive-manufactured $AlCoCrFeNi_{2.1}$ high-entropy alloys. *J. Alloys Compd.* **2025**, *1010*, 178032. [CrossRef]
31. Erdogan, A.; Sunbul, S.E.; Icin, K.; Doleker, K.M. Microstructure, wear and oxidation behavior of AlCrFeNiX (X= Cu, Si, Co) high entropy alloys produced by powder metallurgy. *Vacuum* **2021**, *187*, 110143. [CrossRef]
32. Yang, T.; Deng, H.; Wang, Y.; Feng, J.; Meng, Q.; Qi, J.; Wei, F.; Hu, Z.; Meng, D.; Sui, Y. Study on the Microstructure and Mechanical Properties of $AlCr_{1.6}Fe_xNi_{(3.2\text{-}X)}Si_{0.2}$ High-Entropy Alloys. *SSRN* **2024**, 4705447. [CrossRef]
33. Yue, K.; Yang, X.; Wang, L.; Su, L.; Xu, Q.; Xu, Z.; Cheng, C.; Wang, Y.; Wang, Z.; Chen, Z. Effect of Ti content on spinodal decomposed microstructure and properties of AlCoCrFeNiTix high-entropy alloy coatings prepared by laser cladding. *J. Mater. Res. Technol.* **2025**, *34*, 1120–1129. [CrossRef]

34. Li, S.; Lv, J.; Xu, K.; Lou, M.; Hu, X.; Xiao, X.; Liu, S.; Chang, K. Manipulation of the spinodal decomposition behavior in a multicomponent system. *Scr. Mater.* **2022**, *219*, 114893. [CrossRef]
35. Wei, X.; Zhang, L.; Zhang, C.; Li, G. A sunflower-like eutectic microstructure in the $AlCrFeMnNi_2$ high-entropy alloy with excellent compressive mechanical properties. *Mater. Lett.* **2023**, *339*, 134107. [CrossRef]
36. Ma, Y.; Wang, Q.; Jiang, B.B.; Li, C.L.; Hao, J.M.; Li, X.N.; Dong, C.; Nieh, T.G. Controlled formation of coherent cuboidal nanoprecipitates in body-centered cubic high-entropy alloys based on $Al_2(Ni,Co,Fe,Cr)_{14}$ compositions. *Acta Mater.* **2018**, *147*, 213–225. [CrossRef]
37. Seidman, D.N.; Marquis, E.A.; Dunand, D.C. Precipitation strengthening at ambient and elevated temperatures of heat-treatable Al (Sc) alloys. *Acta Mater.* **2002**, *50*, 4021–4035. [CrossRef]
38. Nembach, E. Precipitation hardening caused by a difference in shear modulus between particle and matrix. *Phys. Status Solidi (A)* **1983**, *78*, 571–581. [CrossRef]
39. Jansson, B.; Melander, A. On the critical resolved shear stress from misfitting particles. *Scr. Metall.* **1978**, *12*, 497–498. [CrossRef]
40. Courtney, T.H. *Mechanical Behavior of Materials*; Waveland Press: Long Grove, IL, USA, 2005.
41. Hudok, D. *Properties and Selection: Irons, Steels, and High-Performance Alloys*; Metals Handbook; ASM Press: Washington, DC, USA, 1990; Volume 1, pp. 200–211.
42. Wasilewski, R. Elastic constants and youngs modulus of NiAl. *AIME Met Soc. Trans.* **1966**, *236*, 455–457.
43. Melander, A.; Persson, P.Å. The strength of a precipitation hardened AlZnMg alloy. *Acta Metall.* **1978**, *26*, 267–278. [CrossRef]
44. Hong, T.; Freeman, A. Effect of antiphase boundaries on the electronic structure and bonding character of intermetallic systems: NiAl. *Phys. Rev. B* **1991**, *43*, 6446. [CrossRef]
45. Martin, J.W. Particle strengthening of metals and alloys. *Mater. Sci. Technol.* **1997**, *13*, 705.

metals

MDPI

Article

Microscopic Modeling of Interfaces in Cu-Mo Nanocomposites: The Case Study of Nanometric Metallic Multilayers

Abdelhafid Akarou, Florence Baras * and Olivier Politano *

Laboratoire Interdisciplinaire Carnot de Bourgogne ICB UMR 6303, Université Bourgogne Europe, CNRS, F-21000 Dijon, France; abdelhafid.akarou@u-bourgogne.fr

* Correspondence: florence.baras@u-bourgogne.fr (F.B.); olivier.politano@u-bourgogne.fr (O.P.)

Abstract: Nanocomposites composed of Cu and Mo were investigated by means of molecular dynamics (MD) simulations to study the incoherent interface between Cu and Mo. In order to select an appropriate potential capable of accurately describing the Cu-Mo system, five many-body potentials were compared: three Embedded Atom Method (EAM) potentials, a Tight Binding Second Moment Approximation (TB-SMA) potential, and a Modified Embedded Atom Method (MEAM) potential. Among these, the EAM potential proposed by Zhou in 2001 was determined to provide the best compromise for the current study. The simulated system was constructed with two layers of Cu and Mo forming an incoherent *fcc*-Cu(111)/*bcc*-Mo(110) interface, based on the Nishiyama–Wassermann (NW) and Kurdjumov–Sachs (KS) orientation relationships (OR). The interfacial energies were calculated for each orientation relationship. The NW configuration emerged as the most stable, with an interfacial energy of 1.83 J/m^2, compared to 1.97 J/m^2 for the KS orientation. Subsequent simulations were dedicated to modeling Cu atomic deposition onto a Mo(110) substrate at 300 K. These simulations resulted in the formation of a dense layer with only a few defects in the two Cu planes closest to the interface. The interfacial structures were characterized by computing selected area electron diffraction (SAED) patterns. A direct comparison of theoretical and numerical SAED patterns confirmed the presence of the NW orientation relationship in the nanocomposites formed during deposition, corroborating the results obtained with the model *fcc*-Cu(111)/*bcc*-Mo(110) interfaces.

Keywords: EAM; TB-SMA; MEAM; nanocomposite; nanometric metallic multilayers; Immiscible metals; Cu/Mo interface; Molecular dynamics simulations

1. Introduction

Recently, thin films have gained significant importance in advanced technologies, including protective coatings (acting as diffusion or thermal barriers and providing wear or corrosion resistance), biosensors, plasmonic devices, optical coatings, electrically conductive layers, and thin-film photovoltaic cells. Among these materials, nanometric metallic multilayers (NMMs), also known as layered composites or nanolaminates, have emerged as a particularly promising class of thin films. These structures are composed of hundreds of nanometric layers made of two metallic elements, with individual layer thicknesses typically ranging from 4 to 100 nm. Compared to monolithic films, NMMs offer superior properties due to their abundant interfaces and nanometric layered architecture. Recently, their unique physical and chemical properties have attracted increasing attention [1].

In this context, our study focuses on a specific type of NMMs composed of immiscible metals, referred to as NMMIs (Nanometric Metallic Multilayers of Immiscible Metals). These systems offer a wide range of mechanical, optical, magnetic, and radiation tolerance properties that can be tailored ("materials on demand") to specific applications by controlling the combined metal ratio, layer thickness, and texture. For instance, the family of NMMIs combining one element with high thermal and electrical conductivity with a refractory element that has a low coefficient of thermal expansion is promising for next-generation multi-material devices (e.g., heat sinks in electronic devices). However, their thermal stability remains a significant challenge, as high temperatures can profoundly alter their layered microstructure, thereby limiting their durability and potential applications [2]. The structural instability is closely related to interface and grain boundary energies and is further amplified by the significant stresses present in the multilayers. Although numerous experimental studies have investigated the structural stability of NMMIs under thermal treatments (e.g., Cu-Ag [3,4], Cu-W [5–7], Cu-Mo [8], Hf-Ti [9], Cu-Ag/Fe [10], Ag-Ni [11,12]), the precise mechanisms underlying this thermal instability remain poorly understood.

A starting explanation can be provided by simulations at the atomic scale. During the last decades, MD simulations have proven instrumental in exploring various aspects of NMMs. Previous studies examined reactive metallic multilayers like Al-Ni and Al-Ti, investigating their self-propagating exothermic reactions [13,14], reaction front propagation [15], and the role of different microstructures, such as amorphous and crystalline grains, on combustion behavior [16–18]. Recently, approaches focused on the thermal stability of immiscible metal nanolaminates. MD simulations of Ag-Ni multilayers investigated the kinetic pathways accountable for the degradation of NMMIs to nanocomposites (NCs) [19]. The effect of in-plane strain on thermal degradation kinetics of NMMs into NCs was investigated in Cu-W coupling experiments and DFT calculations [7,20].

As a case study, this work employs molecular dynamics (MD) to investigate Cu-Mo NMMIs, with particular emphasis on characterizing the interface between Cu and Mo. Here, the complete immiscibility between the metals prevents intermetallic phase formation and/or chemical intermixing, as observed in reactive NMMS. Therefore, other mechanisms are responsible for the thermal instability of the laminated architecture and need to be identified. Molecular dynamics requires reliable interatomic potentials, which describe the interactions between atoms in a system. Five many-body potentials fitted for Cu and Mo binary systems from the literature were benchmarked to select the most appropriate one to investigate the Cu/Mo interface.

Epitaxy relationships between *bcc* and *fcc* phases have been the topic of many works over the years, with the development of pioneering models for the ferrite-austenite interface in iron alloys. Two primary orientation relationships have been identified: the Nishiyama–Wassermann (NW) relationship [21,22], where $[1\bar{1}0]_{fcc} \parallel [100]_{bcc}$, and the Kurdjumov–Sachs (KS) relationship [23], where $[1\bar{1}0]_{fcc} \parallel [1\bar{1}1]_{bcc}$. These orientation relationships describe the atomic alignment at the interface and significantly influence the structural and energetic stability of the system. To date, few studies have been dedicated to the Cu/Mo interface despite the remarkable properties of these composite systems. Cui et al. [24] and Yeom et al. [8] observed using transmission electron microscopy (TEM) the Cu{111}//Mo{110} fiber texture. Nevertheless, they do not mention any preferential orientation relationships. In Cu-W [6] and Cu-Nb [25,26] systems, TEM observations reported Cu{111}, W{110} and Nb{110} textures with a Cu$\{111\}\langle 10\bar{1}\rangle \parallel$ W$\{110\}\langle 00\bar{1}\rangle$ and Cu$\{111\}\langle 110\rangle \parallel$ Nb$\{110\}\langle 111\rangle$ orientation relationships, corresponding to the NW and KS orientation relationships, respectively. Additionally, small regions with NW orientation were observed in the Cu-Nb system. On the other hand, for these two last systems, Bodlos et al. [27] using atomistic simulations reported a preferential NW orientation relationship. To determine the preferential epitaxial orientation in Cu-Mo nanometric metallic multilayers (NMMs), interfacial analyses were carried out for both KS and NW configurations. For this purpose, we created two simulation boxes, one corresponding to a KS orientation relationship and the other to a NW orientation relationship, and calculated the interfacial energy in both cases to identify the most stable configuration. To further investigate the orientation relationships predicted with the selected potential, we performed simulations of copper deposition onto a molybdenum substrate. The resulting deposits were analyzed and characterized using selected area electron diffraction (SAED) patterns.

2. Methods

The molecular dynamics simulations were performed using the LAMMPS (22 aug 2018 version) software package [28] using the five many-body potentials listed in Table 1. Three of them were based on the Embedded Atom Method (EAM), which is widely used for modeling metals [29]. One was developed by Gong et al. based on ab initio calculations that were performed to predict the structures, lattice constants, and cohesive energies of the metastable $Cu_{75}Mo_{25}$, $Cu_{50}Mo_{50}$, and $Cu_{25}Mo_{75}$ phases, as well as the elastic behavior of the pure Cu and Mo phases [30]. The two others were proposed by Zhou et al. in 2001 [31] and 2004 [32] in the context of simulating multilayers sputtering processes and properties. They are part of a 16-element normalized potential database that enables calculations of alloys. They were fitted to basic material properties such as lattice constants, elastic constants, bulk moduli, vacancy formation energies, sublimation energies, and heats of solution. They were successively used to study nanometric films such as CoFe/Cu/CoFe [31], CoFe/NiFe [32], Au/NiFe [33], and Cu/NiCo/Cu [34]. We also considered the Tight Binding in the Second Moment Approximation (TB-SMA) potential developed by M. A. Karolewski [35]. This potential was specifically developed to study sputtering simulations involving 26 *fcc* and *bcc* metals. Finally, a second-nearest-neighbor modified embedded-atom method (2NN MEAM) potential proposed by Wang et al. was also tested [36]. To our knowledge, this potential has not been used to study multilayers until now but was applied to investigate the effect of pressure on the enthalpy of mixing of *fcc*/*bcc* binary solid solutions [37]. For each individual potential, we computed microscopic, thermal, and mechanical properties of pure Cu and Mo with LAMMPS [28]. All quantities were then compared with experimental or ab initio data to assess their range of validity. The best compromise was found for the two pure elements with Zhou's 2001 potential, which was selected to investigate the Cu/Mo interface.

The computational efficiency of each potential was evaluated by running an NPT simulation for 100 ps using a simulation box containing 66,564 atoms. The performance, measured in nanoseconds per day (ns/day) on 20 cores, is presented in Figure S1. As expected, EAM and TB-SMA potentials exhibit better performances, with the Gong potential being the most efficient at about 22 ns/day on 20 cores. On the benchmarked system, the MEAM potential proposed by Wang et al. is about 27 times slower than the Gong potential.

Table 1. List of the benchmarked Cu-Mo potentials.

Potential	Ref	Type	Notation	Performance (ns/day)
Gong et al.	[30]	EAM	Gong	20.51
Zhou et al. 2001	[31]	EAM	Zhou.01	8.66
Zhou et al. 2004	[32]	EAM	Zhou.04	9.71
M. A. Karolewski et al.	[35]	TB-SMA	TB-SMA	12.77
J. Wang et al.	[36]	MEAM	MEAM	0.76

3. Results

3.1. Physical Properties of the Pure Elements in the System

3.1.1. Thermal Expansion

The variation of the lattice parameter with temperature and the coefficient of thermal expansion (CTE) were first determined. To this end, we performed 600 picoseconds NPT runs (600,000 timesteps) for temperatures ranging from 50 K to 1200 K in steps of 50 K using simulation boxes containing 15 × 15 × 15 lattice units. The box sizes were averaged over the last 100 ps in order to determine the lattice parameter at the target temperature. A fourth-order polynomial equation $a(T) = A + BT + CT^2 + DT^3 + ET^4$ was used to fit the lattice parameter as a function of temperature for the two elements. As shown in Figure S2 in the Supplementary Material (SM), the five potentials correctly predict an increase in the lattice parameter with temperature. The coefficients A, B, C, D, and E are given in the Tables S3 and S4 in the Supplementary Material (SM) for Cu and Mo, respectively. The percentage variation of $\Delta L/L_0 = (a(T) - a_0)/a_0$ as a function of temperature is plotted in Figure 1 along with the experimental values reported by Touloukian [38]. Here, a_0 is the lattice parameter at 293 K.

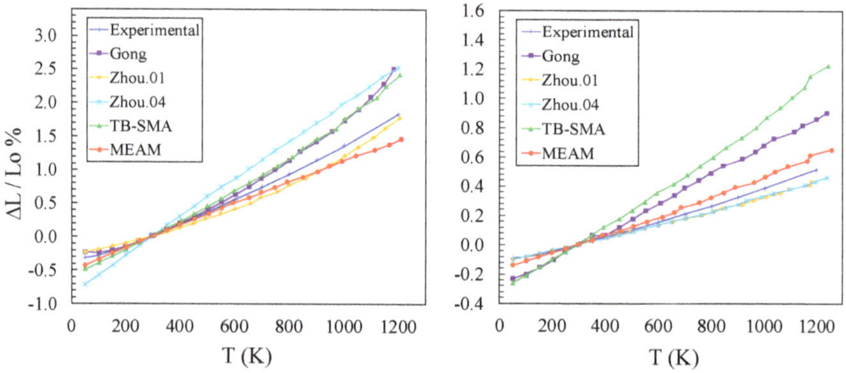

Figure 1. Comparison between the MD results obtained for Cu and Mo with experimental results from Touloukian [38].

For copper, Gong, TB-SMA, and MEAM potentials underestimate the value of $\Delta L/L_0$ at high temperatures, while the Zhou.04 potential shows underestimation even at low temperatures. The Zhou.01 potential provides more accurate values of $\Delta L/L_0$ across the entire temperature range. In the case of Mo, Gong and TB-SMA potentials similarly underestimate the value of $\Delta L/L_0$ across all temperature ranges. The Zhou.01 and Zhou.04 potentials also show underestimation at high temperatures but offer more accurate predictions at temperatures below 1200 K. The MEAM potential gives a more precise estimation across all temperature ranges. Therefore, the Zhou.01 potential emerges as the most reliable for both copper and molybdenum in the temperature range of interest in our studies (0 to 1200 K). The coefficient of thermal expansion was computed using $\alpha = \frac{1}{a(T)}\frac{da(T)}{dT}$ (see Figure 2) and compared to the experimental values available in [38]. For copper, the MEAM, Zhou.04, and TB-SMA potentials generally underestimate the CTE across the entire temperature range. The Zhou.01 potential offers a slightly better trend and provides the closest approximation across the entire temperature range. For molybdenum, the TB-SMA and Gong potentials overestimate the CTE, while the Zhou.01, Zhou.04, and MEAM potentials are closer to the experimental measurement. The Zhou.01 potential emerges as the most reliable model, closely matching the experimental data across the entire temperature range.

The difference in thermal expansion between copper and molybdenum will be the source of thermal stress in NMMIs. Thus, the calculation of the ratio of these two quantities will help to understand the presence of internal stress measured in an NMMIs at finite temperature (see Figure 3). Experimentally, the thermal expansion of copper is about three times higher than that of molybdenum in the given temperature range. The Zhou.01 potential provides the closest match to the experimental data. The Gong potential improves at higher temperatures but still underestimates the ratio. The Zhou.04 potential consistently overestimates the ratio, and the MEAM potential significantly underestimates it. Thus, the Zhou.01 potential appears to be the most reliable for predicting the thermal expansion ratio between copper and molybdenum across the temperature range.

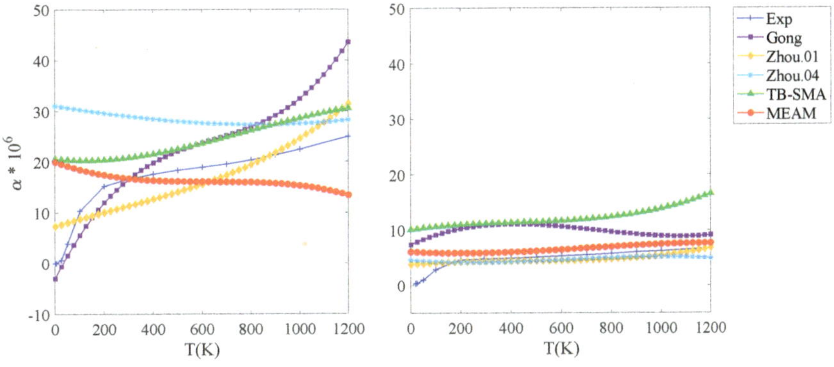

Figure 2. The evolution of the coefficient of thermal expansion as a function of temperature of copper and molybdenum.

3.1.2. Melting Temperature

The melting temperature was determined using the hysteresis method. The simulation box was heated from 300 K up to a target temperature well above the melting point, using a procedure similar to that used to calculate thermal expansion coefficients. Subsequently, we initiated a cooling process with a constant temperature ramp. The potential energy per atom was plotted as a function of the temperature (see Figure S3 in Supplementary Material). The bulk melting temperature, T_m, is estimated with $T_m = T_+ + T_- - \sqrt{T_+T_-}$, where T_+ represents the temperature at which the solid melted during the heating process, and T_- is the temperature at which the liquid solidified during the cooling process [39]. The melting temperatures obtained for Cu and Mo are listed in Table 2. The closest calculated melting points were 2778 K for Mo with the MEAM and 1332 K for Cu with the TB-SMA. The other values are relatively distant from the experimental values. The melting temperature is generally not considered in the fitting of potentials. This often results in significant discrepancies with experimental data, as observed here.

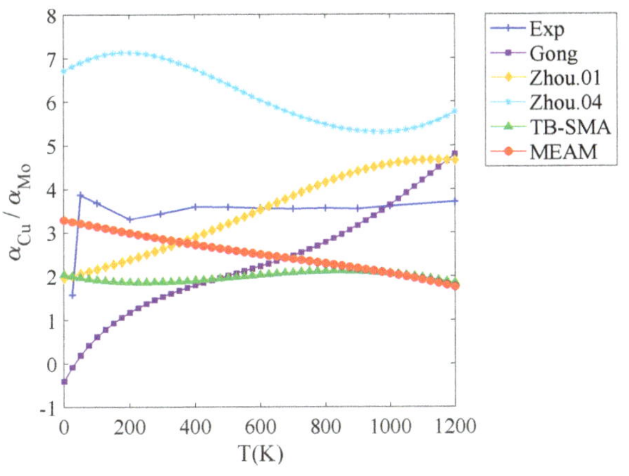

Figure 3. Evolution of the ratio of the thermal expansion coefficient of Cu and Mo as a function of temperature.

Table 2. Melting temperatures obtained with the hysteresis method for Cu and Mo.

T_m (K)	Cu		Mo	
	This Work	Experimental	This Work	Experimental
Gong	1085		3143	
Zhou.01	1131		3404	
Zhou.04	1153	1358	3420	2896
TB-SMA	1332		2236	
MEAM	1572		2778	

3.1.3. Elastic Constants and Elastic Moduli

Elastic constants were computed by imposing a deformation, ε_{ij}, on the simulation box and measuring the resulting stress σ_{ij}. To respect the assumption of linear elasticity, the size of the simulation box was deformed from −0.1% up to 0.1% at 0 K. The elastic constants were directly computed using $\sigma_{ij} = C_{ij}\varepsilon_{ij}$. The results are reported in Table 3. The inability of the TB-SMA potential to model the elastic properties of Mo (*bcc* lattice) is attributed to its lack of angular dependence [35]. The calculations were repeated at finite temperatures up to 1100 K (see Figure S5 in Supplementary Material (SM)). Using these elastic constants along with the Voigt-Reuss-Hill approximation, which is an averaging scheme, the anisotropic single-crystal elastic constants can be converted into isotropic polycrystalline elastic moduli [40,41]. The Young modulus E, Poisson ratio v, shear modulus G, and bulk modulus K are calculated using the elastic constants was computed at 0 K and at finite temperature up to 1100 K. The bulk modulus is expressed in terms of the elastic constants:

$$K = \frac{C_{11} + 2C_{12}}{3}. \tag{1}$$

The upper (Voigt approximation) and lower bounds (Reuss approximation) of shear moduli read

$$G_V = \frac{C_{11} - C_{12} + 3C_{44}}{5} \tag{2}$$

and

$$G_R = \frac{5C_{44}(C_{11} - C_{12})}{4C_{44} + 3(C_{11} - C_{12})}. \tag{3}$$

Based on the Hill empirical average [40,41], we introduced the isotropic elastic shear modulus as follows:

$$G = \frac{G_V + G_R}{2}. \tag{4}$$

It is then straightforward to find Poisson's ratio, v, and Young's modulus, E, from the well-known isotropic relations as follows:

$$v = \frac{3K - 2G}{2(3K + G)} \tag{5}$$

and

$$E = \frac{9GK}{3K + G}. \tag{6}$$

Table 3. The values of the elastic constants C_{ij}, Young modulus E, Poisson ratio v, shear modulus G, and bulk modulus K of Cu and Mo at 0 K.

Material	Method	C_{11} (GPa)	C_{12} (GPa)	C_{44} (GPa)	E (GPa)	G_V (GPa)	G_R (GPa)	G (GPa)	K (GPa)	v
Cu	Experimental	170 [42]	122.5 [42]	75.8 [42]	128.31	54.98	40.39	47.69	138.33	0.35
	Gong	170.03	122.50	75.80	128.34	54.99	40.41	47.70	138.34	0.35
	Zhou.01	182.84	121.15	70.64	135.79	54.72	46.59	50.66	141.71	0.34
	Zhou.04	169.62	122.13	75.74	128.21	54.94	40.38	47.66	137.96	0.35
	TB-SMA	176.03	121.71	83.63	141.99	61.04	45.66	53.35	139.82	0.33
	MEAM	176.13	124.90	81.76	137.67	59.30	43.56	51.43	141.98	0.34
Mo	Experimental	464.7 [43]	161.5 [43]	108.9 [43]	322.20	125.98	122.73	124.35	262.57	0.30
	Gong	500.18	196.94	108.92	327.55	126.00	122.75	124.37	298.02	0.32
	Zhou.01	456.59	166.55	113.11	323.64	125.87	124.03	124.95	263.23	0.30
	Zhou.04	456.56	166.55	113.11	323.63	125.87	124.02	124.94	263.22	0.30
	TB-SMA	182.84	173.35	123.76	121.07	76.15	11.22	43.69	176.51	0.39
	MEAM	474.79	160.51	115.10	336.13	131.92	128.89	130.40	265.27	0.29

The results obtained at finite temperature are presented in Figure 4 for Cu and in Figure 5 for Mo. The symbols are the calculated points, and the dotted lines are the corresponding fourth-order polynomial fits. Regarding the elastic constants at 0 K, the Gong and Zhou.04 potentials are the most accurate for copper, closely matching the experimental values for all three elastic constants (C_{11}, C_{12}, and C_{44}). Additionally, the Zhou.01 potential shows results that are very close to the experimental values. For molybdenum, the Zhou.01, Zhou.04, and MEAM potentials provide the best approximation to the experimental values. With respect to the variation of elastic properties with temperature and by comparing the simulation results with the experimental data [44,45], the Zhou.01 potential appears as the most reliable for predicting the elastic properties of both materials across the temperature range. Calculations carried out on pure metals indicate that the Zhou.01 potential provides the best compromise for most of the quantities calculated. This potential was therefore adopted to simulate the interface in the Cu-Mo bilayer system.

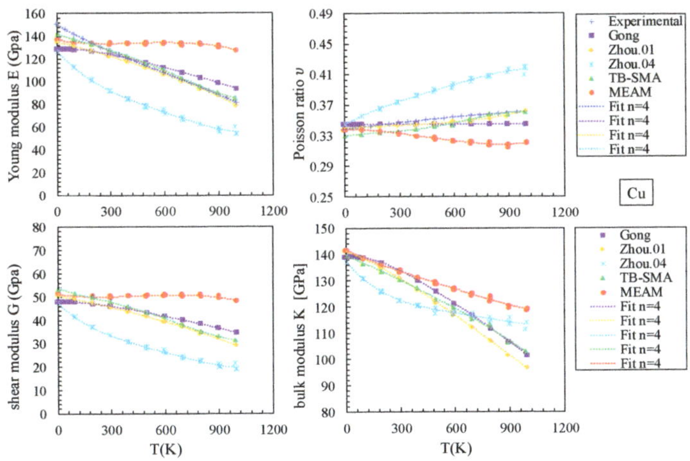

Figure 4. The variation of the young modulus E, Poisson ratio v, shear modulus G, and bulk modulus K of Cu as a function of temperature.

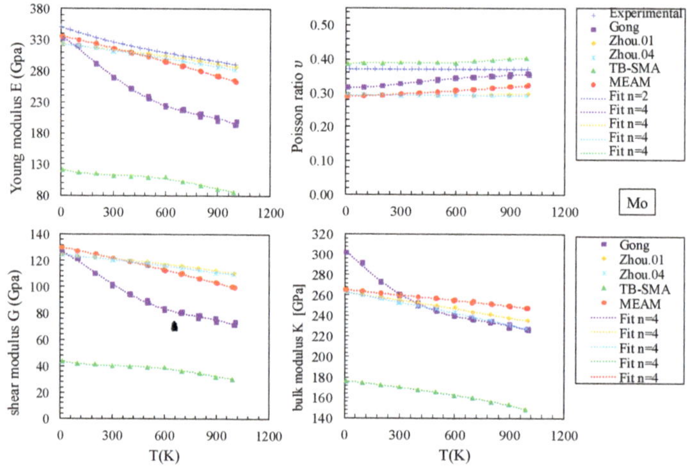

Figure 5. The variation of the young modulus E, Poisson ratio υ, shear modulus G, and bulk modulus K of Mo as a function of temperature.

3.2. *Modeling of the Cu/Mo Interface*

3.2.1. Energy, Atomic Volume and Local Atomic Environment

Bimetallic interfaces are categorized based on their crystal structure and the degree of lattice mismatch between the two metals [1]. This lattice misfit is commonly defined as $f = a(A)/a(B)$, where $a(A)$ and $a(B)$ are the lattice parameters of metals A and B, respectively. When the two metals in contact share the same crystal structure, coherent interfaces occur for $f < 10\%$, semi-coherent interfaces are observed when f ranges from 10% to 20% and incoherent interfaces for $f > 20\%$. For A and B having different crystalline structures, incoherent interfaces are observed regardless of the lattice misfit. In the case of Cu-Mo, the interface is incoherent with a lattice misfit of 15%. For the interface between *fcc*(111) and *bcc*(110) crystals, two types of orientation relationships have been identified during epitaxial growth experiments. One is the Nishiyama–Wassermann (NW) relationship [21,22] with $[1\bar{1}0]_{fcc} \parallel [100]_{bcc}$, and the other is the Kurdjumov–Sachs (KS) relationship [23] with $[1\bar{1}0]_{fcc} \parallel [1\bar{1}1]_{bcc}$. To determine the preferential epitaxial orientation in Cu-Mo NMMs using Zhou0.1, we created two simulation boxes: one corresponding to KS and the other to NW orientation relationship. As shown in Figure 6, the initial configuration includes a Mo layer and a Cu layer. The typical size of the simulation box is $L_x = 8.755$ nm, $L_y = 11.05$ nm, and $L_z = 9.30$ nm, fitted with 99% occupation in the x-direction for Cu and Mo in the KS box and 99% occupation in the x-direction for Cu in the NW box. The KS box contained 22 atomic planes of Cu and 21 of Mo along the z-direction, as shown in Figure 6 (KS). Along the lateral directions, the simulation box consisted of 34 atomic planes of Cu and 32 of Mo along the x-direction, and 50 atomic planes of Cu and 43 of Mo along the y-direction. The box was composed of 66,296 atoms (37,400 Cu and 28,896 Mo atoms). The NW box contained 22 atomic planes of Cu and 21 of Mo along the z-direction, as shown in Figure 6 (NW). Along the lateral directions, the simulation box consisted of 34 atomic planes of Cu and 28 of Mo along the x-direction, and 50 atomic planes of Cu and 50 of Mo along the y-direction. The box was composed of 66,800 atoms (37,400 Cu and 29,400 Mo atoms).

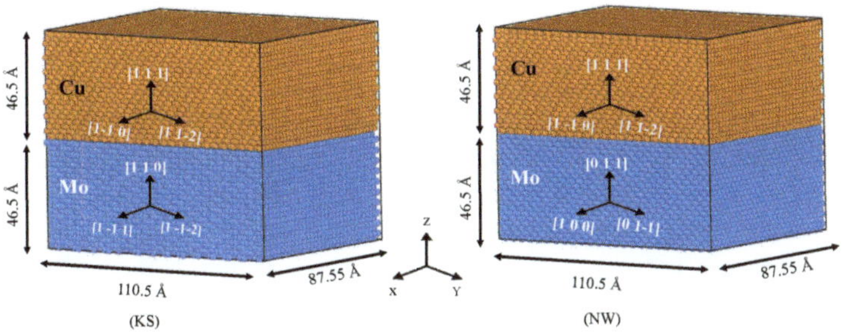

Figure 6. KS and NW boxes. Mo and Cu atoms are represented by blue and orange spheres, respectively.

We minimized the energy of the system by allowing the atom positions and box size to change using a conjugate gradient procedure. We then computed with the OVITO software the partial radial distribution function, $g_{ij}(r)$, where i and j represent Cu or Mo atoms and r denotes the atomic distance [46]. The radial distribution function (RDF) is expressed as:

$$g(r) = c_i^2 g_{ii}(r) + 2c_i c_j g_{ij}(r) + c_j^2 g_{jj}(r), \tag{7}$$

where c_i and c_j are the concentrations of i and j, respectively. Figure 7 shows the partial radial distribution functions $g(r)$ computed in 2D for the first atomic planes of Mo and Cu at the interface. Vertical lines marked with a cross ($*$) indicate the theoretical distances of the first neighbors in a bulk system, while those marked with a diamond (♦) correspond to the second neighbors. The computed peaks align well with the theoretical distances of the first and second neighbors in the bulk. For the first Mo plane neighboring the interface, the peaks are located precisely at the theoretical positions ($\frac{\sqrt{3}a_0}{2}$, a_0, and $\sqrt{2}a_0$), confirming that the local structure remains consistent with a bulk configuration. For the first Cu plane neighboring the interface, the peaks also align with the expected positions (a_0, $\sqrt{2}a_0$, and $\sqrt{\frac{3}{2}}a_0$), but they exhibit slight broadening. This broadening can be attributed to the misfit, i.e., the difference in lattice parameters between Mo and Cu, which introduces a slight distortion in the atomic structure of Cu at the interface. In contrast, the Mo atoms keep their bulk configuration.

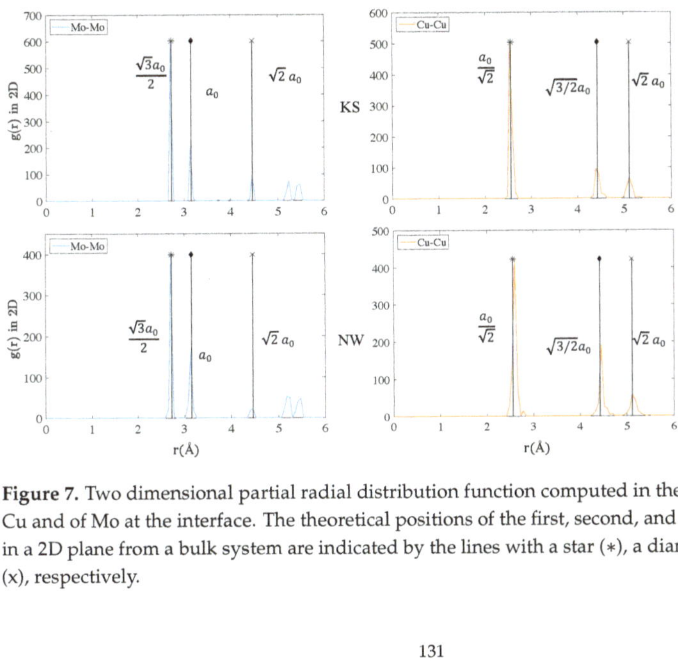

Figure 7. Two dimensional partial radial distribution function computed in the first atomic plane of Cu and of Mo at the interface. The theoretical positions of the first, second, and third neighbor peaks in a 2D plane from a bulk system are indicated by the lines with a star ($*$), a diamond (♦), and a cross (x), respectively.

Snapshots of the first Mo and Cu planes adjacent to the interface are shown in Figure 8 for KS orientation and in Figure 9 for NW. Atoms were colored according to different indicators such as potential energy per atom, volume per atom, and local environment (i.e., polyhedral template matching analysis). We recovered the typical Moiré fringes observed in [27] for the *fcc*(111) and *bcc*(110) interface. For KS and NW configurations, the potential energy per atom is higher at the interface compared to bulk atoms, reflecting the strain caused by lattice mismatch. However, the KS interface (Figure 8) exhibits larger variations in energy. For Mo, the potential energy ranges from −6.08 eV/atom to −6.37 eV/atom, while for Cu it varies from −3.23 eV/atom to −3.54 eV/atom. By comparison, the NW interface (Figure 9) shows a narrower energy range, with Mo spanning from −6.18 eV/atom to −6.40 eV/atom and Cu ranging from −3.30 eV/atom to −3.57 eV/atom. These smaller variations in the NW case suggest a better accommodation of the lattice mismatch, resulting in a more uniform energy distribution across the interface.

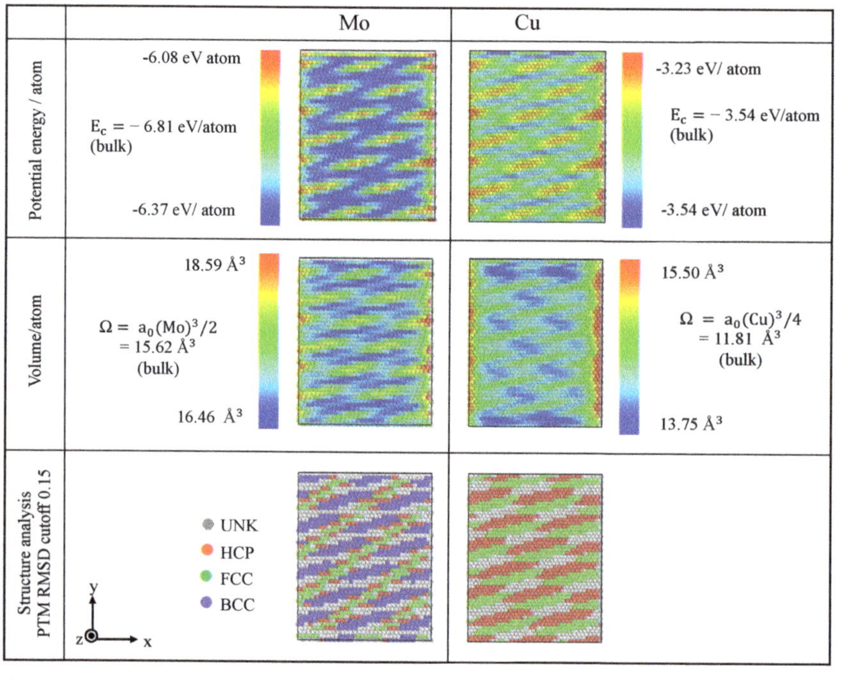

Figure 8. Energy, atomic volume and local atomic environment of KS box. For direct comparison, cohesive energy and volume per atom for bulk systems are also reported.

The atomic volume further highlights differences between the two configurations. In the KS case, Mo atoms exhibit a volume ranging from 16.46 Å^3 to 18.59 Å^3, deviating significantly from the bulk value of 15.62 Å^3, while Cu atoms show volumes between 13.75 Å^3 and 15.50 Å^3, compared to a bulk value of 11.81 Å^3. In contrast, the NW interface shows smaller deviations: Mo volumes range from 15.52 Å^3 to 17.76 Å^3, and Cu volumes range from 13.27 Å^3 to 14.60 Å^3. These reduced distortions in the NW configuration indicate a more stable atomic arrangement, with fewer local compressions and expansions caused by the interface. Additionally, structural analysis using PTM (with a cutoff of 0.15 RMSD) reinforces these observations. In the KS configuration, the alternating Moiré fringe patterns are more pronounced (more regions of disordered atomic arrangements (*unk*)), reflecting stronger lattice mismatch and interfacial strain. The NW configuration, on the other hand, exhibits a more gradual transition between the *fcc*(Cu) and *bcc*(Mo) phases, with a reduced strain distribution across the interface. This further supports the idea that the NW orientation relationship better accommodates the differences in lattice parameters between Mo and Cu.

Overall, the NW interface exhibits smaller variations in potential energy and atomic volume compared to the KS interface, reflecting reduced strain and more favorable interfacial stability. These results suggest that the NW configuration provides a more energetically and structurally stable interface for the Cu-Mo system.

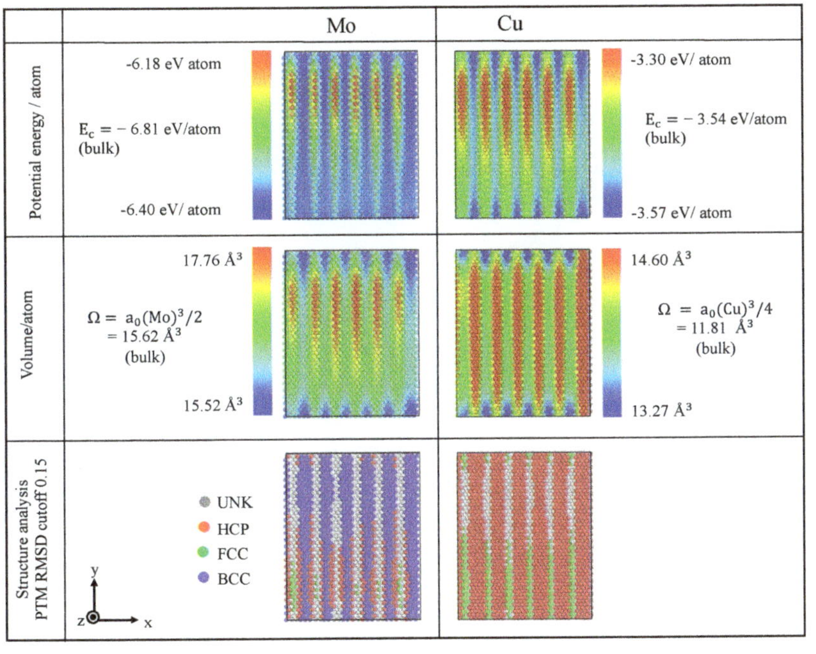

Figure 9. Energy, atomic volume and local atomic envirronement of NW box. For direct comparison, cohesive energy and volume per atom for bulk systems are also reported.

3.2.2. Energy of the Interface

The interfacial region can be clearly identified according to the average potential energy of atoms in each plane adjacent to the interface after relaxation. Figure 10 depicts the excess potential energy compared to bulk energy in each atomic plane parallel to the interface. The potential energies of the planes near the interface significantly differ from those in the bulk region (i.e., the region far away from the interface, in the middle of the layer). The interface energy γ_i is given by:

$$\gamma_i = \frac{1}{A}\left[\sum_{\ell}\left(E_{\ell}^{\mathrm{Cu}} - E_{0,\ell}^{\mathrm{Cu}}\right) + \left(E_{\ell}^{\mathrm{Mo}} - E_{0,\ell}^{\mathrm{Mo}}\right)\right] \tag{8}$$

where E is the actual potential energy of plane ℓ, $E_{0,\ell} = N_{\ell}^{\mathrm{Cu}}E_0^{\mathrm{Cu}} + N_{\ell}^{\mathrm{Mo}}E_0^{\mathrm{Mo}}$ is the reference bulk energy, with N_{ℓ}^{Cu} representing the number of Cu atoms in plane ℓ, N_{ℓ}^{Mo} representing the number of Mo atoms in plane ℓ, and A being the interface area. The interface energy evaluated with Equation (8) is $\gamma_i = 1.97\,\mathrm{J/m^2}$ for an interface with Kurdjumov-Sachs (KS) orientation $[1\bar{1}0]_{fcc} \parallel [1\bar{1}1]_{bcc}$, and $\gamma_i = 1.83\,\mathrm{J/m^2}$ for an interface with Nishiyama-–Wassermann (NW) orientation $[1\bar{1}0]_{fcc} \parallel [100]_{bcc}$. Although there is limited information in the literature regarding the interfacial energy of Cu-Mo systems, some studies have reported data that agree well with our observations. Experimentally, Yeom et al. [8] estimated (111)Cu/(110)Mo interface energy in a range between 0.5 and 2.0 J/m² at 600 °C, in good agreement with our values. Other studies were conducted using an ab initio approach by Li et al. [47] with three other interface orientations, namely (111)Cu/(111)Mo, (110)Cu/(110)Mo, and (100)Cu/(100)Mo. Despite the difference of orientations, these results fall within the same order of magnitude, with values of 3.51, 4.11, and 3.37 J/m², respectively. For Cu/W and Cu/Nb, Bodlos et al. [27] computed the interfacial (111)/(110) energy using DFT and EAM calculations, showing that NW is more stable by around 0.15 eV, again corroborating the present findings. From an energetic perspective, the NW orientation proves to be more stable than the KS orientation.

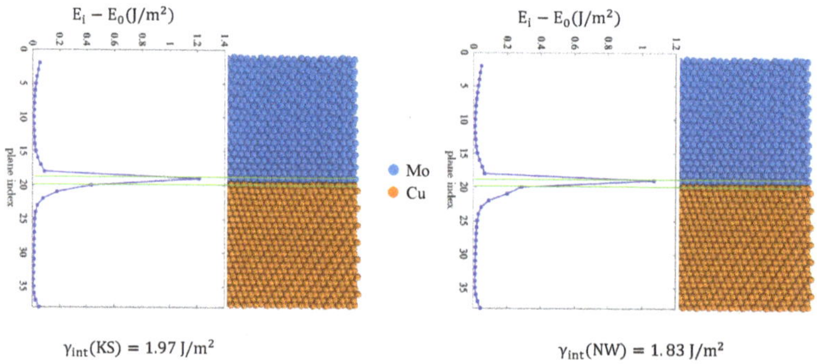

Figure 10. Excess of potential energy at the interface at 0 K.

3.2.3. MD Simulation of the Growth of Sputtered Cu Film on Mo Substrate

The calculations at 0 K presented above revealed that the NW orientation is energetically more stable. To validate at finite temperature this theoretical finding and investigate the naturally occurring orientation relationship, deposition simulations were carried out using the Zhou.01 potential. Figure 11 illustrates the MD model for the deposition of Cu atoms onto the Mo (110) and (011) surfaces. The simulation domain measured $21.5a \times 21.8a \times 27a$ (where $a = 3.15$ Å, the lattice constant of Mo). The substrate itself had dimensions of $21.5a \times 21.8a \times 11.1a$. The z-axis represents the film growth direction. For Mo (110), the crystal orientations were $x = [1\bar{1}1]$, $y = [1\bar{1}\bar{2}]$, and $z = [110]$, whereas for Mo (011), we used $x = [100]$, $y = [01\bar{1}]$, and $z = [011]$. Periodic boundary conditions were applied along the x- and y-axes, while the z-axis was subject to non-periodic and fixed boundary conditions. The incident Cu atoms originate from the region between the incident plane and the virtual plane (see Figure 11), with a velocity of −3.43 Å/ps (i.e., $E_c = 0.039$ eV) and a deposition rate of 1 atom per 100 ps. A virtual plane allows us to reflect Cu atoms that eventually reach the top of the box. Mo atoms located at the bottom of the box (i.e., $z < 2.3$ nm) were held fixed while the others (i.e., 2.3 nm $< z <$ 3.5 nm) were thermalized at 300K using a Langevin thermostat with a damping parameter equal to 1 ps.

After deposition, the atomic ordering of the first three deposited Cu layers was investigated. Figure 12 shows the structural analysis of Cu after deposition on Mo(110) at 300 K, focusing on the first three atomic layers at the interface. The $g(r)$ curve in 2D (on the left) shows the distribution of distances between the nearest neighbors for the first three Cu layers and the last Mo layer (the first Mo layer in contact with Cu). In the first Cu layer, the snapshots (see images a–d in Figure 12) show a mixture of *fcc* (green atoms) and *bcc* (blue atoms) structures, indicating that Cu locally adopts a *bcc* structure in certain regions under the strong influence of the Mo substrate, along with some regions exhibiting an *hcp* (stacking fault) structure (in red). The white color represents atoms whose structure could not be identified. In the second Cu layer, the peaks begin to align with the theoretical positions for *fcc*(111), but with a slightly larger full width at half maximum, reflecting residual strain as the structure transitions completely from *bcc* (first layer) to *fcc* (third layer). Finally, in the third Cu layer, the peaks correspond well to the theoretical positions for *fcc*(111), indicating the presence of the stable *fcc* structure. We observed initially a layer-by-layer (2D) growth mode for the first six copper planes. This growth mode ensures initial adherence consistent with the substrate structure. After that stage, the mechanism changed and as shown in Figure 12d, the formation of islands is observed with the presence of 3D growth. However, as growth progresses, small islands appear in the upper layers, signaling a transition from two-dimensional to three-dimensional (3D) growth. In summary, the results demonstrate that Cu deposited on Mo(110) is initially influenced by the *bcc* structure of the substrate in the first atomic layer, before transitioning to its stable *fcc*(111) structure by the third layer. This transition, combined with the shift from a 2D to a 3D growth mode, highlights the complex interactions between the substrate and the deposited material, influencing both local structure and overall growth mode. Note that similar simulations were conducted using the Gong potential. With that potential, the computed interface energy is also lower for the NW ($\gamma_i = 1.54\,\text{J/m}^2$) orientation compared to the KS ($\gamma_i = 1.61\,\text{J/m}^2$). Nevertheless, the deposition of Cu on Mo subtrates never resulted in a clear orientation relationship as obtained with Zhou.01.

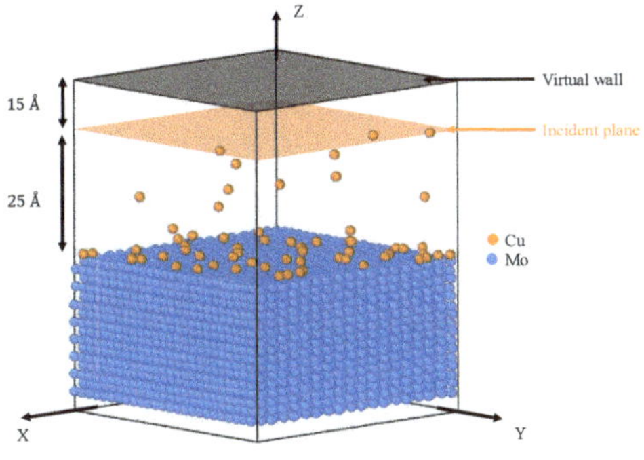

Figure 11. MD model of the deposition of sputtered Cu atoms onto Mo surface.

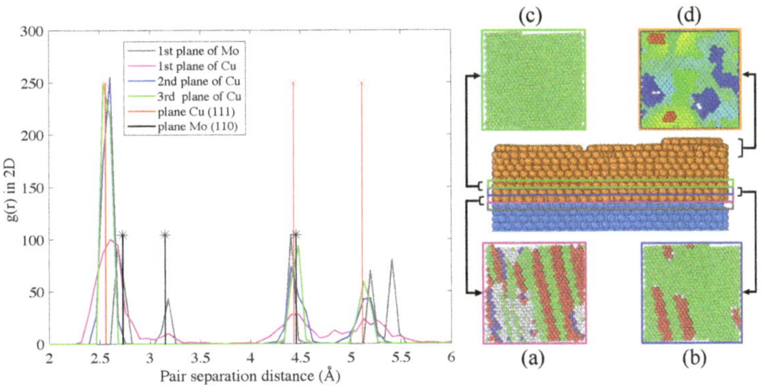

Figure 12. Two-dimensional radial distribution functions computed for the first three planes of Cu and the Mo plane located at the interface. The positions of the peaks computed in Cu(111) and Mo(110) planes in bulk systems are presented for comparison by lines with a cross (x) and a star (∗), respectively. Side views of the final configurations from simulations at 300 K. Thin slices of Cu atoms at different heights are shown on (**a**–**d**). In snapshots (**a**–**c**), the atoms are colored according to their local environment (i.e., *fcc* (green), *bcc* (blue), *hcp* (red), and *unk* (gray)), whereas in snapshot (**d**), the atoms are colored according to their relative height.

3.2.4. SAED Patterns

To identify the preferred orientation relationship adopted by the system, it was essential to perform detailed analyses following the deposition. Given that such analyses are often complex and unclear in real space, we opted for a more precise approach. Selected area electron diffraction (SAED) patterns, typically obtained through TEM experiments, provide a reliable method to study orientation relationships.

For an *fcc*(111)/*bcc*(110) interface, there are three variants of the Nishiyama–Wassermann (NW) orientation relationship and six variants of the Kurdjumov–Sachs (KS) orientation relationship. To establish a theoretical basis, SAED patterns were prepared for all these variants (see Figures S6 and S7 in Supplementary Materials (SM)). Figure 13 presents the necessary diffraction images used to construct the theoretical SAED patterns for the two systems introduced earlier, which correspond to one variant of KS (KS1 see Figure S6 in Supplementary Materials (SM)) and one of NW (NW1 see Figure S7 in Supplementary Materials (SM)).

The calculated diffraction pattern of *bcc*-Mo with zone axis along the [110] direction, shown in Figure 13a, corresponds to the low surface orientation of solid Mo [110] and with zone axis along the [011] direction, shown in Figure 13b, corresponds to the low surface orientation of solid Mo [011]. Cu was observed to grow in the [111] direction. The calculated diffraction pattern of *fcc*-Cu is depicted in Figure 13c. Each point of the reciprocal lattice corresponds to an observable peak (hkl). The unit cell of the reciprocal lattice is a rectangle with $l = d = 0.317$ Å^{-1} and $L = \sqrt{2}d = 0.449$ Å^{-1} for Mo and an hexagon with $L = 2d\sqrt{2} = 0.0.782$ Å^{-1} for Cu where $d = 1/a$ Å^{-1}, and a is the lattice parameter in real space. Some representative points are indexed in Figure 13. To construct the KS theoretical diffraction pattern, we rotated the diffraction pattern of *bcc*-Mo with the zone axis along [110] by an angle of 144.74°. Then, we rotated the diffraction pattern of Cu by 180° to align the two directions along x ($x = [1\bar{1}0]$ for Cu and $x = [1\bar{1}1]$ for Mo), and to construct the NW theoretical diffraction pattern, we rotated the diffraction pattern of Cu by 180° to align the two directions along x ($x = [1\bar{1}0]$ for Cu and $x = [100]$ for Mo). Not all points contribute in an equal manner to the diffraction pattern.

Next, the computational method developed in [48] was applied to produce a virtual electron diffraction pattern directly from atomistic simulations. The SAED pattern was computed for perfect KS and perfect NW boxes, as represented in Figure 6, to validate our theoretical patterns. The blue square in Figures 14 and 15, defined by the Mo peaks, and the orange hexagon, defined by the Cu peaks, are drawn as guides to identify the structure. The virtual SAED pattern is reproduced with striking similarity.

Figure 16a presents the calculated SAED patterns after Cu deposition on Mo, with orientations $x = [100]$, $y = [01\bar{1}]$, and $z = [011]$. In Figure 16b, the theoretical SAED of NW1 is superimposed on the SAED of deposition, revealing a perfect alignment between the patterns. Figure 17a illustrates the SAED patterns obtained after Cu deposition on Mo for $x = [1\bar{1}1]$, $y = [1\bar{1}\bar{2}]$, and $z = [110]$. In Figure 17b, the superposition of the theoretical SAED of KS1 with the SAED of the deposition highlights a shift of 5.26°, corresponding to the difference between the KS1 and NW2 orientation relationships (see Figure S8 in Supplementary Materials (SM)) The NW orientation relationship is always obtained, regardless of the orientations of Mo(110), confirming that the system favors the NW relationship in agreement with previous observations.

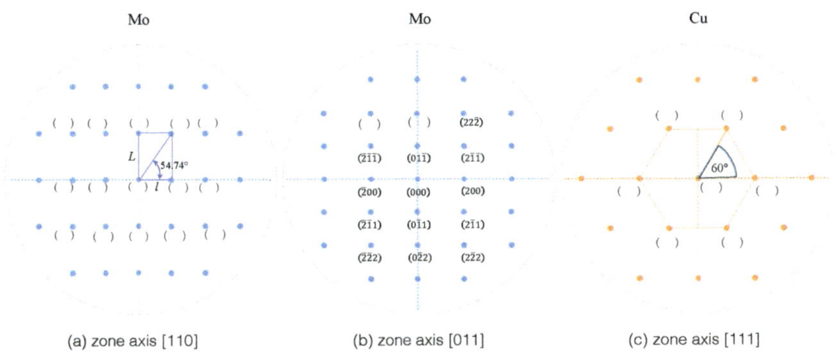

Figure 13. (**a**,**b**) Calculated diffraction pattern of *bcc*-Mo substrate with zone axis aligned along the [011] and [110] directions. (**c**) Calculated diffraction pattern of *fcc*-Cu with zone axis aligned along the [111]-direction.

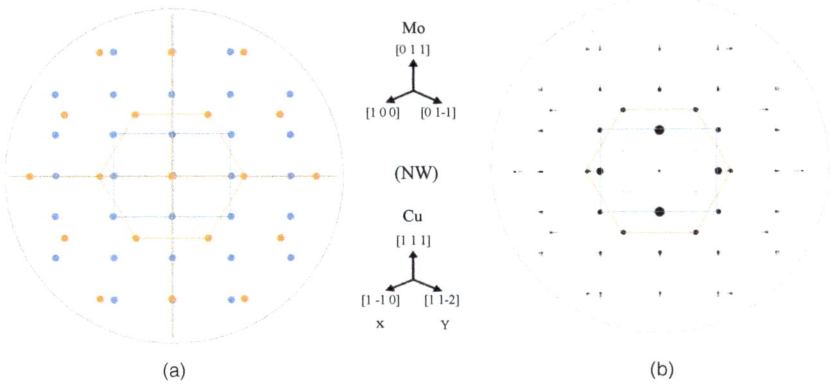

Figure 14. SAED for Nishiyama–Wassermann (NW) relationship. (**a**) Theoretical diffraction patterns. (**b**) Virtual diffraction patterns. Mo and Cu peaks are represented in blue and orange, respectively.

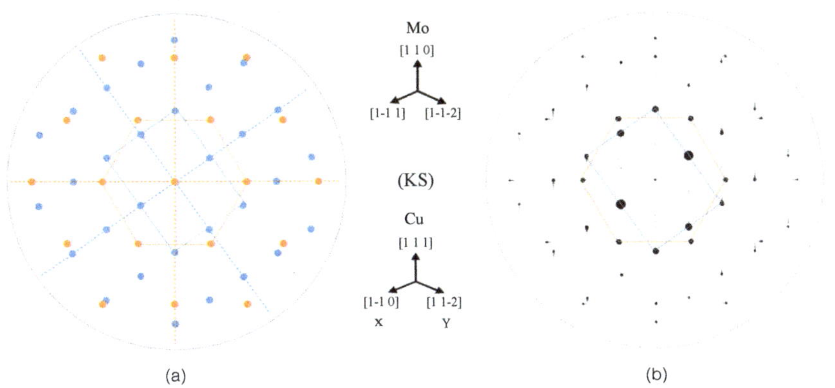

Figure 15. SAED for Kurdjumov–Sachs (KS) relationship. (**a**) Theoretical diffraction patterns. (**b**) Virtual diffraction patterns. Mo and Cu peaks are represented in blue and orange, respectively.

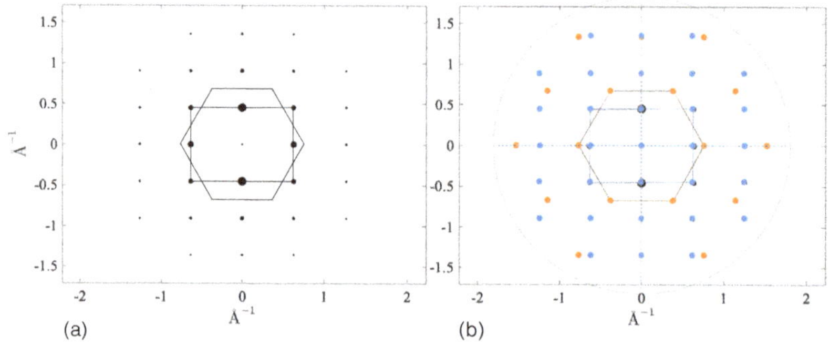

Figure 16. (**a**) SAED patterns computed for the full system at the end of deposition simulation. The black square defined by the Mo peaks, and the black hexagon, defined by the Cu peaks, are drawn as guides to identify the structure. (**b**) superposition of theoretical and virtual SAED patterns (deposition simulation). Theoretical Mo and Cu peaks are represented in blue and orange, respectively.

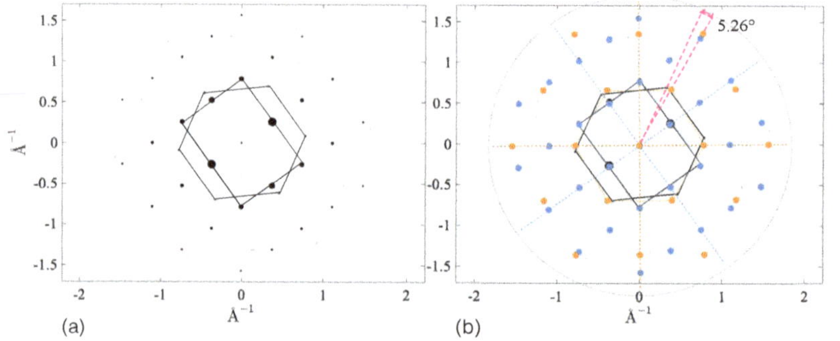

Figure 17. (**a**) SAED patterns computed for the full system at the end of deposition simulation. The black square defined by the Mo peaks, and the black hexagon, defined by the Cu peaks, are drawn as guides to identify the structure. (**b**) superposition of theoretical and virtual SAED patterns (deposition simulation). Theoretical Mo and Cu peaks are represented in blue and orange, respectively.

4. Conclusions

By means of molecular dynamics simulations, we studied the Cu/Mo interface in layered nanocomposite systems. Since there are only a few MD studies dedicated to this binary system, we first had to select an appropriate interatomic potential capable of simulating both metals as well as the interface orientation relationship. The comparison of the computed physical properties of the pure elements (i.e., lattice parameters, cohesive energy, thermal expansion, coefficient of thermal expansion, melting temperature, elastic constants, and moduli) with the experimental data led us to select the Zhou.01 potential.

Experimentally, nanometric Cu-Mo multilayers are obtained by magnetron-sputtering. Following this elaboration step, annealing at high temperature is often performed to reduce internal stresses. Degradation of these multilayers, revealing structural instability, is observed at 600 °C [8]. One of the causes of this instability is the interface between Cu and Mo. To investigate the possible source of instability, we conducted a study of the Cu/Mo interface by MD. Experimentally, Mo grows with a (110) orientation along the z-axis, and Cu grows with a (111) orientation along the z-axis, resulting in an *fcc*(111)/*bcc*(110) interface with a fiber texture, as Cui et al. [24] and Yeom et al. [8] have shown using transmission electron microscopy (TEM). Nevertheless, they were not able to characterize the preferential orientation relationships (either NW or KS). Both configurations were investigated by MD to determine the most stable one. The interface energy is lower for the NW orientation relationship compared to KS. This was confirmed by simulations of Cu deposition on Mo (110). The SAED diffraction patterns computed after deposition perfectly match the theoretical SAED calculated for NW. These results clearly show that the system favors the NW orientation relationship. Similar results were found experimentally by Moszner et al. [6] and numerically by Bodlos et al. [27] for the Cu-W system, which also has an incoherent interface and a lattice misfit very close to that of Cu-Mo (15%). This study helped to characterize the Cu/Mo interface that plays a key role in the nanocomposite's stability. It represents a first step towards understanding the lack of thermal stability of the Cu-Mo system at the atomic level, thereby limiting its potential applications for the development of high-performance nanocomposite materials.

Supplementary Materials: The following supporting information can be downloaded at: https://www.mdpi.com/article/10.3390/met15030282/s1, Figure S1: Performance of potentials with different numbers of cores; Figure S2: Evolution of the lattice parameter of Cu and Mo as a function of temperature; Figure S3: Evolution of the potential energy per atom (E_p/atom) as a function of temperature (T) during two processes: heating and cooling; Figure S4: Two types of deformations were applied to compute the elastic constants: uniaxial deformation for C_{11} and C_{12} (left) and shear deformation for C_{44} (right); Figure S5: Variation of the elastic constant of Cu (left) and Mo (right) as a function of temperature; Figure S6: SAED theoretical diffraction patterns of KS variants; Figure S7: SAED theoretical diffraction patterns of NW variants; Figure S8: Comparison between a KS variant and an NW variant, showing the small angle difference of 5.26°; Table S1: MD calculations of lattice parameters a for pure Cu and Mo using the five selected potentials, compared with selected literature references; Table S2: MD calculations of cohesive energies E_c for pure Cu and Mo using the five selected potentials, compared with selected literature references; Table S3: The coefficient A, B, C, D, and E for the fit $a(T) = A + BT + CT^2 + DT^3 + ET^4$ of the evolution of the lattice parameter of Cu as a function of temperature; Table S4: The coefficient A, B, C, D, and E for the fit $a(T) = A + BT + CT^2 + DT^3 + ET^4$ of the evolution of the lattice parameter of Mo as a function of temperature; Table S5: The elastic constants of Cu; Table S6: The elastic constants of Mo.

Author Contributions: Conceptualization, F.B. and O.P.; Methodology, F.B. and O.P.; Software, A.A. and O.P.; Validation, A.A., F.B. and O.P.; Formal analysis, A.A., F.B. and O.P.; Investigation, A.A.; Data curation, A.A. and O.P.; Writing – original draft, A.A.; Writing – review & editing, A.A., F.B. and O.P.; Visualization, A.A.; Supervision, F.B. and O.P.; Project administration, O.P.; Funding acquisition, O.P. All authors have read and agreed to the published version of the manuscript.

Funding: This research was funded by [Agence Nationale de la Recherche] grant number [ANR-23-CE09-0002-01].

Data Availability Statement: The original contributions presented in this study are included in the article/Supplementary Material. Further inquiries can be directed to the corresponding authors.

Acknowledgments: The use of computational facilities at the Computing Center of the Université Bourgogne Europe PSIUN-CCUB, is gratefully acknowledged. We also thank Jolanta Janczak-Rusch and Helena Zapolsky for fruitful discussions.

Conflicts of Interest: The authors declare no conflicts of interest.

References

1. Sáenz-Trevizo, A.; Hodge, A.M. Nanomaterials by design: A review of nanoscale metallic multilayers. *Nanotechnology* **2020**, *31*, 292002. [CrossRef] [PubMed]
2. Lewis, A.C.; Josell, D.; Weihs, T.P. Stability in thin film multilayers and microlaminates: The role of free energy, structure, and orientation at interfaces and grain boundaries. *Scr. Mater.* **2003**, *48*, 1079–1085. [CrossRef]
3. Knoedler, H.L.; Lucas, G.E.; Levi, C.G. Morphological stability of copper-silver multilayer thin films at elevated temperatures. *Metall. Mater. Trans.* **2003**, *34A*, 1043–1054. [CrossRef]

4. Ma, Y.J.; Wei, M.Z.; Sun, C.; Cao, Z.H.; Meng, X.K. Length scale effect on the thermal stability of nanoscale Cu/Ag multilayers. *Mater. Sci. Eng. A 2017*, *686*, 142–149. [CrossRef]
5. Druzhinin, A.V.; Cancellieri, C.; Jeurgens, L.P.H.; Straumal, B.B. The effect of interface stress on the grain boundary grooving in nanomaterials: Application to the thermal degradation of Cu/W nano-multilayers. *Scr. Mater.* **2021**, *199*, 113866. [CrossRef]
6. Moszner, F.; Cancellieri, C.; Chiodi, M.; Yoon, S.; Ariosa, D.; Janczak-Rusch, J.; Jeurgens, L.P.H. Thermal stability of Cu/W nano-multilayers. *Acta Mater.* **2016**, *107*, 345–353. [CrossRef]
7. Troncoso, J.F.; Lorenzin, G.; Cancellieri, C.; Turlo, V. Explaining the effect of in-plane strain on thermal degradation kinetics of Cu/W nano-multilayers. *Scr. Mater.* **2024**, *242*, 115902. [CrossRef]
8. Yeom, J.; Lorenzin, G.; Ghisalberti, L.; Cancellieri, C.; Janczak-Rusch, J. The thermal stability and degradation mechanism of Cu/Mo nanomultilayers. *Sci Technol Adv Mater.* **2024**, *25*, 2357536 . [CrossRef]
9. Appleget, C.D.; Riano, J.S.; Hodge, A.M. An Overview of Nano Multilayers as Model Systems for Developing Nanoscale Microstructures. *Materials* **2022**, *15*, 382. [CrossRef]
10. Niu, T.; Zhang, Y.; Cho, J.; Li, J.; Wang, H.; Zhang, X. Thermal stability of immiscible Cu-Ag/Fe triphase multilayers with triple junctions. *Acta Mater.* **2021**, *208*, 116679. [CrossRef]
11. Schweitz, K.O.; Ratzke, K.; Foord, D.; Thomas, P.J.; Greer, A.L.; Geisler, H.; Chevallier, J.; Bottiger, J. The microstructural development of Ag/Ni multilayers during annealing. *Phil. Mag. A* **2000**, *80*, 1867–1877. [CrossRef]
12. Schweitz, K.O.; Bottiger, J.; Greer, A.L.; Thomas, P.J.; Foord, D.T. The mechanism of degradation of Ag/Ni multilayers deposited at different temperatures. *Philos. Mag. A* **2001**, *81*, 1–10. [CrossRef]
13. Baras, F.; Turlo, V.; Politano, O.; Vadchenko, S.G.; Rogachev, A.S.; Mukasyan, A.S. SHS in Ni/Al Nanofoils: A Review of Experiments and Molecular Dynamics Simulations. *Adv. Eng. Mater.* **2018**, *20*, 1800091. [CrossRef]
14. Baloochi, M.; Shekhawat, D.; Riegler, S.S.; Matthes, S.; Glaser, M.; Schaaf, P.; Bergmann, J.P.; Gallino, I.; Pezoldt, J. Influence of Initial Temperature and Convective Heat Loss on the Self-Propagating Reaction in Al/Ni Multilayer Foils. *Materials* **2021**, *14*, 7815. [CrossRef]
15. Politano, O.; Baras, F. Molecular dynamics simulations of self-propagating reactions in Ni–Al multilayer nanofoils. *J. Alloys Compd.* **2015**, *652*, 25–29. [CrossRef]
16. Politano, O.; Baras, F. Reaction front propagation in nanocrystalline Ni/Al composites: A molecular dynamics study. *J. Appl. Phys.* **2020**, *128*, 215301. [CrossRef]
17. Politano, O.; Rogachev, A.S.; Baras, F. Molecular Dynamics Studies in Nanojoining: Self-Propagating Reaction in Ni/Al Nanocomposites. *J. Mater. Eng. Perform.* **2021**, *30*, 3160–3166. [CrossRef]
18. Schwarz, F.; Spolenak, R. Molecular dynamics study of the influence of microstructure on reaction front propagation in Al–Ni multilayers. *Appl. Phys. Lett.* **2021**, *119*, 133901. [CrossRef]
19. Baras, F.; Politano, O.; Li, Y.; Turlo, V. A Molecular Dynamics Study of Ag-Ni Nanometric Multilayers: Thermal Behavior and Stability. *Nanomaterials* **2023**, *13*, 2134. [CrossRef]
20. Lorenzin, G.; Troncoso, J.F.; Liyanage, M.; Druzhinin, A.V.; Jeurgens, L.P.H.; Cancellieri, C.; Turlo, V. Experimental and ab initio derivation of interface stress in nanomultilayered coatings: Application to immiscible Cu/W system with variable in-plane stress. *Appl. Surf. Sci.* **2024**, *661*, 159994. [CrossRef]
21. Wassermann, G. Einflu der—Umwandlung eines irreversiblen Nickelstahls auf Kristallorientierung und Zugfestigkeit. *Arch. Eisenhüttenwesen* **1933**, *6*, 347–351. [CrossRef]
22. Nishiyama, Z. X-ray investigation of the mechanism of the transformation from face centered cubic lattice to body centered cubic. *Sci. Rep. Tohoku Univ.* **1934**, *23*, 637.
23. Kurdjumow, G.; Sachs, G. Über den Mechanismus der Stahlhärtung. *Z. Phys.* **1930**, *64*, 325–343. [CrossRef]
24. Cui, Y.; Derby, B.; Li, N.; Mara, N.A.; Misra, A. Suppression of shear banding in high-strength Cu/Mo nanocomposites with hierarchical bicontinuous intertwined structures. *Mater. Res. Lett.* **2018**, *6*, 184. [CrossRef]
25. Yu-Zhang, K.; Embury, J.D.; Han, K.; Misra, A. Transmission electron microscopy investigation of the atomic structure of interfaces in nanoscale Cu-Nb multilayers. *Philos. Mag.* **2008**, *88*, 2559–2567. [CrossRef]
26. Monclús, M.A.; Karlik, M.; Callisti, M.; Frutos, E.; Lorca, J.L.; Polcar, T.; Molina-Aldareguí, J.M.A. Microstructure and mechanical properties of physical vapor deposited Cu/W nanoscale multilayers: Influence of layer thickness and temperature. *Thin Solid Films* **2014**, *571*, 275–282. [CrossRef]
27. Bodlos, R.; Fotopoulos, V.; Spitaler, J.; Shluger, A.L.; Romaner, L. Energies and structures of Cu/Nb and Cu/W interfaces from density functional theory and semi-empirical calculations. *Materialia* **2022**, *21*, 101362. [CrossRef]
28. Plimpton, S. Fast Parallel Algorithms for Short-Range Molecular Dynamics. *J. Comp. Phys.* **1995**, *117*, 1. [CrossRef]
29. Daw, M.S.; Baskes, M.I. Embedded-atom method: Derivation and application to impurities, surfaces, and other defects in metals. *Phys. Rev. B* **1984**, *29*, 6443–6453. [CrossRef]
30. Gong, H.R.; Kong, L.T.; Liu, B.X. Metastability of an immiscible Cu-Mo system calculated from first-principles and a derived n-body potential. *Phys. Rev. B* **2004**, *69*, 024202. [CrossRef]

31. Zhou, X.; Johnson, R.A.; Wadley, H.N.G. Misfit-energy-increasing dislocations in vapor-deposited CoFe/NiFe multilayers. *Phys. Rev. B* **2004**, *69*, 144113. [CrossRef]
32. Zhou, X.; Johnson, R.A.; Wadley, H.N.G. Atomic scale structure of sputtered metal multilayers. *Acta Mater.* **2004**, *49*, 4005–4015. [CrossRef]
33. Zhou, X.W.; Wadley, H.N.G. Misfit dislocations in gold/Permalloy multilayers. *Philos. Mag.* **2004**, *84*, 193–212. [CrossRef]
34. Wadley, H.; Zhou, X.; Butler, W.H. *Atomic Assembly of Magnetoresistive Multilayers*; Springer: Berlin/Heidelberg, Germany, 2008; Volume 109, pp. 497–559. [CrossRef]
35. Karolewski, M.A. Tight-binding potentials for sputtering simulations with fcc and bcc metals, Radiation Effects and Defects in Solids. *Radiat. Eff. Defects Solids* **2001**, *153*, 239–255. [CrossRef]
36. Wang, J.; Lee, B.-J. Second-nearest-neighbor modified embedded-atom method interatomic potential for V-M (M=Cu, Mo, Ti) binary systems. *Comput. Mater. Sci.* **2021**, *188*, 110177. [CrossRef]
37. Oh, S.-H.; Lu, X.-G.; Chen, Q.; Lee, B.-J. Pressure dependence of thermodynamic interaction parameters for binary solid solution phases: An atomistic simulation study. *Calphad* **2021**, *75*, 102342. [CrossRef]
38. Touloukian, Y.S.; Kirby, R.K.; Taylor, R.E.; Desai, P.D. Thermophysical properties of matter-the tprc data series. In *Thermal Expansion Metallic and Elements and Alloys*; Plenum Press: New York, NY, USA, 1975; Volume 12.
39. Luo, F.; Chen, X.; Cai, L.; Ji, G. Solid–Liquid Interfacial Energy and Melting Properties of Nickel Under Pressure from Molecular Dynamics. *J. Chem. Eng. Data Data* **2010**, *55*, 5149–5155. [CrossRef]
40. Rassoulinejad-Mousavi, S.M.; Mao, Y.; Zhang, Y. Evaluation of copper, aluminum, and nickel interatomic potentials on predicting the elastic properties. *J. Appl. Phys.* **2016**, *119*, 244304. [CrossRef]
41. Hill, R. The Elastic Behaviour of a Crystalline Aggregate, Proceedings of the Physical Society. *Sect. A* **1952**, *65*, 349. [CrossRef]
42. Simmons, G.; Wang, H.F. Single Crystal Elastic Constants and Calculated Aggregate Properties. In *A Handbook*; 2nd ed.; M.I.T. Press: Cambridge, MA, USA, 1971 .
43. Bujard, P.; Sanjinés, R.; Walker, E.; Ashkenazi, J.; Peter, M. Elastic constants in Nb-Mo alloys from zero temperature to the melting point: Experiment and theory. *J. Phys. Met. Phys.* **1981**, *11*, 775–786. [CrossRef]
44. ASM Handbook Committee. *Properties and Selection: Nonferrous Alloys and Special-Purpose Materials*; ASM International: New York, NY, USA, 1952. [CrossRef]
45. Simon, N.J.; Drexler, E.S.; Reed, R.P. *Properties of Copper and Copper Alloys at Cryogenic Temperatures*; NIST Monograph: Boulder, CO, USA, 1992. [CrossRef]
46. Stukowski, A. Visualization and analysis of atomistic simulation data with OVITO-the open visualization tool. Model. *Simul. Mater. Sci. Eng.* **2009**, *18*, 015012. [CrossRef]
47. Li, Z.; Ding, P.; Wang, R.; Wang, C.; Li, Q.; Sun, B.; Zhang, Y. First-principles investigation on the effects of alloying elements on Cu/Mo interface. *Chem. Phys. Lett.* **2023**, *832*, 140895. [CrossRef]
48. Coleman, S.P.; Spearot, D.E.; Capolungo, L. Virtual diffraction analysis of Ni [010] symmetric tilt grain boundaries. *Model. Simul. Mater. Sci. Eng.* **2013**, *21*, 055020. [CrossRef]

metals

MDPI

Article

Characterizing the Behavior and Microstructure of Cu-La_2O_3 Composite Processed via Equal Channel Angular Pressing

Lenka Kunčická * and Radim Kocich

Department of Metallurgical Technologies, Faculty of Materials Science and Technology, VŠB Technical University of Ostrava, 17. listopadu 2172-15, 708 00 Ostrava, Czech Republic; radim.kocich@vsb.cz
* Correspondence: lenka.kuncicka@vsb.cz; Tel.: +420-596-991-265

Abstract: Cu-based alloys and composites are popular to prepare electroconductive parts. However, their processing can be challenging, especially in case of composites strengthened with oxides. To save the necessary time and costs, numerical simulations can be of help when determining the deformation behaviour of (newly introduced) materials. The study presents a combined method of strengthening of Cu by adding 5 wt.% of La_2O_3 particles and performing shear-based deformation by equal channel angular pressing (ECAP). The effects of the method on the microstructure, mechanical properties, and thermal stability of the composite are examined both numerically and experimentally. The results showed that the La_2O_3 addition caused the maximum imposed strain to be higher for the composite than for commercially pure Cu, which led to the development of subgrains and shear bands within the microstructure, and a consequent increase in microhardness. The numerical predictions revealed that the observed differences could be explained by the differences in the material plastic flow (comparing the composite to commercially pure Cu). The work hardening supported by the addition of La_2O_3 led to a significant increase in stress and punch load during processing, as well as contributed to a slight increase in deformation temperature in the main deformation zone of the ECAP die. Certain inhomogeneity of the parameters of interest across the processed workpiece was observed. Nevertheless, such inhomogeneity is typical for the ECAP process and steps prospectively leading to its elimination are proposed.

Keywords: copper; composite; equal channel angular pressing; finite element method; microstructure

1. Introduction

Cu of a high purity (commercially pure, CP) is a highly versatile conductive material, but features relatively low strength [1]. Advantageous ways of increasing the mechanical properties of Cu include application of various heat, deformation, and thermomechanical treatments [2–5], as well as additions of strengthening/reinforcing elements to create alloys [6,7] and composites [8,9]. As for the first mentioned way, optimized treatments enhancing the mechanical properties of Cu can advantageously be carried out by methods of intensive and severe plastic deformation (IPD, e.g., [10–12], and SPD, e.g., [13–16]). Such methods impart strengthening by introducing nucleation of dislocations and formation of substructure, resulting in grain refinement and consequently to improvement of the mechanical properties [17,18]. As for the second mentioned way, besides the alloying element typically added to fabricate the commonly known Cu-based alloys (e.g., bronzes and brasses [19,20]), additions of specific elements can improve the properties of Cu. For

https://doi.org/10.3390/met15040368

example, small amounts of Ag, Cd, or Fe were reported to improve the durability of Cu at elevated temperatures [21,22]. Cu-based composites are very advantageous for achieving specific combinations of properties. Both laminated composites, consisting of (multiple) layers of Cu + other elements (e.g., Mg [23–25], Al [26–28], Nb [29–31], Ti [32–34], combinations of elements [35,36], various alloys [37], graphene [38], etc.), and powder-based Cu composites, usually containing strengthening particles (carbon [39–42], various oxides, such as Al_2O_3 [43–45], or Y_2O_3 [46–49]), are widely researched.

The presented study introduces a Cu-based composite consisting of Cu + 5 wt.% of La_2O_3 particles. Cu is mainly used for electroconductive components, operating typically in contact with other (such) parts. Therefore, such parts can be subject to not only electrical erosion, but also wear by abrasion and adhesion [50]. For this reason, it is of a high importance to increase the mechanical (as well as tribological) properties of Cu-based electroconductive materials. This can advantageously be done by adding stable and resistant reinforcing particles, such as the La_2O_3. The La_2O_3 oxide was, similar to the Y_2O_3 one, documented to enhance the mechanical properties, especially hardness and elastic modulus, without having significantly negative effects on the electric conductivity [51]. Moreover, adding La_2O_3 to Cu was shown to be highly supportive of grain refinement, which is advantageous from the viewpoint of introducing another strengthening, by the Hall–Petch effect [52]. Nevertheless, contrary to Y_2O_3, the effects of La_2O_3 additions on the overall behavior of Cu have not been sufficiently studied yet (either experimentally or via numerical simulations).

The processing method selected herein is equal channel angular pressing (ECAP), a widely used and researched SPD method suitable for processing of relatively large billets (see, e.g., [53,54]). The positive effects of ECAP on the microstructures and properties of CP Cu and selected Cu-based alloys were documented. For example, Heidari et al. [55] increased the lifetime of a tool electrode Cu by ECAP, and Volokitina et al. [56] subjected CP Cu to ECAP under cryogenic conditions and achieved double the effectivity of room temperature ECAP as regards the increase in the mechanical properties. Guo et al. [57] subjected a Cu1Cr0.2Si alloy to four pass ECAP (route Bc [58]) and achieved an increase in its tensile strength by 47% (up to 564 MPa), whereas Chu et al. [59] used combinations of ECAP, rolling, and heat treatments to impart nano-sized structural features within a CuCrZr alloy to increase its tensile strength up to 730 MPa, while maintaining the electric conductivity of 75.5% IACS (International Annealed Copper Standard). Several types of Cu-based composites were also subjected to ECAP and other ECAP-based methods; for example, Cu-Al laminates were prepared by ECAP [60], ECAP + drawing [61], or TCAP (twist channel angular pressing) [62].

Numerical simulations using the finite element method (FEM) advantageously provide deeper insights into the production processes, as they enable to predict processing parameters, such as distribution of temperature, strain, stress, forming forces and their developments, etc. [63,64]. They can be advantageously used to study the microstructures and behaviors of metallic materials during processing by SPD methods, such as the ECAP process applied herein. For example, Borodin et al. [65] used FEM to predict the development of microstructural features within CuCrZr alloys during ECAP and multidirectional forging, Bratov et al. [66] used modelling to predict dislocations and grain size development within Cu and Al subjected to ECAP, Aal [67] numerically studied the deformation behavior of pure Al when deformed by ECAP, and a combination of ECAP and direct extrusion, Hongyu et al. [68] predicted the quality of shape control of a square Al tube during ECAP, Vafaeenezhad et al. [69] assembled a numerical model to analyze the effect of severe plastic deformation on wear of Al-7075 alloy deformed by ECAP, and Ghosh et al. [70] used computer simulations to investigate the deformation behavior of EN

AW 7075 alloy during ECAP processing through dies with different outer corner angles. Nevertheless, to successfully assemble a numerical simulation, numerous input data and boundary conditions have to be set. Therefore, when proposing a new material, or examining a material the properties of which are not widely known, its behavior needs to be thoroughly studied at first to gain input data for the numerical modelling.

The study presented herein features a combined method of strengthening of CP Cu by adding La_2O_3 oxide particles and performing SPD by ECAP. The results were acquired both experimentally and via numerical simulations; the simulations were also performed for CP Cu to provide in-depth insight into the effects of La_2O_3 addition on the behavior of the composite. As characterized in the following section, the study starts with the investigation of the mechanical properties of the as-cast Cu-La_2O_3 composite. The acquired data were then advantageously used to assemble the numerical simulation to predict the behavior of the composite during the ECAP processing. Along with examining the mechanical properties, the experimentally prepared composite billets were subjected to detailed investigations of microstructures and investigation of their thermal stability (as electroconductive parts can experience temperature fluctuations). Subsequently, the numerical simulations were focused on assessing the material plastic flow and stress–strain conditions during processing, as well as on the processing parameters, such as deformation temperature and punch load. Finally, the acquired results were put in correlation to characterize the enhancement resulting from the performed processing steps.

2. Materials and Methods

2.1. Experimental

The used materials were electro-conductive CP Cu (impurities of, in wt.%, 0.002 O, 0.002 Zn, and 0.015 P), and La_2O_3 particles. The initial Cu was a commonly available as-cast electro-conductive Cu rod (UCB Technometal, s.r.o., Loděnice u Berouna, Czech Republic), while the La_2O_3 powder of 99.99% purity was obtained from Thermo Scientific Chemicals Inc. (Waltham, MA, USA). The initial composite material was cast by vacuum induction melting (VIM) to a 200 mm long billet with the diameter of 20 mm. Subsequently, the billet was machined to acquire 100 mm long workpieces with the cross-section of 10×10 mm^2 for room temperature ECAP processing, which was performed via a single pass through a die with the bending angle of 110°. In order to confirm robustness of the experiment and repeatability of the results, all the experimental works, from casting to investigations, were doubled with another billet. As no significant discrepancies in the mechanical properties, microstructures, and overall behaviors of the billets were observed, the results from one billet are presented to keep the manuscript within a reasonable length.

Mechanical properties of the as-cast composite, as well as ECAP-processed composites, were at first examined using uniaxial compression testing (ASTM D7012-23 standard [71]) at room temperature (20 °C) and strain rate of 0.01 s^{-1}. This particular strain rate was chosen based on the fact that the experimental ECAP processing was performed using a hydraulic press featuring very low strain rates, so the acquired data could subsequently be used to reliably perform the numerical simulations. However, higher strain rate values of 0.1 s^{-1} and 10 s^{-1} were also applied in order to experimentally characterize the deformation behavior of the prepared composite at a wider range of thermomechanical conditions, as well as to assess its thermal stability. The testing was carried out using a Gleeble 3800-GTC thermal-mechanical simulator equipped with a Hydrawedge II testing unit (both Dynamic System Inc., Poestenkill, NY, USA), see, e.g., [72–74]. The specimens were of a cylindrical shape with the dimensions of 10 mm in diameter and 15 mm in length. Each specimen was deformed by uniaxial compression to the final true strain of 1.0. To reduce friction on the contact surfaces between the anvils and the specimen and to protect the anvils, nickel-based lubricant with tantalum washers were used. The Gleeble 3800 equipment and testing samples of the above-described dimensions were also used for the experiments investigating the thermal stability of the ECAP-processed Cu-La_2O_3 composite. For this purpose, the heat treatment regime of 250° C/30 min was selected and applied on the ECAP-processed composite samples.

To experimentally acquire data on the mechanical properties for subsequent detailed comparison of the deformation behavior of the composite with the results of numerical simulations, Vickers microhardness HV0.05 was measured along two intersecting diagonals across the cross-sections of both the as-cast and ECAP-processed billets (see Figure 1). The load for the measurements was 50 g, and indent load time was 10 s (ASTM E384 standard [75], FM ARS 900 equipment by Future Tech, Spectrographic Limited, Leeds, UK).

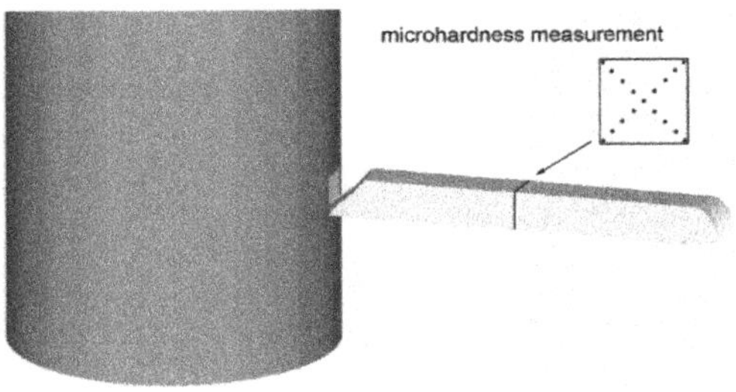

Figure 1. Depiction of locations for microhardness measurements.

Finally, samples for scanning electron microscopy (SEM) (electron backscatter diffraction, EBSD), were prepared from both, the as-cast and ECAP-processed billets (SEM by Tescan Orsay Holding a.s., Brno, Czech Republic). The billets were cut longitudinally along the axis to prepare samples, which were mechanically ground, and finally polished via an OPS (Oxide Polishing Suspension, by Struers s.r.o, Ostrava, Czech Republic). EBSD scans were acquired both from relatively larger areas to provide overviews of the microstructures, as well as from detailed views on the microstructures with 0.05 µm scan step to closely investigate the development of substructure. The results were evaluated with the limit values of 5° for low angle grain boundaries (LAGB), and 15° for high angle grain boundaries (HAGB).

2.2. Numerical Simulation

The numerical simulation of the ECAP process was assembled using the Forge software (version NxT®, Transvalor S.A., Biot, France), see, e.g., [76,77]. The comparative simulation of CP Cu was carried out using data files for 99.97% pure Cu already included in the material database of the software. The material behavior of the prepared composite was defined using the data acquired via the above-mentioned compression testing procedure. The data needed to be implemented into the software to assess the behavior of the composite. The material file of the prepared composite was filled in with the following parameters: Young's modulus E of 110 MPa, Poisson's ratio ν of 0.3, specific heat capacity c of 435 $J\cdot kg^{-1}\cdot K^{-1}$, and density ρ, of 8100 $kg\cdot m^{-3}$. The modification also consisted in adjusting the material constants of the Hensel–Spittel rheology law [78], see Equation (1),

$$\sigma = A \cdot e^{m_1 \cdot T} \cdot \varepsilon^{m_2} \cdot \dot{\varepsilon}^{m_3} \cdot e^{m_4/\varepsilon} \cdot (1+\varepsilon)^{m_5 \cdot T} \cdot e^{m_7 \cdot \varepsilon} \cdot \dot{\varepsilon}^{m_8 \cdot T} \cdot T^{m_9} \quad (1)$$

where T (°C), ε (-), $\dot{\varepsilon}$ (s^{-1}), and σ (MPa) represent the deformation temperature, true strain, strain rate, and true flow stress, respectively. The constants originally determined for the 99.97% pure Cu were recalculated for the examined Cu-La_2O_3 composite using regression analysis of the experimentally acquired data. The resulting constants, together with the original constants, are summarized in Table 1. When compared to the Hensel–Spittel model for the pure Cu, the model assembled for the Cu-La_2O_3 composite differed significantly in several key constants (see Table 1). These differences point to and explain the differences in the results predicted for the Cu and Cu-La_2O_3 composite, as further discussed in Section 3. Fundamental parameters affecting the material behavior are the coefficients influencing the strain hardening/softening processes, i.e., m_2, m_4, m_5, and m_7 (constants related to true strain). In the material database of the used Forge NxT® software, the constants m_5 and m_7 are equal to zero for pure Cu. However, for the Cu-La_2O_3 composite, these constants were considered to be non-zero in order to achieve the most reliable fit of the experimental data during the regression analysis.

Table 1. Material constants for 99.97% pure Cu and Cu-La_2O_3 composite.

Parameter	Cu 99.97	Cu-La_2O_3
A	411.19	318.87
m_1	−0.00121	−0.01241
m_2	0.21554	−0.10732
m_3	0.01472	0.00346
m_4	−0.00935	−0.01617
m_5	0	0.11678
m_7	0	−0.90557
m_8	0	0.00054
m_9	0	−0.11201

The accuracy of the assembled model was verified by the root mean squared error *RMSE* (5.042 MPa), and Pearson's correlation coefficient *R* (0.998), calculated as follows (Equations (2) and (3)):

$$RMSE = \sqrt{\frac{1}{n} \cdot \sum_{i=1}^{n} (E_i - M_i)^2} \tag{2}$$

$$R = \frac{\sum_{i=1}^{n} (E_i - \overline{E}) \cdot (M_i - \overline{M})}{\sqrt{\sum_{i=1}^{n} (E_i - \overline{E})^2 \cdot \sum_{i=1}^{n} (M_i - \overline{M})^2}} \tag{3}$$

In Equations (2) and (3), E_i (MPa) and M_i (MPa) represent the i-th experimental and modelled true flow stress, respectively, $\overline{E}$ (MPa) and $\overline{M}$ (MPa) represent the corresponding arithmetic means, and $i = [1, n] \subset \mathbb{N}$, where n is the number of datapoints.

The above-mentioned favorable values of *RMSE* and *R* parameters are further supported by the histogram, showing the distribution of residuals of the Hensel–Spittel rheology law (Equation (1)), depicted in Figure 2. As can be observed, the model deviations from the experimental data were mostly lower than 5 MPa (a negligible number of residues exceeded the limit of 10 MPa, with the maximum of 30 MPa). However, further analysis of the residuals showed that such high deviations were exclusively associated with very low true strain values (below ~0.025). This corresponds to the very beginning of the deformation process and, therefore, does not fundamentally affect the accuracy of the calculations.

The assembly of the room temperature ECAP process for the numerical simulation, i.e., channel (die), workpiece (billet), and plunger, is depicted in Figure 3. The initial dimensions of the workpiece were $10 \times 10 \times 100$ mm^3, the dimensions of the plunger were $10.05 \times 10.05 \times 110$ mm^3, and the cross-section of the channel was 10.1×10.1 mm^2, with the bending angle of 110°, all corresponding to the real experimental setup. The entire die was a cylinder with the diameter of 300 mm and height of 300 mm. The workpiece was considered as a deformable body, while both the tools (die, plunger) were considered as rigid bodies. Correspondingly, the die and plunger were surface-meshed, while the workpiece, defined with tetrahedron elements, was volume-meshed. The mesh of the workpiece consisted of a total of 13,128 nodes connecting 61,271 elements, while the meshes of the plunger and channel consisted of 2120 and 4652 nodes connecting 4236 and 9304 elements, respectively. To increase the efficiency and reduce the time required to calculate the simulations, anisotropic mesh was chosen, i.e., finer meshes were selected for the workpiece, and die locations of primary interest (Figure 3). As for the die, the channel part, being in a direct contact with the workpiece, was finely meshed, similar to the lower part of the plunger. Based on our previous experience with numerical simulations of SPD processes, the re-meshing procedure was activated.

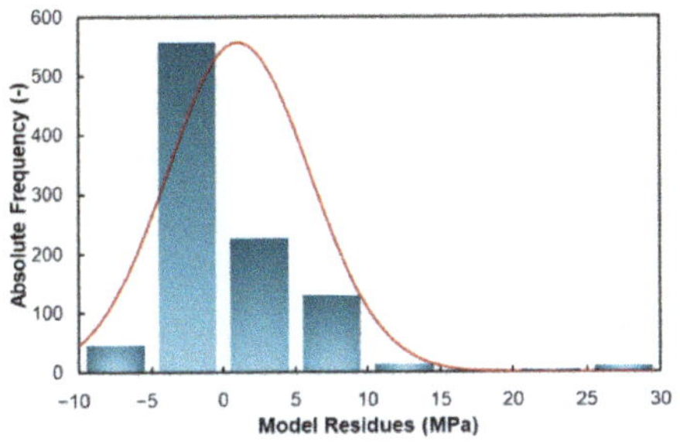

Figure 2. Distribution of Hensel–Spittel model residues.

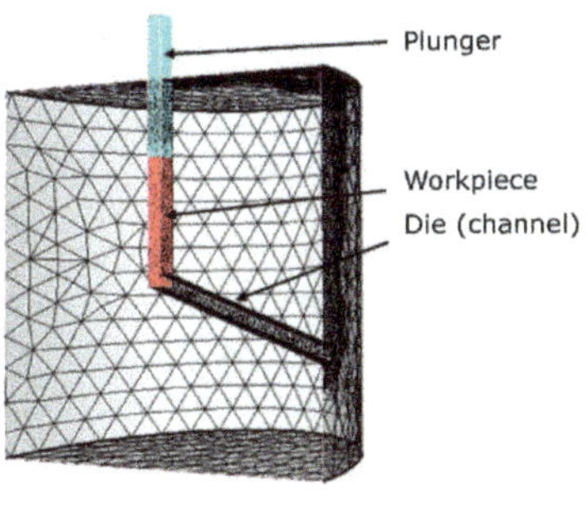

Figure 3. Graphical depiction of a cut through the assembly for ECAP processing.

The initial temperature of the workpiece, as well as of the die, plunger, and the ambient environment, was 20 °C. The temperature conditions were based on the conditions of the real ECAP process, which was performed at room temperature, without any preheating. The thermal exchange between the equipment, i.e., steel die and plunger, and the workpiece was determined by a heat transfer coefficient α of 10 000 $W{\cdot}m^{-2}{\cdot}K^{-1}$, and tool effusivity E of 11,764 $J{\cdot}m^{-2}{\cdot}K^{-1}{\cdot}s^{-1/2}$ (interface with the rigid die). The surrounding environment was set as air, and the thermal exchange between the assembly and the environment was characterized by a heat transfer coefficient α of 10 $W{\cdot}m^{-2}{\cdot}K^{-1}$, which corresponds to the assumption of free convection in the surroundings. Considering the fact that the ECAP process is a bulk forming process, and taking into account our previous experience, the combined (Coulomb and Tresca) friction law was selected. The friction between the tools and the workpiece was determined using a friction coefficient μ of 0.05 and friction factor m of 0.1. The movement of the plunger was controlled by a hydraulic press with the speed rate of 1 $mm{\cdot}s^{-1}$.

In the following section, the results of the numerical simulations performed for the Cu-La_2O_3 composite and, for comparison, for CP Cu are represented as graphical depictions of the distributions of the parameters of interest, i.e., temperature, effective strain, material plastic flow, stress state, and punch load. For the purposes of more detailed investigations of selected parameters, a monitoring plane was created along the longitudinal cross-sectional cut through the workpiece, and three individual sensors were located on this plane. Figure 4a shows that sensor 1 was located on one (top) workpiece surface, sensor 2 was located in the axis (i.e., middle of workpiece thickness), and sensor 3 was located at the opposite (bottom) surface of the extruded workpiece. Furthermore, a superimposed grid defined with square cells with the dimensions of 0.5×0.5 mm^2 was created in the monitored plane to observe the material plastic flow (Figure 4b).

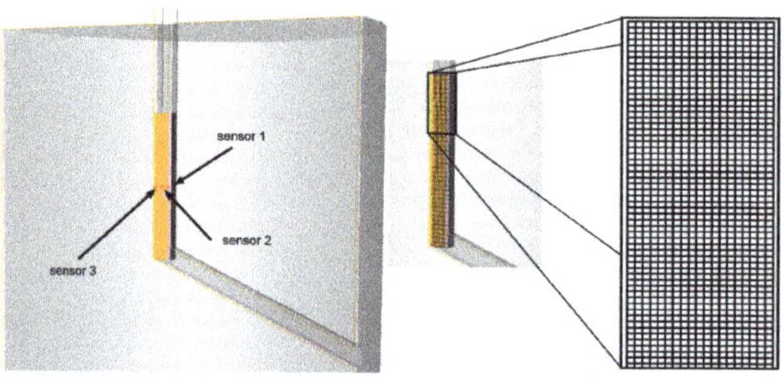

(a) (b)

Figure 4. Axial cutting plane with localization of: sensors (**a**), superimposed grid (**b**).

3. Results and Discussions

3.1. Experimental Analysis

3.1.1. Mechanical Behavior

The results of the compression tests performed for both the Cu-La_2O_3 composites (as-cast and ECAP-processed) at the strain rate of 0.01 s^{-1} are depicted as the true flow stress–true strain curves in Figure 5. The figure shows that the maximum flow stress, achieved at the true strain of 1.0, was 340 MPa for the as-cast composite, and reached to 382 MPa for the ECAP-processed one. The difference in the maximum values was thus approximately 13%. Nevertheless, the behaviors of both the material states during the loading differed significantly. The as-cast composite exhibited a steep increase in the flow stress up to the value of ~170.4 MPa (true strain of 0.37). After reaching this value, the increase continued, but in a more gradual manner. On the other hand, the ECAP-processed composite exhibited a rapid increase in the flow stress up to the value of ~352.5 MPa (true strain of 0.055). With continuing deformation, the material exhibited more or less steady state like behavior. The rapid increase in flow stress observed at the very beginning of deformation for both the investigated material states can be attributed to rapid work hardening due to the presence of oxide particles acting as effective barriers for dislocations movement (introducing dislocations pinning and pile-ups) [17]. The acquired results can be compared to those acquired by Zhang et al. [79], who investigated the compressive strengths of Cu with additions of Y_2O_3 of 4, 7, and 10 wt.%. The compressive yield strengths achieved herein were 274, 438, and 613 MPa, respectively. Considering that the composite features presented herein were 5 wt.% of La_2O_3, the strengthening effect can be considered comparable to that of Y_2O_3. However, adding La_2O_3 is more favorable as it is less costly and generally features better room temperature conductivity than Y_2O_3 [80]. The very high initial increase in flow stress observed for the ECAP-processed composite studied herein can be attributed to the deformation history, i.e., to the ECAP-processing introducing numerous lattice defects, substructure development, and accumulation of dislocations (primarily at subgrain boundaries and oxide particles, as further discussed). It also confirms the highly favorable effect of La_2O_3 on dislocations pinning during deformation.

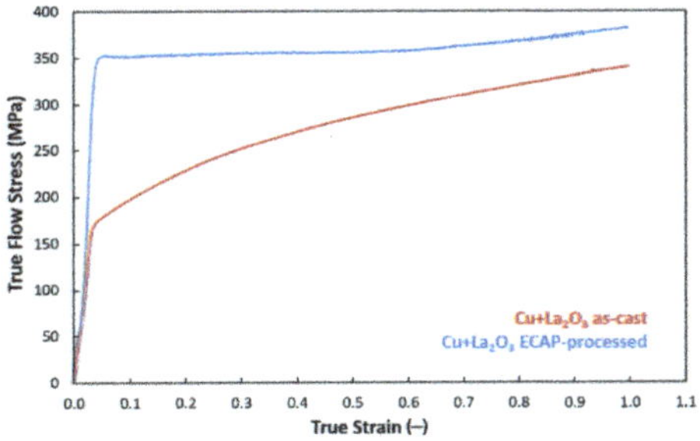

Figure 5. Flow stress curves for Cu-La_2O_3 composites, as-cast and ECAP-processed.

The fact that the microstructures and mechanical properties of SPD-processed materials are strongly dependent on temperature is known [81–83]. In this respect, thermal stability of the ECAP-processed composite under various thermomechanical conditions was further investigated. Figure 6a depicts the true flow stress–true strain curves acquired at the strain rates of 0.1 s^{-1} and 10 s^{-1} for the ECAP-processed composite, while Figure 6b depicts the true flow stress–true strain curves at identical strain rates for the ECAP-processed composite further subjected to the 250 °C/30 min heat treatment.

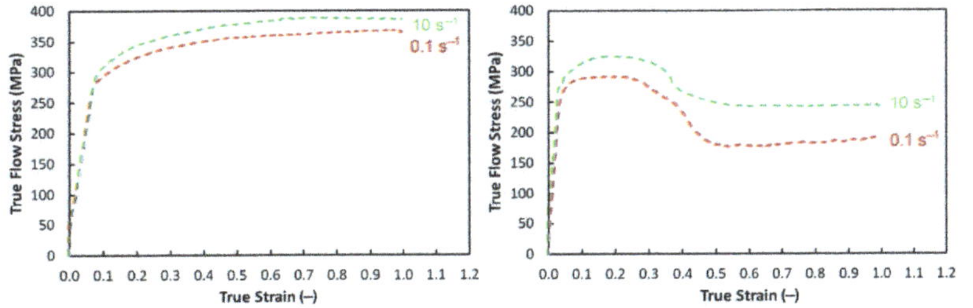

Figure 6. Flow stress curves of ECAP-processed composite: with no heat treatment (**a**), and after 250 °C/30 min heat treatment (**b**).

As can be seen in Figure 6a, the samples taken from the ECAP-processed composite with no heat treatment exhibited very similar deformation behaviors regardless the applied strain rate. This points to a relatively high stability of the ECAP-processed composite at room temperature. The fact that higher values of flow stress were achieved for the higher strain rate of 10 s^{-1} corresponds to: (i) the general hypothesis applicable for deformation processing of the majority of metallic alloys, (ii) the presence of La_2O_3 particles. In other words, the increased flow stress was related to: (i) the aggravated conditions for possible development of dynamic restoration processes given by speeding up the process of plastic deformation (i.e., work hardening dominated over dynamic restoration) [84], as well as (ii) increased frequency of interactions between dislocations and oxide particles during the (substantial) work hardening. As can be seen in Figure 6a, both the flow stress curves exhibited slight but continuous increases after reaching the strain value of 0.1 s^{-1}. In other words, work hardening, supported by the intense dislocations pinning effect of the reinforcing oxide particles, dominated over dynamic softening (recovery) within the ECAP-processed composite under all the examined strain rates.

Figure 6b then documents that the applied heat treatment significantly influenced the deformation behavior of the ECAP-processed composite. The work hardening was significant already at the beginning of testing, and was visibly more progressive than in the case of the ECAP-processed composite with no heat treatment (compare Figure 6a,b, the flow stress curves in the latter exhibited steeper increases). Despite the fact that both the examined material states achieved similar flow stress values at the true strains at which deformation softening occurred, the deformation behaviors differed entirely. After reaching the peak stress, both the flow stress curves exhibited plateaus, followed with flow stress decreases. Such behavior confirmed the occurrence of dynamic softening, i.e., dynamic recrystallization. Although the applied heat treatment temperature of 250 °C was relatively low to trigger dynamic recrystallization, the process was initiated by the accumulated energy (severe strain imposed during ECAP), which lowered the necessary activation temperature [84]. Figure 6b also shows that the flow stress curve acquired for the strain rate of 0.1 s^{-1} exhibited a more significant decrease compared to that acquired for 10 s^{-1}, i.e., softening was more prominent at the lower strain rate. This finding can be attributed to the differences in the mutual effects of work hardening and subsequent heat treatment and can be explained by similar phenomena as above, i.e., by speeding up the deformation process, the conditions for possible development of dynamic restoration were aggravated. It confirms the results acquired for the ECAP-processed composite with no heat treatment, and is in accordance with the above-mentioned hypothesis.

3.1.2. Vickers Microhardness

The average Vickers microhardness, measured for the CP Cu used to fabricate the Cu-La_2O_3 composite, was 46.2 HV0.05. The results of measurements of Vickers microhardness for the cross-sectional samples taken from both the as-cast and ECAP-processed composites are summarized in Figure 7. The measurements were performed along the intersecting diagonals across the cross-sections of the workpieces, as depicted in Figure 1. The as-cast Cu-La_2O_3 composite featured the average microhardness value of 90.1 HV0.05, which represents almost a 100% increase when compared to the original CP Cu. As shown in Figure 7, the diagonal trends across the cross-section of the as-cast composite were more or less random. In other words, the as-cast composite did not feature any clear decreasing/increasing trend in the microhardness across the cross-section, neither did it exhibit any severe local HV drops/increases, which would point to the presence of clusters of oxide particles.

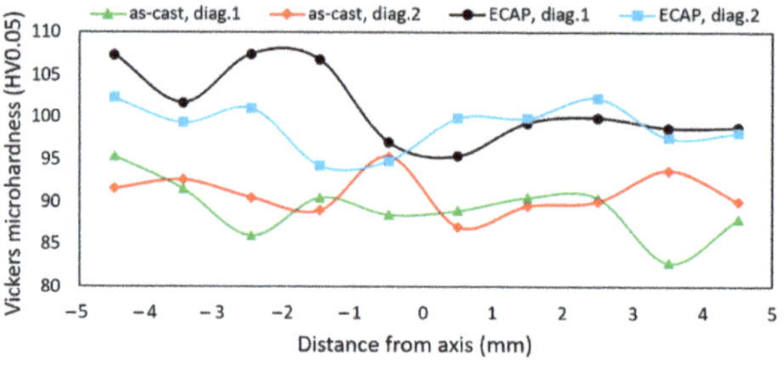

Figure 7. Vickers microhardness measured along intersecting diagonals across cross-sections of as-cast and ECAP-processed composites.

The average Vickers microhardness value for the composite processed via a single pass ECAP increased to 100.6 HV0.05, which is another increase by more than 10% compared to the as-cast composite, and more than 120% increase when compared to the original CP Cu. Contrary to the as-cast composite, the microhardness values measured across the cross-section of the ECAP-processed composite exhibited a decreasing trend towards the bottom edge of the workpiece. In other words, the microhardness exhibited a trend specific for ECAP processing: the composite processed via ECAP exhibited increased microhardness values primarily in its (sub)surface region, in the vicinity of the top edge of the billet [54,85]. This phenomenon, which will further be discussed also in correlation with the results from the numerical simulations, can primarily be attributed to the influence of friction between the (sub)surface material layers and ECAP die. By the effect of friction, the material plastic flow was aggravated in this region, and thus the grains sheared and fragmented significantly therein. This effect was supported even more by the presence of oxide particles pinning the occurring and moving dislocations. These factors resulted in the observed microhardness increase (especially in the top workpiece region). Towards the bulk volume of the ECAP-processed composite, the effect of friction gradually mitigated. Therefore, the microhardness values gradually decreased, then remained more or less constant with no excessive deviations throughout the cross-section.

The phenomenon of cross-sectional inhomogeneity can be reduced by several approaches, one of which is applying multiple ECAP passes [86–88]. Another one is incorporating another feature in the die, such as when applying the twist channel (multi) angular pressing method (TCAP, TCMAP) [62,89]. These approaches were proven to successfully improve the homogeneity of the effective imposed strain across the cross-section of an ECAP-processed workpiece.

3.1.3. Microstructure

The microstructure of the as-cast Cu-La_2O_3 composite was analyzed at first. The Orientation Image Map (OIM) depicted in Figure 8a shows that the microstructure of the as-cast composite consisted of coarse equiaxed grains featuring a major fraction (more than 90%) of HAGB. The arithmetic mean grain size for the as-cast composite (measured as equivalent circle diameter) reached up to 190 μm, and the grains exhibited no dominant preferential orientations. Figure 8b, showing a detailed phase map of the as-cast composite, depicts that the distribution of the La_2O_3 particles within the Cu grains was more or less homogeneous. The homogeneity of the as-cast microstructure was documented by the stress–strain curve (Figure 5) and microhardness (Figure 7) acquired for the as-cast material state, both of which exhibited homogeneous trends with no obvious discrepancies (which could be caused by possibly present inhomogeneities or casting defects).

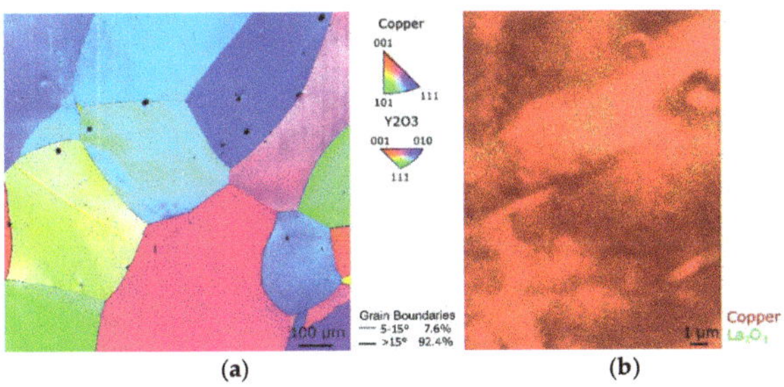

Figure 8. Microstructure of as-cast Cu-La_2O_3 composite: OIM image (**a**); phase map (**b**).

Figure 9a shows the OIM image of the microstructure of the ECAP-processed Cu-La_2O_3 composite. As can be seen, a single ECAP pass imparted substantial changes in the micro and substructure. The grains, which were more or less randomly oriented after casting, exhibited the tendency to form typical <111> | | SD (shear direction) shear texture fiber already after the single pass ECAP; see also the Inverse Pole Figures (IPF) depicting texture orientations and texture intensity in Figure 9b. The fraction of LAGB changed remarkably, too, and increased from a negligible value observed before ECAP-processing up to more than 17%. Both tendencies—to form substructure (increase in the fraction of LAGBs) and texture fibers—can be attributed to the fact that the La_2O_3 particles acted as effective obstacles and hindered the movement of dislocations [17,52]. This hypothesis was confirmed by subsequent detailed observations, which revealed the presence of shear bands within the ECAP-processed Cu grains. Figure 9c depicts the OIM map showing a more detailed view on the microstructure of the ECAP-processed composite, while the OIM image in Figure 9e shows a very close view depicting the substructure. Both the Figures clearly show the presence of dislocation walls and developing subgrains in the shear bands within a single grain affected by the ECAP processing. Sousa et al. [90] previously reported that a single ECAP pass is able to increase the dislocations density within pure Cu to more than 1×10^{12}. Such an increase in dislocations density imparted by ECAP, together with the presence of La_2O_3 particles, non-negligibly resulted in the intense strain hardening, as documented in Sections 3.1.1 and 3.1.2. As revealed by the phase map depicted in Figure 9d, as well as by the very detailed phase map shown in Figure 9f, (the close vicinity of) the shear band boundaries exhibited the presence of (increased volume fractions) the oxide particles. In other words, the analysis confirmed that the presence of La_2O_3 particles supported the development of substructure and shear banding within the composite workpiece during ECAP processing (both of which contributed to increase in the mechanical properties, as discussed above). The experimentally acquired results on microstructures and mechanical properties are further put in correlation with the overall material behavior, which was examined via a numerical simulation (Section 3.2).

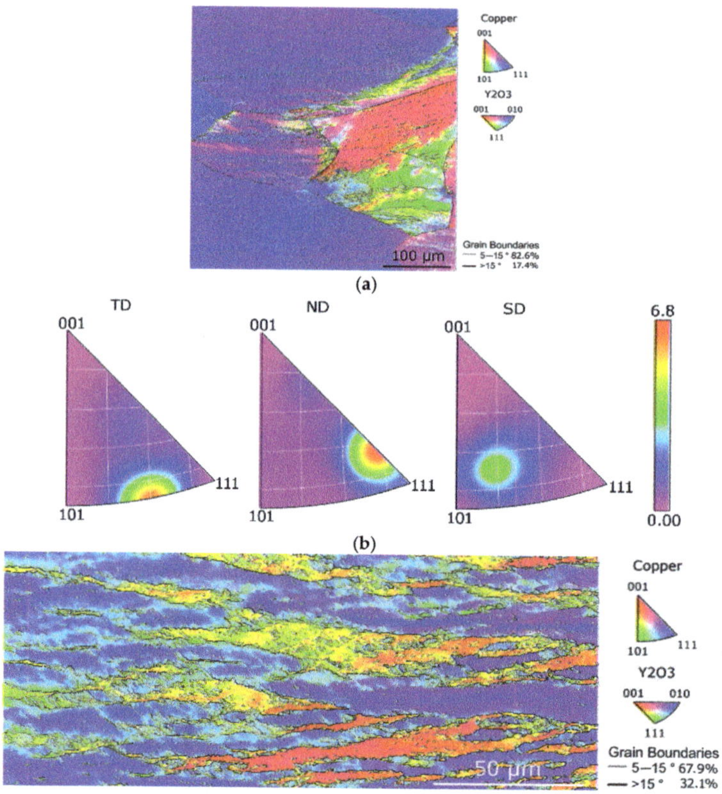

(c)

Figure 9. *Cont.*

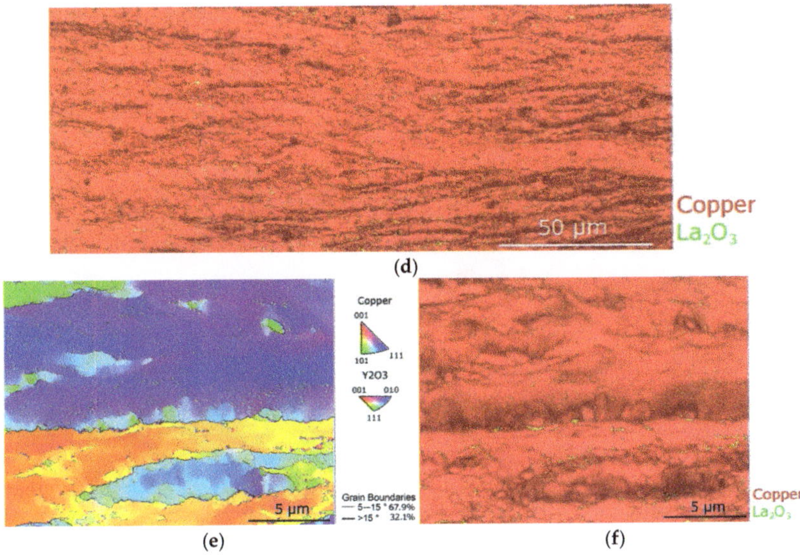

Figure 9. Microstructure of ECAP-processed Cu-La_2O_3 composite: OIM image (**a**); IPF images (**b**). Detailed views on microstructure: OIM image (**c**); phase map (**d**). Close view on substructure: OIM image (**e**); phase map (**f**).

3.2. Numerical Prediction

3.2.1. Deformation Temperature

The predicted developments of deformation temperature within the Cu-La_2O_3 composite and CP Cu during the ECAP process are depicted in Figure 10a. Both the workpieces clearly exhibited slight increases in temperature in the region of the main deformation zone (i.e., passage through the channel bending). For both the investigated materials, this phenomenon can be attributed to the development of deformation heat, see, e.g., [85,89]. However, the Figure also demonstrates that this increase was more intense (approximately by 20%) for the composite with La_2O_3 particles than for the CP Cu. This finding is in accordance with the expected substantial development of dislocations during ECAP processing, and also confirms that the presence of the oxide particles contributed to a more intense work hardening within the composite.

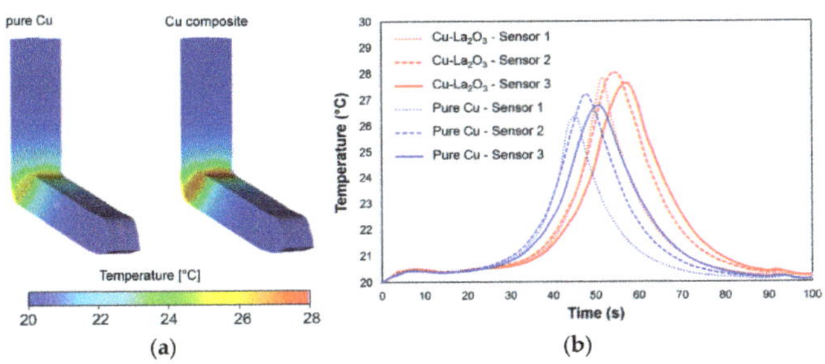

Figure 10. Temperature distribution during ECAP processing: contours (**a**), time development for individual sensors (**b**).

As regards the bulk volume of an extruded workpiece, the highest temperature was detected in sensor 2 (i.e., in the vicinity of workpiece axis) for both the investigated materials; see Figure 10b depicting the developments of temperature in time of extrusion for both the CP Cu and Cu-La_2O_3. This result is in agreement with the localization of the main deformation zone in the ECAP die, the simple shear in which is the most intense, and, simultaneously, temperature transfer into the die is the most aggravated, also due to the occurrence of the dead zone (Section 3.2.5).

3.2.2. Effective Strain

Figure 11a depicts the distributions of the effective imposed strain within both the CP Cu and Cu-La_2O_3 composite during ECAP processing. Similar to the distributions of deformation temperature, both the extruded workpieces exhibited increased imposed strain after passing through the main deformation zone of the die. Figure 11b shows gradual increases in strain over time in the individual sensors for both the materials; such development corresponded to gradual approaching of a workpiece (location with the sensors) to the main deformation zone. Entering of a workpiece (i.e., location with sensors) to the main deformation zone was then accompanied with a rapid increase in the effective imposed strain. Further, as the workpiece passed through the channel bending zone, the effective imposed strain remained more or less constant. However, the differences between both the processed workpieces were evident.

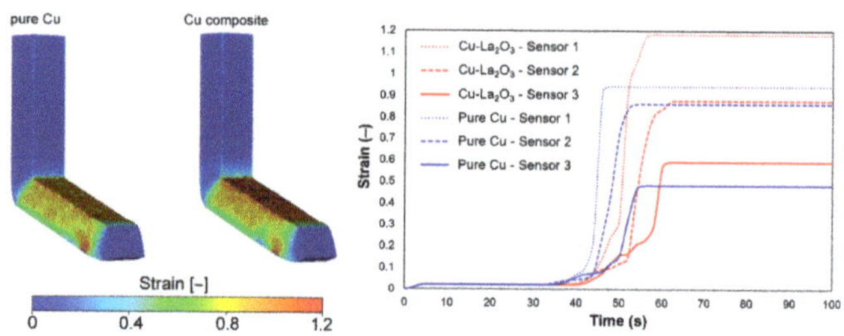

(a) (b)

Figure 11. Distribution of imposed effective strain during ECAP: contours (**a**), time development for individual sensors (**b**).

Primarily, the increase in strain was again greater for the composite than for CP Cu, in all the monitored sensors. In other words, the predicted maxima of the effective imposed strain were higher for the Cu-La_2O_3 composite. This finding was in accordance with: (i) the experimental results and above-mentioned hypotheses on dislocations development and work hardening, and (ii) the fact that the values of the strain hardening/softening constants acquired for the reliable fit of the Hensel–Spittel model (Equation (1)) for the Cu-La_2O_3 composite pointed to a more significant work hardening when compared to CP Cu. Also, during the processing, two interesting phenomena were noticed. The first one was that the increase in the effective strain in time was quicker for the composite than for the CP Cu. In other words, the Cu-La_2O_3 composite exhibited a more rapid increase in the imposed strain in all the examined locations than CP Cu, which was confirmed by steeper growths of the curves at the time of 40–60 s, see Figure 11b. This phenomenon can be attributed to the rapid work hardening of the composite, as also documented by the experimentally acquired data (Sections 3.1.1 and 3.1.2). The second intriguing phenomenon was that sensor 1 for the CP Cu generally exhibited slightly lower increase in the effective imposed strain than sensors 2 and 3. The effective imposed strain in the top (sub)surface region of the composite (sensor 1) thus achieved a higher maximum value (by ~20%) when compared to CP Cu. This phenomenon again points to intense dislocations development and work hardening of the composite. However, it also points to a certain inhomogeneity of the imposed strain across the cross-section of the workpiece.

The cross-sectional inhomogeneity can be explained by the material plastic flow, i.e., the paths the atoms (crystallites) take during passing through the ECAP channel (see also Section 3.2.5 on plastic flow). The strain imposed during ECAP processing generally tends to be the highest along the top surface (sensor 1) of a workpiece, as the paths of the crystallites in the (very) vicinity of the top workpiece–die interface are the shortest [54,85]. Towards the longitudinal axis of a workpiece (sensor 2) the paths of the crystallites lengthen, and are the longest along the bottom workpiece surface, in the (very) vicinity of the bottom workpiece–die interface (sensor 3). For this reason, the effective imposed strain generally tends to be higher along the top surface of a processed workpiece than along its bottom surface. The inhomogeneity can also be affected by friction (i) and (ii) die bending angle. (i) The effect of friction, aggravating the material plastic flow at the workpiece–die interface, diminishes towards the bulk of the processed workpiece [91,92]. (ii) Die angles higher than 90° are generally preferred for processing of materials with low formability, such as Mg-based ones [56,93], or materials for which a high hardening rate is expected, such as the Cu-La_2O_3 composite. On the other hand, increasing the die bending angle generally decreases the maximum imposed strain achievable within a single pass, and increases the cross-sectional inhomogeneity [94,95].

The predicted results went hand in hand with the experimentally observed inhomogeneity in microhardness (Section 3.1.2). Nevertheless, the absolute value of the imposed strain is dependent on the particular material behavior. Note that the difference in the imposed strain between sensor 1 and sensor 2 was substantially higher for the composite than for the CP Cu, Figure 11b. The most probable reason for this was the presence of oxide particles imparting differences in the material plastic flow, as further discussed in detail. The presence of the oxide particles introduced the above-mentioned work hardening, modified the effectivity of the acting simple shear deformation mechanism and consequently changed the plastic flow, i.e., depth of penetration of the imposed strain into the bulk of the workpiece. The comparisons of the predicted effective imposed strain for both the CP Cu and Cu-La_2O_3 composite depicted in Figure 11a,b document that the latter exhibited a higher hardening rate during ECAP processing, although the inhomogeneity typical for materials processed with a single pass ECAP could be observed for both.

3.2.3. Stress Distribution

The predicted distribution of von Mises stress depicted in Figure 12a shows evident differences between the two ECAP-processed materials. The Cu-La_2O_3 composite exhibited the maximum stress of up to 600 MPa when passing through the main deformation zone, which was by up to 50% higher than the maximum stress predicted for the CP Cu. This indicates that the Cu-La_2O_3 composite required a (much) greater punch load to be successfully deformed by the ECAP process than CP Cu (see also Section 3.2.4). This finding supported the experimentally acquired results and confirmed that the addition of La_2O_3 particles has a highly strengthening effect on Cu. When compared to the CP Cu, the Cu-La_2O_3 composite exhibited notable increase in stress, especially in the vicinity of the top workpiece–die interface. Similar to the effective strain distribution, the stress distribution corresponds with the observed cross-sectional inhomogeneity of mechanical properties, as documented in Section 3.1.3.

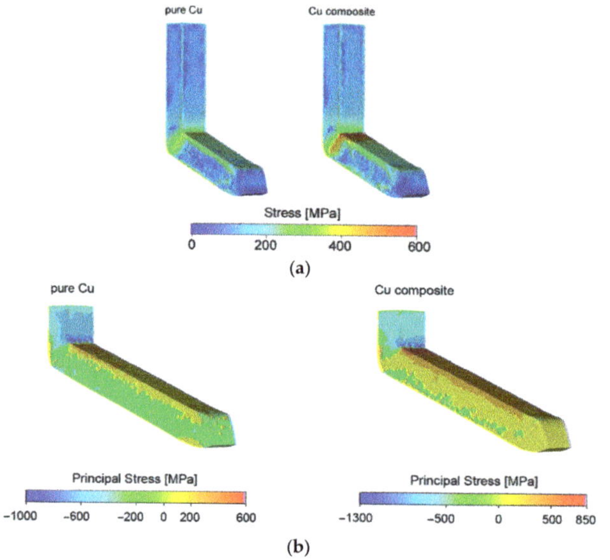

Figure 12. Predicted distribution of: von Mises stress (**a**), residual stress (**b**).

Figure 12b depicts predicted residual stress within the ECAP-processed workpieces. As evident from the Figure, the maximum values of both the tensile and compressive stress were substantially higher (by more than 20%) for the composite than for the CP Cu. Interestingly, the majority of the volume of the ECAP-processed composite exhibited tensile residual stress, while the processed CP Cu featured predominantly compressive residual stress. The presence of tensile stress is related to a higher risk of occurrence of cracks during processing. In this respect, processing of the composite via ECAP can be considered to be more problematic than ECAP processing of CP Cu. With a high probability, the presence of tensile residual stress originated from the presence of a highly effective mechanism of work hardening introduced by the oxide particles.

3.2.4. Punch Load

Among the very important parameters characterizing deformation processing is the force that needs to be applied for the process to be successfully performed. In case of ECAP processing, the punch load parameter is crucial for the choice of a suitable extrusion machine, as well as for determination of punch stability and durability. Figure 13 shows that the necessary load for ECAP processing of the Cu-La_2O_3 composite can be up to 30% higher in comparison with ECAP processing of CP Cu. In addition, two distinct differences can be noticed. First, processing of the composite featured a substantially higher increase in the load force than processing of the CP Cu at the moment of entering/crossing the main deformation zone (time of extrusion ~15 s). Second, contrary to the punch load predicted for the CP Cu, the load predicted for the composite featured a final increase at the very end of processing. The most probable reason for these differences was the presence of the oxide particles and their influence on the work hardening; the oscillations on both the curves are typically related to instability of material plastic flow, which, again, goes hand in hand with work hardening. The punch load is another factor supporting the experimentally acquired results and predicted differences between both the examined materials.

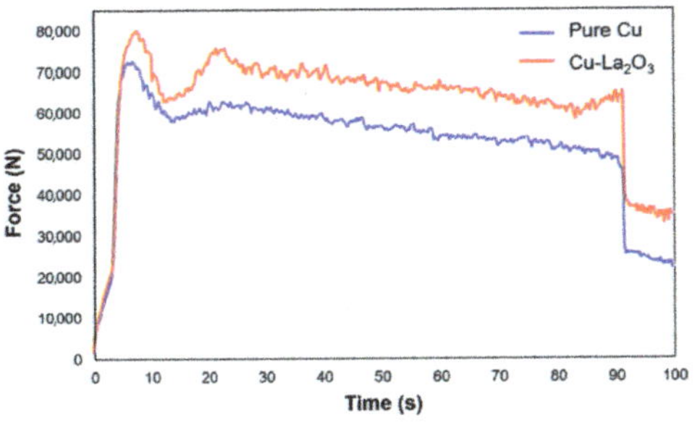

Figure 13. Development of punch load in time during ECAP processing.

3.2.5. Plastic Flow

The (differences in the) material plastic flows of both the processed materials were the primary reasons for/consequences of the majority of the above discussed differences observed during the simulation, as well as experiment. At first, the predicted plastic flow was evaluated using the superimposed grid (see Figure 14a). The grid clearly shows the differences in the intensity of forward material flow of the Cu-La_2O_3 composite and CP Cu. The original square cells of the grids within both the workpieces were deformed by the effect of simple shear. However, the effect of the oxide particles, acting as effective obstacles for dislocations movement and thus aggravating the forward plastic flow, is clearly visible in Figure 14a. In case of CP Cu, a more complicated material plastic flow can be expected as this material features (virtually) no obstacles, such as precipitates and oxide particles (pure metals tend to exhibit more "turbulent" plastic flows than alloys or composites), e.g., [89]. On the other hand, the material plastic flow of the Cu-La_2O_3 composite exhibited the tendency to delay, particularly along the top and bottom interfaces. This was primarily caused by the effect of friction and intense work hardening in these regions. The plastic flow was aggravated already in the main deformation zone (the transverse lines of the grid within the CP Cu evidently moved faster than those within the composite). This phenomenon also directly affected the size of the dead zone, i.e., the zone in the corner of the bending channel, the workpiece in which was not in contact with the channel wall. The dead zone was thus larger when processing CP Cu than the composite, see Figure 14a,b. Increasing the size of the dead zone generally results in a higher heterogeneity of the imposed strain [96]. As can be seen from the comparison of the imposed strains for both the workpieces (Figure 11b), the material plastic flow predicted within this study is in accordance with this hypothesis. Figure 14b then shows that the velocity of the material plastic flow of the composite tended to decelerate in the main deformation zone. The plastic flow vectors in the bulk of the composite changed their directions, particularly in the area of the inner corner of the die channel. In other words, the plastic flow within the composite exhibited a tendency to flow towards the outer corner of the die, contrary to CP Cu. By this effect, the size of the dead zone decreased during ECAP processing of the composite (see Figure 14b).

Considering the results acquired from the experimental investigation and numerical simulation of a vacuum induction cast Cu-La_2O_3 composite processed via a single pass ECAP and their mutual correlation, the conclusion that such a composite could be manufactured by the proposed method in commercial conditions can be drawn. The designed composite could be used to manufacture little durable components, such as specific components in the microelectronics, and their final properties could be tailored by modifying and optimizing the processing conditions, especially when considering wider processing possibilities, such as the application of TC(M)AP or multiple pass ECAP. Nevertheless, in order to tailor and optimize the properties of such a Cu-La_2O_3 composite for application in the microelectronics, the relation of processing conditions and electric conductivity should be further investigated. Our prospective research is going to involve a study of the effects of ECAP processing by multiple passes via various deformation routes on the homogeneity of the imposed strain across the cross-sections of Cu-La_2O_3 composite workpieces, in correlation with the electric properties.

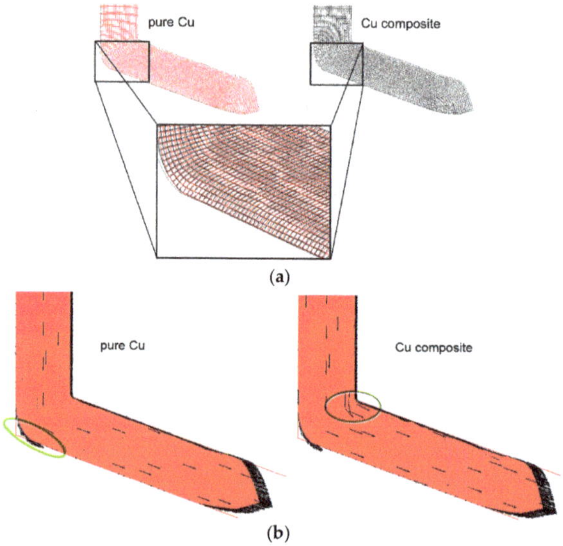

Figure 14. Plastic flow during ECAP: superimposed grids (**a**), vector profiles (**b**).

4. Conclusions

The study involved experimental and numerical examination of an innovative Cu-based composite with 5 wt.% addition of La_2O_3, processed by equal channel angular pressing (ECAP) through a 110° die. The most important results were:

- La_2O_3 hindered movement of dislocations and supported the development of shear bands and substructure during ECAP; their presence promoted formation of <111> | | SD texture.
- ECAP processing caused accumulation of dislocations and consequently increased the maximum flow stress of the composite to 382 MPa, average Vickers microhardness increased by 120% compared to CP Cu.
- the composite exhibited substantial work hardening, which increased the effective imposed strain, but at the expense of tensile residual stress and increased punch load; the presence of La_2O_3 directly influenced the size of the dead zone in the ECAP die.
- the energy accumulated during ECAP decreased the activation energy for dynamic recrystallization and influenced thermal stability of the mechanical properties.

The acquired results document that the ECAP method is highly promising for preparation of Cu-based composites with enhanced mechanical properties. The observed inhomogeneities introduced by the ECAP process can be overcome by optimized additional deformation processing (e.g., using multiple ECAP passes).

Author Contributions: Conceptualization, data curation, investigation, writing—original draft preparation: L.K.; investigation, project administration, funding acquisition, software, writing—review and editing: R.K. All authors have read and agreed to the published version of the manuscript.

Funding: The authors acknowledge the support of project no. 25-16860S by the Czech Science Foundation.

Data Availability Statement: The original data supporting the research are available upon a reasonable request from the corresponding author.

Acknowledgments: We would like to thank Petr Opěla and Josef Walek (Department of Metallurgical Technologies, VŠB-Technical University of Ostrava) for their assistance.

Conflicts of Interest: The authors declare no conflicts of interest. The funders had no role in the design of the study, in the collection, analyses, or interpretation of data, in the writing of the manuscript, or in the decision to publish the results.

References

1. Russell, A.; Lee, K.L. *Structure-Property Relations in Nonferrous Metals*, 1st ed.; John Wiley & Sons, Inc.: Hoboken, NJ, USA, 2005; ISBN 978-0-471-64952-6.
2. Konchakova, I.S.; Kolmakov, A.G.; Gvozdev, A.E. Features of the Influence of Thermal Action on Structural Changes in Copper Conductors. *Inorg. Mater. Appl. Res.* **2021**, *12*, 993–999. [CrossRef]
3. Gilev, V.G.; Bezmaternykh, N.V.; Morozov, E.A. Study of Steel–Copper Pseudo Alloy Microstructure and Microhardness After Laser Heat Treatment. *Met. Sci. Heat Treat.* **2014**, *56*, 262–268. [CrossRef]
4. Chu, Z.Q.; Wei, K.X.; Wei, W.; Alexandrov, I.V.; An, X.L.; Wang, D.D.; Liu, X.K. Simultaneously Enhancing Mechanical Properties and Electrical Conductivity of Cu-0.5%Cr Alloy as 5G Connector Material. *J. Alloys Compd.* **2023**, *948*, 169750. [CrossRef]
5. Kocich, R.; Kunčická, L. Crossing the Limits of Electric Conductivity of Copper by Inducing Nanotwinning via Extreme Plastic Deformation at Cryogenic Conditions. *Mater. Charact.* **2024**, *207*, 113513. [CrossRef]
6. Popovich, A.; Sufiiarov, V.; Polozov, I.; Borisov, E.; Masaylo, D.; Orlov, A. Microstructure and Mechanical Properties of Additive Manufactured Copper Alloy. *Mater. Lett.* **2016**, *179*, 38–41. [CrossRef]
7. Wu, Y.; Zhang, W.; Li, Y.; Yang, F.; Liu, H.; Zou, J.; Jiang, J.; Fang, F.; Ma, A. An Overview of Microstructure Regulation Treatment of Cu-Fe Alloys to Improve Strength, Conductivity, and Electromagnetic Shielding. *J. Alloys Compd.* **2024**, *1002*, 175425. [CrossRef]
8. Kunčická, L.; Kocich, R. Effect of Activated Slip Systems on Dynamic Recrystallization during Rotary Swaging of Electro-Conductive Al-Cu Composites. *Mater. Lett.* **2022**, *321*, 10–13. [CrossRef]
9. Hu, X.; Hu, H.; Lai, R.; Xie, Q.; Zhi, Y. Cyclic Warm Rolling: A Path to Superior Properties in Mo Cu Composites. *Int. J. Refract. Met. Hard Mater.* **2025**, *126*, 106926. [CrossRef]
10. Strunz, P.; Kunčická, L.; Beran, P.; Kocich, R.; Hervoches, C. Correlating Microstrain and Activated Slip Systems with Mechanical Properties within Rotary Swaged WNiCo Pseudoalloy. *Materials* **2020**, *13*, 208. [CrossRef]
11. Kunčická, L.; Kocich, R.; Németh, G.; Dvořák, K.; Pagáč, M. Effect of Post Process Shear Straining on Structure and Mechanical Properties of 316 L Stainless Steel Manufactured via Powder Bed Fusion. *Addit. Manuf.* **2022**, *59*, 103128. [CrossRef]
12. Canelo-Yubero, D.; Kocich, R.; Hervoches, C.; Strunz, P.; Kunčická, L.; Krátká, L. Neutron Diffraction Study of Residual Stresses in a W–Ni–Co Heavy Alloy Processed by Rotary Swaging at Room and High Temperatures. *Met. Mater. Int.* **2022**, *28*, 919–930. [CrossRef]
13. Wang, Z.; Chen, J.; Kocich, R.; Tardif, S.; Dolbnya, I.P.; Kunčická, L.; Micha, J.-S.; Liogas, K.; Magdysyuk, O.V.; Szurman, I.; et al. Grain Structure Engineering of NiTi Shape Memory Alloys by Intensive Plastic Deformation. *ACS Appl. Mater. Interfaces* **2022**, *14*, 31396–31410. [CrossRef] [PubMed]
14. Kunčická, L.; Kocich, R.; Ryukhtin, V.; Cullen, J.C.T.; Lavery, N.P. Study of Structure of Naturally Aged Aluminium after Twist Channel Angular Pressing. *Mater. Charact.* **2019**, *152*, 94–100. [CrossRef]
15. Segal, V. Review: Modes and Processes of Severe Plastic Deformation (SPD). *Materials* **2018**, *11*, 1175. [CrossRef]
16. Arbuz, A.; Lutchenko, N.; Yordanova, R. FEM Method Study of the Advanced ECAP Die Channel and Tool Design. *Modelling* **2025**, *6*, 19. [CrossRef]
17. Verlinden, B.; Driver, J.; Samajdar, I.; Doherty, R.D. *Thermo-Mechanical Processing of Metallic Materials*; Elsevier: Amsterdam, The Netherland, 2007; ISBN 9780080444970.
18. Bodyakova, A.; Tkachev, M.; Raab, G.I.; Kaibyshev, R.; Belyakov, A.N. Regularities of Microstructure Evolution in a Cu-Cr-Zr Alloy during Severe Plastic Deformation. *Materials* **2022**, *15*, 5745. [CrossRef]
19. Sukhov, I.A.; Shafeev, G.A.; Voronov, V.V.; Sygletou, M.; Stratakis, E.; Fotakis, C. Generation of Nanoparticles of Bronze and Brass by Laser Ablation in Liquid. *Appl. Surf. Sci.* **2014**, *302*, 79–82. [CrossRef]
20. Muñoz, A.I.; Antón, J.G.; Guiñón, J.L.; Herranz, V.P. Corrosion Behavior and Galvanic Studies of Brass and Bronzes in Aqueous Lithium Bromide Solutions. *Corrosion* **2002**, *58*, 560–569. [CrossRef]
21. Zhou, M.; Geng, Y.; Zhang, Y.; Ban, Y.; Li, X.; Jia, Y.; Liang, S.; Tian, B.; Liu, Y.; Volinsky, A.A. Enhanced Mechanical Properties and High Electrical Conductivity of Copper Alloy via Dual-Nanoprecipitation. *Mater. Charact.* **2023**, *195*, 112494. [CrossRef]

22. Dubskii, G.A.; Vdovin, K.N.; Nefed'ev, A.A.; Egorova, L.G. Thermal Conductivity of Low-Alloy Copper for Molds. *Russ. Metall.* **2018**, *2018*, 252–258. [CrossRef]
23. Tian, W.; Zhang, F.; Han, S.; Chen, X.; Gao, P.; Zheng, K. Analysis of Microstructure and Properties in Cold Rotary Swaged Copper-Clad Magnesium Wires. *Metals* **2023**, *13*, 467. [CrossRef]
24. Volkov, A.Y.; Kalonov, A.A.; Zavalishin, V.A.; Glukhov, A.V.; Komkova, D.A.; Antonov, B.D. The Influence of Interfaces on the Physicomechanical Properties of Cu/Mg Composites. *Phys. Met. Metallogr.* **2020**, *121*, 568–574. [CrossRef]
25. Volkov, A.Y.; Kalonov, A.A.; Komkova, D.A.; Glukhov, A.V. Structure and Properties of Cu/Mg Composites Produced by Hydrostatic Extrusion. *Phys. Met. Metallogr.* **2018**, *119*, 946–955. [CrossRef]
26. Kocich, R.; Kunčická, L. Optimizing Structure and Properties of Al/Cu Laminated Conductors via Severe Shear Strain. *J. Alloys Compd.* **2023**, *953*, 170124. [CrossRef]
27. Rogachev, S.O.; Sundeev, R.V.; Andreev, V.A.; Andreev, N.V.; Tabachkova, N.Y.; Korotkova, N.O. The Microstructure and Conductivity of Copper–Aluminum Composites Prepared by Rotary Swaging. *Phys. Met. Metallogr.* **2022**, *123*, 1193–1200. [CrossRef]
28. Zhang, X.; Yu, Y.; Liu, B.; Ren, J. Mechanical Properties and Tensile Fracture Mechanism Investigation of Al/Cu/Ti/Cu/Al Laminated Composites Fabricated by Rolling. *J. Alloys Compd.* **2019**, *805*, 338–345. [CrossRef]
29. Popova, E.N.; Deryagina, I.L.; Valova-Zaharevskaya, E.G.; Ruello, M.L.; Popov, V.V. Microstructural Features in Multicore Cu–Nb Composites. *Materials* **2021**, *14*, 7033. [CrossRef]
30. Permyakova, I.E.; Glezer, A.M.; Karpov, M.I.; Vnukov, V.I.; Shtansky, D.V.; Gorshenkov, M.V.; Schetinin, I.V. Structure Amorphization and Mechanical Properties of Nanolaminates of the Copper–Niobium System During High-Pressure Torsion. *Russ. Phys. J.* **2018**, *61*, 428–438. [CrossRef]
31. Deryagina, I.L.; Popova, E.N.; Valova-Zaharevskaya, E.G.; Patrakov, E.I. Structure and Thermal Stability of High-Strength Cu–18Nb Composite Depending on the Degree of Deformation. *Phys. Met. Metallogr.* **2018**, *119*, 92–102. [CrossRef]
32. Bokstein, B.S.; Vnukov, V.I.; Golosov, E.V.; Karpov, M.I.; Kolobov, Y.R.; Kolesnikov, D.A.; Korzhov, V.P.; Rodin, A.O. Structure and Diffusion Processes in Laminated Composites of a Cu–Ti System. *Russ. Phys. J.* **2009**, *52*, 811–815. [CrossRef]
33. Liu, Z.; Wang, X.; Liu, M.; Liu, Y.; Liu, J.; Ignatov, A.V.; Wang, T. Microstructure and Mechanical Behavior of Ti/Cu/Ti Laminated Composites Produced by Corrugated and Flat Rolling. *Trans. Nonferrous Met. Soc. China* **2022**, *32*, 2598–2608. [CrossRef]
34. Xu, Q.; Zhang, X.; Han, J.; He, X.; Kvanin, V.L. Combustion Synthesis and Densification of Titanium Diboride–Copper Matrix Composite. *Mater. Lett.* **2003**, *57*, 4439–4444. [CrossRef]
35. Mali, V.I.; Bataev, A.A.; Maliutina, I.N.; Kurguzov, V.D.; Bataev, I.A.; Esikov, M.A.; Lozhkin, V.S. Microstructure and Mechanical Properties of Ti/Ta/Cu/Ni Alloy Laminate Composite Materials Produced by Explosive Welding. *Int. J. Adv. Manuf. Technol.* **2017**, *93*, 4285–4294. [CrossRef]
36. Trykov, Y.; Gurevich, L.; Pronichev, D.; Trunov, M. Investigation of the Rupture of Ti/Steel Laminated Composite with Soft Interlayers. *FME Trans.* **2016**, *44*, 16–21. [CrossRef]
37. Xu, W.; Qi, J.; Zhang, Y.; Zeng, G.; Zhou, Y.; Ou, H.; Zhou, H.; Wu, Y.; Yang, Y. Effect of Annealing Treatment on Electromagnetic Shielding Effectiveness of Double-Layer FeSiBCuNb/Cu Composite Strips. *J. Mater. Sci. Mater. Electron.* **2023**, *34*, 376. [CrossRef]
38. Sheinerman, A.G.; Gutkin, M.Y. Multiple Cracking in Deformed Laminated Metal-Graphene Composites. *Compos. Struct.* **2018**, *191*, 113–118. [CrossRef]
39. Zhao, Y.; Zhang, Y.; Wang, W.; Yue, S.; Li, T. A Novel Approach to Achieve High Strength, High Plasticity and High Conductivity of Graphene/Cu Composites with Graphene Core-Shell Structure. *J. Alloys Compd.* **2024**, *1002*, 175207. [CrossRef]
40. Wei, K.X.; Jia, F.L.; Wei, W.; Zhou, H.R.; Chu, F.Q.; Du, Q.B.; Alexandrov, I.V.; Hu, J. Flexible Nanotwinned Graphene/Copper Composites as Thermal Management Materials. *ACS Appl. Nano Mater.* **2020**, *3*, 4810–4817. [CrossRef]
41. Akmal, J.; Salmi, M. Additive Manufacturing of Self-Sensing Parts through Material Extrusion. *Virtual Phys. Prototyp.* **2024**, *19*, e2321200. [CrossRef]
42. Koltsova, T.S.; Bobrynina, E.V.; Larionova, T.V.; Salynova, M.A.; Tolochko, O.V. Structure and Properties of Copper-Based Composite with Different Types of Carbon Nanostructures. *Diam. Relat. Mater.* **2022**, *124*, 108933. [CrossRef]
43. Han, K.; Goddard, R.E.; Toplosky, V.; Niu, R.; Lu, J.; Walsh, R. Alumina Particle Reinforced Cu Matrix Conductors. *IEEE Trans. Appl. Supercond.* **2018**, *28*, 1–5. [CrossRef]
44. Grigoreva, T.F.; Talako, T.L.; Devyatkina, E.T.; Vosmerikov, S.V.; Ancharov, A.I.; Tsybulya, S.V.; Vityaz, P.A.; Lyakhov, N.Z. Modifying Copper with Alumina during a Mechanically Stimulated Reaction. *Phys. Met. Metallogr.* **2023**, *124*, 74–79. [CrossRef]
45. Volkova, I.R.; Tyryshkina, L.E.; Volochaev, M.N.; Zaloga, A.N.; Shabanova, K.A.; Ovchinnikov, A.V.; Lyamkin, A.I. The Effect of Aluminum-Oxide Powders on the Structure and Properties of Copper Electrodeposited Composite Coatings. *Prot. Met. Phys. Chem. Surf.* **2023**, *59*, 71–75. [CrossRef]

46. Ma, B.; Ding, H.; Jiang, F.; Hishinuma, Y.; Luo, L.; Zhang, Y.; Wang, J.; Sheng, X.; Noto, H.; Liu, J.; et al. Effect of Process Control Agent on the Synthesis of Cu-Y_2O_3 by Mechanical Alloying. *Nucl. Mater. Energy* **2024**, *38*, 101599. [CrossRef]
47. Huang, F.; Wang, H.; Yang, B.; Liao, T.; Wang, Z. Pinning Effect of Y_2O_3 Network on Copper Grain Growth during High Temperature Annealing. *Mater. Res. Express* **2018**, *5*, 056520. [CrossRef]
48. Sobahi, N.; El-Kady, O.A.; Nouh, F.; Banoqitah, E.; Ghoname, R.; Moustafa, E.B.; Elsheikh, A. Impact of Nano-Y_2O_3 on the Physical, Microstructure, and Mechanical Characteristics of Cu Composite Fabricated via Powder Metallurgy. *J. Alloys Compd.* **2025**, *1010*, 177042. [CrossRef]
49. Huang, F.; Wang, H.; Yang, B.; Liao, T.; Wang, Z. Uniformly Dispersed Y_2O_3 Nanoparticles in Nanocrystalline Copper Matrix via Multi-Step Ball Milling and Reduction Process. *Mater. Lett.* **2019**, *242*, 119–122. [CrossRef]
50. Zheng, R.G.; Zhan, Z.J.; Wang, W.K. Wear Behavior of Cu-La_2O_3 Composite. *Rare Met. Mater. Eng.* **2010**, *39*, 293–295.
51. Li, X.; Wang, Y.; Yang, P.; Han, T.; Shi, X.; He, K.; Zu, C. Effect of Y_2O_3/La_2O_3 on Structure and Mechanical Properties of Li_2O–Al_2O_3–SiO_2 Glass. *J. Non. Cryst. Solids* **2022**, *596*, 121847. [CrossRef]
52. Zheng, R.G.; Zhan, Z.J.; Liang, B.; Wang, W.K. Tribological Behavior of La_2O_3 Particulate Reinforced Copper Matrix Composites. *Adv. Mater. Res.* **2010**, *150–151*, 979–983. [CrossRef]
53. Kocich, R.; Kursa, M.; Macháčková, A. FEA of Plastic Flow in AZ63 Alloy during ECAP Process. *Acta Phys. Pol. A* **2012**, *122*, 581–587.
54. Kunčická, L.; Kocich, R. Comprehensive Characterisation of a Newly Developed Mg-Dy-Al-Zn-Zr Alloy Structure. *Metals* **2018**, *8*, 73. [CrossRef]
55. Heidari, S.; Afsari, A.; Ranaei, M.A. Increasing Wear Resistance of Copper Electrode in Electrical Discharge Machining by Using Ultra-Fine-Grained Structure. *Trans. Indian Inst. Met.* **2020**, *73*, 2901–2910. [CrossRef]
56. Volokitina, I.E. Evolution of the Microstructure and Mechanical Properties of Copper under ECAP with Intense Cooling. *Met. Sci. Heat Treat.* **2020**, *62*, 253–258. [CrossRef]
57. Guo, T.; Wang, J.; Wu, Y.; Tai, X.; Jia, Z.; Ding, Y. Fabricate of High-Strength and High-Conductivity Cu–Cr–Si Alloys through ECAP-Bc and Aging Heat Treatment. *Materials* **2020**, *13*, 1603. [CrossRef]
58. Zhilyaev, A.P.; Oh-ishi, K.; Raab, G.I.; McNelley, T.R. Influence of ECAP Processing Parameters on Texture and Microstructure of Commercially Pure Aluminum. *Mater. Sci. Eng. A* **2006**, *441*, 245–252. [CrossRef]
59. Chu, Z.; Pan, X.; Wei, W.; Wei, K.; Alexandrov, I.V.; An, X.; Wang, D.; Liu, X. High Strength and Electrical Conductivity of Nanostructured Cu–1Cr–0.1Zr Alloy Processed by Multi–Stage Deformation and Aging. *J. Mater. Res. Technol.* **2024**, *29*, 2051–2060. [CrossRef]
60. Eivani, A.R.; Mirzakoochakshirazi, H.R.; Jafarian, H.R. Investigation of Joint Interface and Cracking Mechanism of Thick Cladding of Copper on Aluminum by Equal Channel Angular Pressing (ECAP). *J. Mater. Res. Technol.* **2020**, *9*, 3394–3405. [CrossRef]
61. Volokitina, I.; Sapargaliyeva, B.; Agabekova, A.; Syrlybekkyzy, S.; Volokitin, A.; Nurshakhanova, L.; Nurbaeva, F.; Kolesnikov, A.; Sabyrbayeva, G.; Izbassar, A.; et al. Increasing Strength and Performance Properties of Bimetallic Rods during Severe Plastic Deformation. *Case Stud. Constr. Mater.* **2023**, *19*, e02256. [CrossRef]
62. Kocich, R.; Kunčická, L. Development of Structure and Properties in Bimetallic Al/Cu Sandwich Composite during Cumulative Severe Plastic Deformation. *J. Sandw. Struct. Mater.* **2021**, *23*, 4252–4275. [CrossRef]
63. Macháčková, A.; Kocich, R.; Bojko, M.; Kunčická, L.; Polko, K. Numerical and Experimental Investigation of Flue Gases Heat Recovery via Condensing Heat Exchanger. *Int. J. Heat Mass Transf.* **2018**, *124*, 1321–1333. [CrossRef]
64. Fiedler, T.; Löffler, R.; Bernthaler, T.; Winkler, R.; Belova, I.V.; Murch, G.E.; Öchsner, A. Numerical Analyses of the Thermal Conductivity of Random Hollow Sphere Structures. *Mater. Lett.* **2009**, *63*, 1125–1127. [CrossRef]
65. Borodin, E.N.; Morozova, A.; Bratov, V.; Belyakov, A.; Jivkov, A.P. Experimental and Numerical Analyses of Microstructure Evolution of Cu-Cr-Zr Alloys during Severe Plastic Deformation. *Mater. Charact.* **2019**, *156*, 109849. [CrossRef]
66. Bratov, V.; Borodin, E.N. Comparison of Dislocation Density Based Approaches for Prediction of Defect Structure Evolution in Aluminium and Copper Processed by ECAP. *Mater. Sci. Eng. A* **2015**, *631*, 10–17. [CrossRef]
67. Abd EL AAL, M.I. 3D FEM Simulations and Experimental Validation of Plastic Deformation of Pure Aluminum Deformed by ECAP and Combination of ECAP and Direct Extrusion. *Trans. Nonferrous Met. Soc. China* **2017**, *27*, 1338–1352. [CrossRef]
68. Hongyu, W.; Jie, W.; Fei, T.; Ting, L.; Juncai, S.; Xiangwei, K.; Dianhua, Z.; Wen, P.; Shunhu, Z. Analysis and Simulation for Shape Control Effects of Square Aluminum Tube during ECAP. *Int. J. Adv. Manuf. Technol.* **2020**, *108*, 3859–3870. [CrossRef]
69. Vafaeenezhad, H.; Chegini, M.; Kalaki, A.; Serajian, H. Micromechanical Finite Element Simulation of Low Cycle Fatigue Damage Occurring During Sliding Wear Test of ECAP-Processed AA7075 Alloy. *Met. Mater. Int.* **2024**, *30*, 143–166. [CrossRef]
70. Ghosh, A.; Das, K.; Eivani, A.R.; Mohammadi, H.; Vafaeenezhad, H.; Murmu, U.K.; Jafarian, H.R.; Ghosh, M. Development of Mechanical Properties and Microstructure for Al–Zn–Mg–Cu Alloys through ECAP after Optimizing the Outer Corner Angles through FE Modeling. *Arch. Civ. Mech. Eng.* **2023**, *23*, 78. [CrossRef]
71. *ASTM D7012-23*; Standard Test Methods for Compressive Strength and Elastic Moduli of Intact Rock Core Specimens under Varying States of Stress and Temperatures. ASTM International: West Conshohocken, PA, USA, 2023.

72. Kunčická, L.; Kocich, R. Deformation Behaviour of Cu-Al Clad Composites Produced by Rotary Swaging. *IOP Conf. Ser. Mater. Sci. Eng.* **2018**, *369*, 012029. [CrossRef]
73. Wang, Z.; Zhang, Y.; Liogas, K.; Chen, J.; Vaughan, G.B.M.; Kocich, R.; Kunčická, L.; Uzun, F.; You, Z.; Korsunsky, A.M. In Situ Synchrotron X-Ray Diffraction Analysis of Two-Way Shape Memory Effect in Nitinol. *Mater. Sci. Eng. A* **2023**, *878*, 145226. [CrossRef]
74. Churyumov, A.Y.; Kazakova, A.A. Prediction of True Stress at Hot Deformation of High Manganese Steel by Artificial Neural Network Modeling. *Materials* **2023**, *16*, 1083. [CrossRef] [PubMed]
75. *ASTM E384*; Standard Test Method for Microindentation Hardness of Materials. ASTM International: West Conshohocken, PA, USA, 2022.
76. Sajjad, M.; Agirre, J.; Plata, G.; Lozares, J.; Mendiguren, J. Characterization of the Heat Transfer Coefficient at near Solidus Forming Condition Using Columnar Pressing Test. *Int. J. Adv. Manuf. Technol.* **2024**, *135*, 721–733. [CrossRef]
77. Dhondapure, P.; Rajakrishnan, N.; Nayak, S.; Champliaud, H.; Morin, J.-B.; Jahazi, M. Influence of Deformation Path on Microstructure Evolution during the Open Die Forging of Large Size Ingot of High Strength Steel: Experiments and FE Analysis. *Int. J. Adv. Manuf. Technol.* **2024**, *134*, 3733–3750. [CrossRef]
78. Ghat, M.; Mohamed, A.; Afify, A.S. Physical Modelling of Ti–6Al–4V Alloy above β Transus at High Temperatures (1010–1150 °C) and High Strain Rates Using Garofalo and Hensel–Spittel Laws. *High Temp.* **2022**, *60*, 479–484. [CrossRef]
79. Zhang, X.-H.; Yuan, H.-X.; Huang, F.; Wang, H.; Yang, B.; Zeng, L.-F.; Li, X.-H.; Huang, Y.-J.; Ahmad, T. Enhanced Strength–Plasticity Synergy of Copper Composites by Designing Uniformly Dispersed Yttria Nanoparticles and a Heterogeneous Grain Structure. *Rare Met.* **2024**, *43*, 6704–6716. [CrossRef]
80. Tallan, N.M.; Vest, R.W. Electrical Properties and Defect Structure of Y_2O_3. *J. Am. Ceram. Soc.* **1966**, *49*, 401–404. [CrossRef]
81. Chuvil'deev, V.N.; Shadrina, I.S.; Nokhrin, A.V.; Kopylov, V.I.; Bobrov, A.A.; Gryaznov, M.Y.; Shotin, S.V.; Tabachkova, N.Y.; Chegurov, M.K.; Melekhin, N.V. An Investigation of Thermal Stability of Structure and Mechanical Properties of Al-0.5Mg–Sc Ultrafine-Grained Aluminum Alloys. *J. Alloys Compd.* **2020**, *831*, 154805. [CrossRef]
82. Qi, Y.; Kosinova, A.; Kilmametov, A.R.; Straumal, B.B.; Rabkin, E. Stabilization of Ultrafine-Grained Microstructure in High-Purity Copper by Gas-Filled Pores Produced by Severe Plastic Deformation. *Scr. Mater.* **2020**, *178*, 29–33. [CrossRef]
83. Arivu, M.; Hoffman, A.; Duan, J.; Poplawsky, J.; Zhang, X.; Liou, F.; Islamgaliev, R.; Valiev, R.; Wen, H. Comparison of the Thermal Stability in Equal-Channel-Angular-Pressed and High-Pressure-Torsion-Processed Fe–21Cr–5Al Alloy. *Adv. Eng. Mater.* **2023**, *25*, 2300756. [CrossRef]
84. Humphreys, F.J.; Hetherly, M.; Rollett, A.; Rohrer, G.S. *Recrystallization and Related Annealing Phenomena*, 2nd ed.; Elsevier Ltd.: Oxford, UK, 2004; ISBN 008-044164-5.
85. Kocich, R.; Greger, M.; Macháčková, A. Finite Element Investigation of Influence of Selected Factors on ECAP Process. In Proceedings of the METAL 2010: 19th International Metallurgical and Materials Conference, Roznov pod Radhostem, Czech Republic, 18–20 May 2010; Tanger Ltd.: Greensboro, NC, USA, 2010; pp. 166–171.
86. Zhao, X.; Yang, X.; Liu, X.; Wang, X.; Langdon, T.G. The Processing of Pure Titanium through Multiple Passes of ECAP at Room Temperature. *Mater. Sci. Eng. A* **2010**, *527*, 6335–6339. [CrossRef]
87. Zhilyaev, A.P.; Swisher, D.L.; Oh-ishi, K.; Langdon, T.G.; McNelley, T.R. Microtexture and Microstructure Evolution during Processing of Pure Aluminum by Repetitive ECAP. *Mater. Sci. Eng. A* **2006**, *429*, 137–148. [CrossRef]
88. Sureshkumar, P.; Ganesan, L.; UmaRani, C.; Stalin, B.; Sasikumar, C.; Rajan, S.T.K.; Borek, W. Effect of Strain Rate on Fractography Texture Descriptor of AA6063/$(Si_3N_4)x/(Cu(NO_3)_2)y$ (x = 12%, y = 2–6%)Composite after Multiple ECAP Passes: Second Order Statistical Texture Analysis Conjunction with Regression Analysis. *J. Mater. Res. Technol.* **2023**, *23*, 2750–2783. [CrossRef]
89. Kocich, R.; Kunčická, L.; Macháčková, A. Twist Channel Multi-Angular Pressing (TCMAP) as a Method for Increasing the Efficiency of SPD. *IOP Conf. Ser. Mater. Sci. Eng.* **2014**, *63*, 012006. [CrossRef]
90. de Sousa, T.G.; Sordi, V.L.; Brandão, L.P. Dislocation Density and Texture in Copper Deformed by Cold Rolling and Ecap. *Mater. Res.* **2017**, *21*, e20170515. [CrossRef]
91. Medeiros, N.; Lins, J.F.C.; Moreira, L.P.; Gouvêa, J.P. The Role of the Friction during the Equal Channel Angular Pressing of an IF-Steel Billet. *Mater. Sci. Eng. A* **2008**, *489*, 363–372. [CrossRef]
92. Agwa, M.A.; Ali, M.N.; Al-Shorbagy, A.E. Optimum Processing Parameters for Equal Channel Angular Pressing. *Mech. Mater.* **2016**, *100*, 1–11. [CrossRef]
93. Lin, P.-C.; Li, J.-Y.; Chen, H.-J.; Lin, K.; Chen, M.-J.; Lin, K.-M.; Lin, H.-C. Indirect Enhancement of ALD Thin-Film Properties Induced by the ECAP Modification of an As-Extruded Mg-Ca Alloy. *Micromachines* **2024**, *15*, 1006. [CrossRef]
94. Agarwal, K.M.; Singh, P.; Dixit, S.; Meshcheryakova, T.S.; Soloveva, O.V.; Solovev, S.A.; Kaushal, K. Optimization of die design parameters in ECAP for sustainable manufacturing using response surface methodology. *Int. J. Interact. Des. Manuf.* **2024**, *18*, 2903–2910. [CrossRef]

95. Skryabina, N.; Aptukov, V.; Romanov, P.; Fruchart, D.; De Rango, P.; Girard, G.; Grandini, C.; Sandim, H.; Huot, J.; Lang, J.; et al. Microstructure Optimization of Mg-Alloys by the ECAP Process Including Numerical Simulation, SPD Treatments, Characterization, and Hydrogen Sorption Properties. *Molecules* **2019**, *24*, 89. [CrossRef]
96. Zhilyaev, A.P.; Langdon, T.G. Microhardness and EBSD Microstructure Mapping in Partially-Pressed Al and Cu through 90° ECAP Die. *Mater. Res.* **2013**, *16*, 586–591. [CrossRef]

metals

MDPI

Article

Unveiling High-Pressure Behavior of Sc_3AlC MAX Phase: A Comprehensive Theoretical Study on Structural, Mechanical, Dislocation, and Electronic Properties

Junping Xi [1,*], Zhipeng Wang [2,*], Linkun Zhang [3], Li Ma [4] and Pingying Tang [4]

1 Department of Mechanical and Automotive Engineering, Science and Technology College of Hubei University of Arts and Science, Xiangyang 441025, China
2 State Key Laboratory of Advanced Design and Manufacturing Technology for Vehicle, Hunan University, Changsha 410082, China
3 State Key Laboratory for Powder Metallurgy, Central South University, Changsha 410083, China
4 Guangxi Key Laboratory of Functional Information Materials and Intelligent Information Processing, Nanning Normal University, Nanning 530001, China
* Correspondence: xjphndx@163.com (J.X.); wzp1205@hnu.edu.cn (Z.W.)

Abstract: The structural, mechanical, dislocation, and electronic properties of the Sc_3AlC MAX phase under applied pressure are investigated in detail using first-principles calculations. Key parameters, including lattice parameter ratios, elastic constants, Young's modulus, bulk modulus, shear modulus, brittle-to-ductile behavior, Poisson's ratio, anisotropy, Cauchy pressure, yield strength, Vickers hardness, and energy factors, are systematically analyzed as a function of applied pressure. The results demonstrate that the Sc_3AlC MAX phase exhibits remarkable mechanical stability within the pressure range of 0 to 60 GPa. Notably, applied pressure markedly improves its mechanical properties, such as resistance to elastic, bulk, and shear deformations. The B/G ratio suggests a tendency toward ductile behavior with increasing pressure, and the negative Cauchy pressure indicates the directional characteristics of interatomic bonding in nature. Vickers hardness and yield strength increase under pressures of 0 to 10 GPa and then decrease sharply above 50 GPa. High pressure suppresses dislocation nucleation due to the increased energy factors, along with twinning deformation. Furthermore, electronic structure analysis confirms that high pressure enhances the interatomic bonding in the Sc_3AlC MAX phase, while the enhancement effect is not substantial. This study offers critical insights for designing MAX phase materials for extreme environments, advancing applications in aerospace and electronics fields.

Keywords: MAX phase; first-principles calculations; mechanical properties; energy factor; high pressure

1. Introduction

Ceramic materials are valued for their high hardness, as well as excellent corrosion and oxidation resistance. Nevertheless, their inherent brittleness restricts their use in structural applications. On the other hand, metallic materials exhibit superior toughness despite inherently inferior corrosion and oxidation resistance [1,2]. In order to combine the advantages of both, a class of materials known as cermets has emerged. Specifically, MAX-phase ceramics (transition metal carbides/nitrides) have attracted significant attention due to their unique layered atomic structure and an exceptional combination of physicochemical properties. As a result, these materials synergistically integrate high hardness and strength with excellent electrical conductivity and thermal stability, thereby positioning them as promising candidates for diverse advanced engineering applications [3,4]. Furthermore, MAX phases exhibit outstanding plasticity at elevated temperatures, alongside high fracture toughness, thermal shock resistance, and stiffness. Consequently, these attributes render them suitable for applications in high-temperature components, wear-resistant coatings, and corrosion-resistant systems [5–9].

As a typical MAX-phase cermet, Sc_3AlC exhibits unique performance advantages, especially in the aerospace industry. Owing to its high specific strength, excellent oxidation resistance, and superior corrosion resistance [10,11], Sc_3AlC is well-suited for manufacturing high-performance aerospace engine components (e.g., engine components under hypersonic flight loads), as well as structural materials for aircraft. In the field of electronics, the semiconductor properties and outstanding electrical performance of Sc_3AlC [12] facilitate its application in the development of advanced electronic devices, including field-effect transistors and solar cells, thereby driving advancements in electronic technology. Compared to mainstream materials, such as titanium alloys (e.g., Ti-6Al-4V) and nickel-based superalloys in aerospace, or silicon and gallium arsenide in electronics, Sc_3AlC offers distinct advantages. Titanium alloys provide excellent strength-to-weight ratios but suffer from inferior oxidation resistance at high temperatures, requiring protective coatings that increase costs and complexity. Nickel-based superalloys, while robust at high temperatures, are denser and less corrosion-resistant in harsh environments. Sc_3AlC, with a density comparable to titanium alloys (~4.5 g/cm^3) and superior oxidation resistance, reduces the need for coatings and enhances durability in oxidative environments [10,11]. In electronics, silicon is cost-effective but lacks the mechanical robustness of Sc_3AlC, which can integrate structural and functional roles in next-generation devices. Other MAX phases, such as Ti_2AlC or Ti_3AlC_2 [9], share similar advantages but may have lower specific strength or electrical conductivity compared to Sc_3AlC due to differences in composition and bonding [4,6]. Despite its promising properties, the widescale implementation of Sc_3AlC faces several obstacles, primarily the high cost and limited availability of scandium, which is a rare and expensive element compared to titanium or other transition metals used in MAX phases like Ti_2AlC or Ti_3AlC_2 [9]. Synthesis challenges also pose barriers, as producing high-purity Sc_3AlC requires precise control to avoid defects or impurities that could degrade its mechanical and electrical properties. Additionally, scalable manufacturing processes, such as cost-effective powder metallurgy or additive manufacturing, are not yet fully optimized for scandium-based MAX phases, limiting their commercial viability [3]. However, recent progress is addressing these challenges. Advances in synthesis techniques, such as spark plasma sintering and reactive hot pressing, have improved the quality and affordability of MAX phases like Ti_3SiC_2 and Ti_2AlC, which are now used in applications such as turbine blades, heat exchangers, and protective coatings [3,9]. Similar methods have been successfully applied to Sc_3AlC, with recent studies reporting high-purity synthesis via reactive sintering at reduced temperatures, enhancing feasibility for industrial-scale production [12]. These developments, coupled with the growing industrial adoption of other MAX phases in aerospace, suggest a promising trajectory for Sc_3AlC in high-value applications.

Currently, research on Sc_3AlC MAX-phase materials has made significant progress. From a structural perspective, both experimental studies and theoretical calculations have confirmed that its crystal structure is of the anti-perovskite type, with a space group of *Pm-3m* [13–15]. Nevertheless, research on Sc_3AlC under high-pressure conditions is still limited. Owing to limitations in high-pressure experimental techniques, the available data on its mechanical properties and electronic structure evolution under extreme pressure are scarce, thereby hindering the acquisition of comprehensive and precise information. Consequently, current studies are still constrained in fully elucidating the complex effects of high pressure on its structural stability and electronic characteristics. Several critical scientific issues require further investigation. These include systematic investigation of the evolution of mechanical properties, particularly the variation of elastic modulus and yield strength under high pressure; exploration of the mechanisms by which high pressure influences its electronic structure, including changes in band structure and electronic density of states; and analysis of the phase transition behavior of Sc_3AlC under extreme conditions and its impact on overall performance. Therefore, this study aims to systematically investigate the structural, mechanical, dislocation, and electronic properties of the Sc_3AlC MAX phase under high pressure. Addressing these critical issues will not only contribute to a deeper understanding of the high-pressure behavior of the Sc_3AlC MAX phase, but also provide valuable guidance for extreme environment applications.

2. Methodology

First-principles calculations based on density functional theory (DFT) are executed using the Vienna ab initio Simulation Package (VASP) [16,17] to systematically investigate the electronic structure and mechanical properties of Sc_3AlC. The ion-electron interactions are modeled using the Projected Augmented-Wave (PAW) method [18] to ensure accurate electronic structure modeling. The generalized gradient approximation (GGA) with the Perdew-Burke-Ernzerhof (PBE) functional is adopted to describe the exchange-correlation potential [19]. The calculations, performed within the DFT framework, utilize a cubic Sc_3AlC structure (space group *Pm-3m*). A plane-wave basis set with a kinetic energy cutoff of 550 eV is applied to expand the electronic wavefunctions. The Brillouin zone is sampled using an $11 \times 11 \times 11$ Γ-centered *k*-point mesh to ensure high-precision *k*-space integration [20]. Convergence criteria are set to 0.01 eV/Å for atomic forces and 10^{-6} eV/atom for total energy. To evaluate pressure-dependent behavior, applied pressures range from 0 to 60 GPa. Total energy calculations under compressive strains are conducted to assess mechanical properties and high-pressure stability. The electronic structure is evaluated through density of states (DOS) calculation. The results are employed to examine element-specific contributions (Sc, Al, C) to the electronic structure. The electronic configurations—Sc ($3d^1$ $4s^2$), Al ($3s^2$ $3p^1$), and C ($1s^2$ $2s^2$ $2p^2$)—critically influence bonding characteristics and material stability. Figure 1 depicts the cubic crystal structure of the Sc_3AlC unit cell.

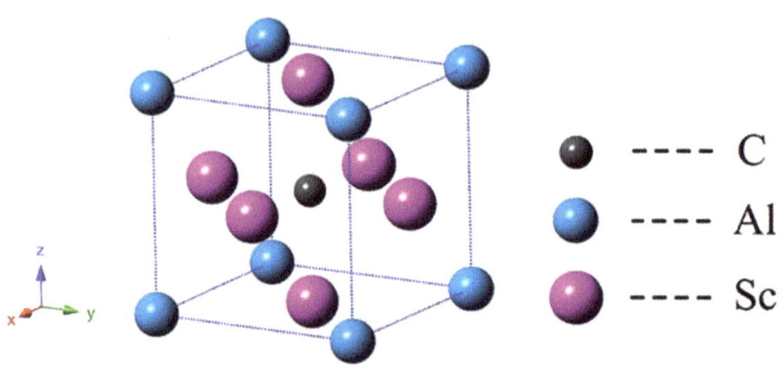

Figure 1. Structural model of Sc_3AlC MAX phase with $a_0 = 4.500$ Å.

3. Results and Discussions

3.1. Structure Properties and Stability

The equilibrium crystal structure of Sc_3AlC in the MAX phase is determined through first-principles calculations. The total energy-volume dependence is characterized using the Birch-Murnaghan equation of state [21], providing a precise description of energy variations as a function of volume. The equilibrium volume is evaluated by systematically varying the volume from 0.9 V_0 to 1.1 V_0, where V_0 denotes the equilibrium volume under the condition of $P = 0$ and $T = 0$. Figure 2 demonstrates that the structure reaches its minimum total energy ($E_t = -34.548438$ eV) at an equilibrium volume of $V_0 = 91.13$ Å^3. The equilibrium lattice constant (a_0) is calculated as 4.500 Å. These results are derived under the condition of $P = 0$ and $T = 0$, confirming that the calculated values reflect the ideal equilibrium structure of Sc_3AlC in the MAX phase. Table 1 illustrates that the calculated results show good agreement with prior experimental and theoretical studies [10,12,22–27].

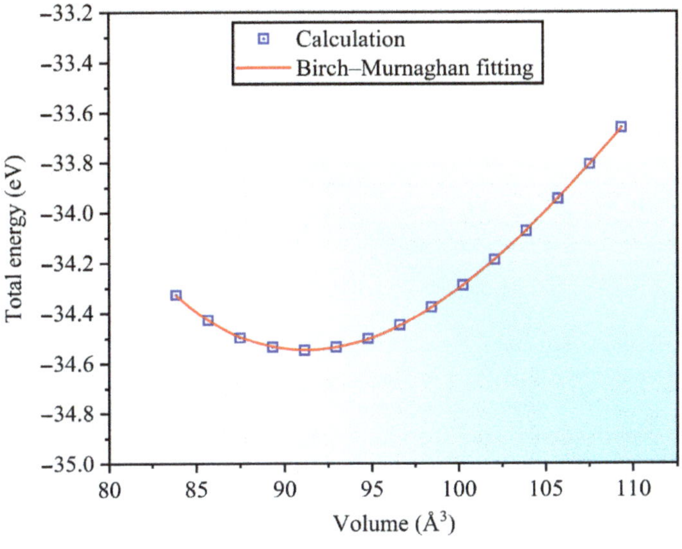

Figure 2. The correlation between total energy and unit cell volume.

Table 1. Lattice parameter of Sc_3AlC MAX phase compared with experimental and theoretical values.

Sc_3AlC	Present	Experimental Values	Theoretical Values
Lattice parameter	4.500	4.498 [22], 4.48 [10], 4.5004 [12]	4.508 [26], 4.480 [25], 4.508 [23], 4.51 [27], 4.51 [24], 4.5119 [12]

To further investigate the effect of high pressure on the crystal structure of Sc_3AlC, this study conducts a series of systematic structural optimization calculations under different pressure conditions, accurately determining the equilibrium lattice constant and volume of Sc_3AlC within the pressure range of 0 to 60 GPa. The corresponding trends are shown in Figure 3. It is clearly observed that as the pressure increases from 0 to 60 GPa, both the lattice constant ratio a/a_0 and the volume ratio V/V_0 exhibit a decreasing trend. This indicates that under high pressure, the interatomic distances within Sc_3AlC continuously decrease, leading to a denser atomic arrangement and enhanced electron interactions.

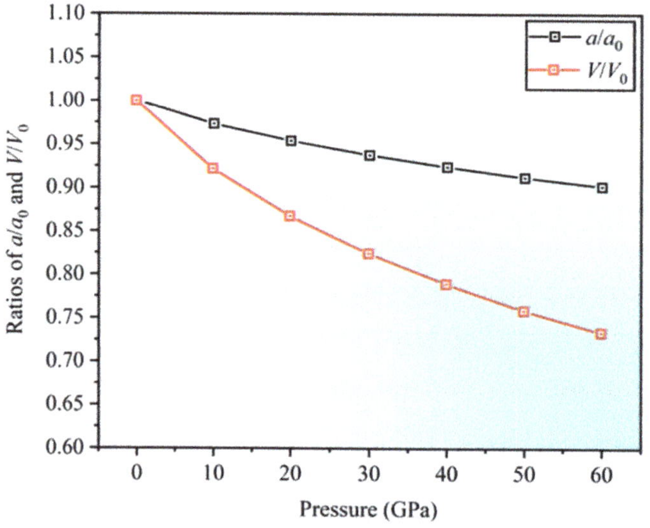

Figure 3. Variation curves depicting V/V_0 and a/a_0 ratios versus applied pressure.

In anisotropic materials, structural stability is intrinsically linked to elastic constants C_{ij}, which indicate the material's ability to resist elastic deformation. Elastic constants C_{ij} not only characterize the material's response to external stresses but also highlight directional variations in mechanical properties. For cubic crystals, especially under high-pressure or large-strain conditions, elastic constants (C_{11}, C_{12}, and C_{44}) are critical in determining the material's resistance to stress-induced deformation. In solid-state physics, the stability of cubic crystals is primarily evaluated based on their elastic constants [28–30]. The stability criteria are defined by specific conditions related to elastic constants, ensuring that the crystal remains stable under diverse external stresses.

$$(C_{11} - C_{12}) > 0, C_{11} > 0, C_{44} > 0, (C_{11} + 2C_{12}) > 0 \quad (1)$$

The computational results obtained in this study demonstrate strong agreement with previous theoretical calculations reported in references [22–27], as evidenced by the comparative analysis presented in Table 2. This consistency between our findings and established theoretical frameworks not only corroborates the validity of our computational approach but also reinforces the reliability and precision of the current research methodology.

Table 2. Calculated results compared with other theoretical data at $T = 0$ and $P = 0$ (Unit: GPa).

Sc_3AlC	Present	Theoretical Data
C_{11}	235.2	220.85 [26], 224.90 [25], 219.20 [23], 220 [24], 225 [27], 220 [22]
C_{12}	45.21	42.15 [26], 40.50 [25], 40.35 [23], 40 [24], 41.5 [27], 43 [22]
C_{44}	80.08	80.12 [26], 80.80 [25], 80.19 [23], 79 [24], 80.3 [27], 79 [22]
B	108.54	101.72 [26], 101.60 [25], 99.98 [23], 99.91 [24], 102.70 [27], 102 [22]
E	203.62	197.03 [26], 199.70 [25], 196.40 [23], 195 [24], 199.27 [27]
G	85.74	83.69 [26], 85.10 [25], 83.80 [23], 83.26 [24], 84.68 [27], 83 [22]
σ	0.16	0.18 [26], 0.17 [25], 0.17 [23], 0.17 [24], 0.18 [27], 0.18 [22]

To explore the effect of high pressure on the elastic properties of Sc_3AlC, the elastic constants are calculated in the pressures range of 0 to 60 GPa. Figure 4 depicts the variation of elastic constants (C_{11}, C_{12}, and C_{44}) as a function of pressure. As pressure increases from 0 to 60 GPa, the elastic constants exhibit a monotonic increase, rising steadily from their initial values. This behavior indicates that Sc_3AlC maintains structural stability under high pressure. Notably, these elastic constants fulfill the stability criterion defined by Equation (1). The results demonstrate that Sc_3AlC retains structural stability under high pressure and exhibits enhanced resistance to structural phase transitions.

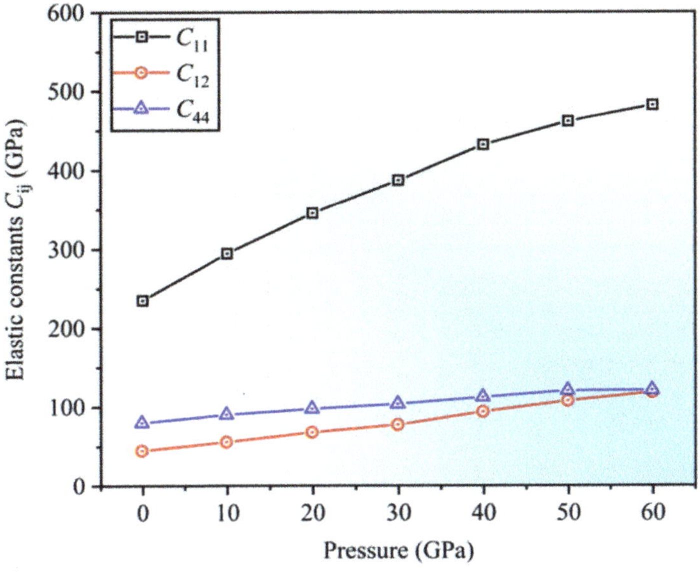

Figure 4. Variation curves depicting elastic constants C_{ij} versus applied pressure.

3.2. Mechanical Properties

The moduli of materials, including bulk modulus B, Young's modulus E, and shear modulus G, play a crucial role in evaluating mechanical properties [31,32]. The magnitude of these moduli is directly related to the material's strength. Materials with high modulus usually have high strength, indicating that they can maintain their shape and resist deformation under external stresses. The Voigt-Reuss-Hill (VRH) method is a commonly used approach to calculate the effective moduli of composite or anisotropic materials [33,34], as with the Equations (2)–(4). It can be used to estimate the relationships between different moduli (such as bulk modulus, Young's modulus, and shear modulus), contributing to evaluate the material's performance under varying stress and deformation conditions. Table 2 presents the computed moduli and compares them with previous studies.

$$B = \frac{1}{3}(C_{11} + 2C_{12}) \tag{2}$$

$$G = \frac{1}{2}(G_V + G_R) \tag{3}$$

$$E = \frac{9BG}{3B + G} \tag{4}$$

In the aforementioned equations, G_V and G_R represent Voigt and Reuss shear modulus, respectively, which are mathematically expressed by Equations (5) and (6):

$$G_V = \frac{(C_{11} - C_{12} + 3C_{44})}{5} \tag{5}$$

$$G_R = \frac{5(C_{11} - C_{12})C_{44}}{4C_{44} + 3(C_{11} - C_{12})} \tag{6}$$

Figure 5 illustrates the variations of bulk modulus (B), Young's modulus (E), and shear modulus (G) within the pressure range of 0 to 60 GPa. As the pressure increases from 0 to 60 GPa, the bulk modulus (B), Young's modulus (E), and shear modulus (G) all exhibit an upward trend. The increase in Young's modulus (E) is the most significant, gradually rising from its initial value to considerably higher values. This indicates that under high pressure, the ability of Sc_3AlC to resist elastic deformation is significantly enhanced, and the atomic bonding strength increases substantially as pressure increases. The bulk modulus (B) also increases steadily, reflecting the gradual enhancement of the ability of Sc_3AlC to resist uniform compression under high-pressure conditions, indicating that the internal structural stability of the crystal improves as the pressure increases. The shear modulus (G) also increases with pressure, although the rate of change is more moderate compared to Young's modulus (E) and bulk modulus (B), suggesting that under high pressure, the ability of Sc_3AlC to resist shear deformation improves, although the rate of change is more moderate. These trends in moduli suggest that the mechanical properties of Sc_3AlC are reinforced under high-pressure conditions.

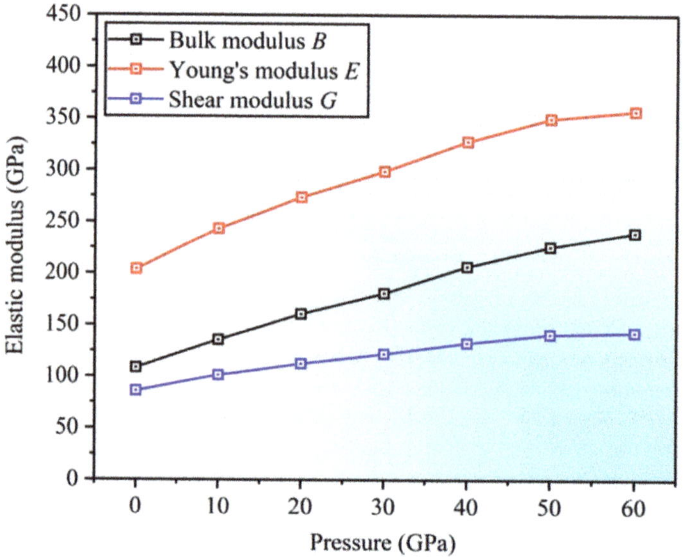

Figure 5. Variation curves depicting Young's modulus, bulk modulus, and shear modulus versus applied pressure.

The brittle-to-ductile transition represents a critical mechanical phenomenon wherein a material undergoes a fundamental transformation in its failure mode, transitioning from brittle fracture characteristics to ductile deformation behavior under applied stress conditions. At certain temperatures and pressures, the plasticity and toughness of a material depend on its crystal structure and the strength of atomic interactions. Specifically, the brittle-to-ductile transition is strongly correlated with the ratio between bulk modulus (B) and shear modulus (G), denoted as B/G [32]. This ratio is an effective indicator for predicting the material's toughness and brittleness behavior under different temperature or stress conditions. The B/G ratio serves as a critical indicator for material toughness, with the threshold value of 1.75 demarcating distinct mechanical behavior;: values exceeding

1.75 typically signify enhanced toughness and ductile characteristics, while ratios below this critical value are indicative of reduced toughness and increased brittleness [35]. As shown in Figure 6, as pressure gradually increases from 0 to 60 GPa, the *B/G* ratio exhibits a continuous upward trend, rising steadily from approximately 1.3 to nearly 1.7. This change indicates that under high-pressure conditions, the ability of Sc_3AlC to resist uniform compression (represented by the bulk modulus *B*) improves more significantly compared to its ability to resist shear deformation (represented by the shear modulus *G*). Within the pressure range of this study, the *B/G* ratio remains below 1.75 but gradually approaches this value with increasing pressure, suggesting that as pressure increases, Sc_3AlC tends to transition towards more toughness mechanical behavior while maintaining relative brittleness characteristics within the 0–60 GPa pressure range.

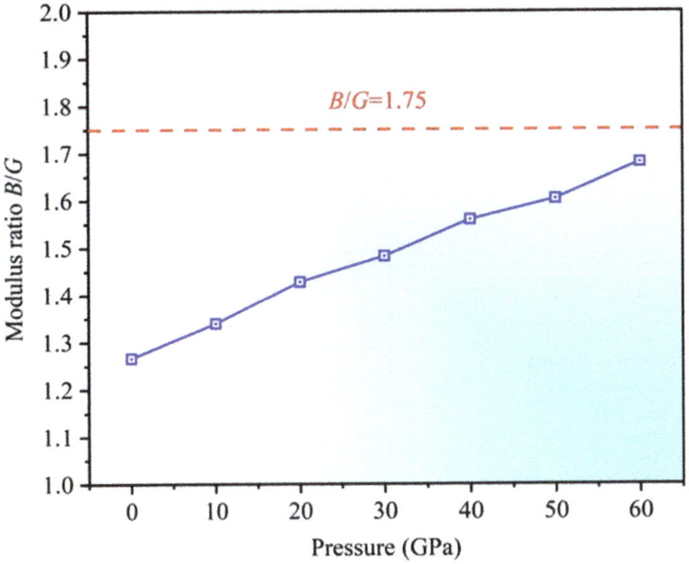

Figure 6. Variation curves depicting *B/G* versus applied pressure.

3.3. Anisotropy

Elastic anisotropy represents a fundamental physical parameter that plays a pivotal role in characterizing and understanding the mechanical behavior of crystalline materials with anisotropic characteristics, particularly in determining their directional dependence of elastic properties. The anisotropy of materials is often characterized using the anisotropy factor *A*, and a greater deviation of anisotropy factor *A* from the one indicates stronger anisotropy [36,37]. The cross-slip pinning model provides a theoretical framework for analyzing dislocation behavior, slip systems, and pinning effects in metallic materials [38]. It is particularly effective in analyzing the mechanisms of plastic deformation and their relationship with anisotropy factor. This model integrates the phenomena of cross-slip and pinning, enabling a more comprehensive understanding of stress transmission and dislocation glide in anisotropic materials. The anisotropy factor for the Sc_3AlC MAX phase can be calculated using the following formulas [39,40]:

$$A_{(100)[001]} = \frac{2C_{44}}{C_{11} - C_{12}} \tag{7}$$

$$A_{(110)[001]} = \frac{C_{44}(C' + 2C_{12} + C_{11})}{C_{11}C' - C_{12}^2} \quad (8)$$

where $C' = (C_{11} + C_{12})/2 + C_{44}$, and both $A_{(100)[001]}$ and $A_{(110)[001]}$ represent the anisotropy factors along the (100)[001] and (110)[001] crystallographic directions, respectively. According to Equations (7) and (8), Figure 7 illustrates the variation trends of anisotropy factors along the (100)[001] direction (labeled as $A_{(100)[001]}$) and the (110)[001] direction (labeled as $A_{(110)[001]}$) under increasing pressure. As the pressure rises from 0 to 60 GPa, both $A_{(100)[001]}$ and $A_{(110)[001]}$ exhibit a declining trend. This suggests that under high pressure, the degree of anisotropy in Sc_3AlC along these two directions gradually increases, and the differences in mechanical properties along these directions increase with increasing pressure. The pressure-dependent variation in the anisotropy factors reflects the rearrangement of atoms, including changes in interatomic distances. These factors lead to the gradual increases in the differences in mechanical responses along different crystallographic directions.

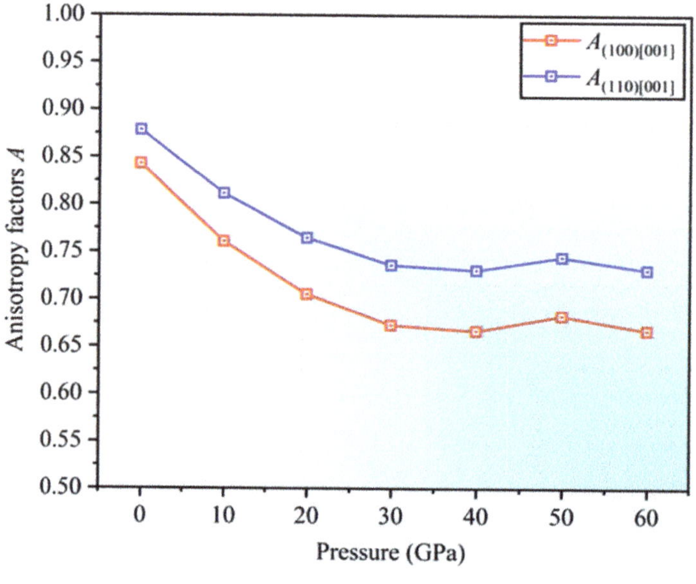

Figure 7. Variation curves depicting anisotropy factors *A* versus applied pressure.

Poisson's ratio σ is a physical quantity that describes the ratio of transverse strain to longitudinal strain when a material is subjected to tension or compression, typically ranging between 0 and 0.5. For different types of materials, Poisson's ratio is closely related to plasticity properties. According to Equations (9) and (10) [39,41], Figure 8 illustrates the variation trends of Poisson's ratio $\sigma_{[001]}$ in the [001] direction and $\sigma_{[111]}$ in the [111] direction under increasing pressure. In general, a large Poisson's ratio typically suggests good plasticity. At the initial stage, the Poisson ratio in both directions of the [001] and [111] is small, and with the increase of pressure, the Poisson ratio in both crystal directions increases, which indicates that the increase of pressure is conducive to improving the plasticity of the material. Furthermore, the value of $\sigma_{[111]}$ consistently exceeds that of $\sigma_{[001]}$

throughout the pressure range, suggesting that under the same pressure conditions, Sc3AlC exhibits better plasticity in the [111] direction compared to the [001] direction.

$$\sigma_{[001]} = \frac{C_{12}}{C_{11} + C_{12}} \tag{9}$$

$$\sigma_{[111]} = \frac{C_{11} + 2C_{12} - 2C_{44}}{2(C_{11} + 2C_{12} + C_{44})} \tag{10}$$

To further elucidate the evolution of the mechanical properties of Sc_3AlC under high pressure, other relevant properties are calculated within the pressure range of 0 to 60 GPa, with the variations illustrated in Figure 9. Figure 9 presents the trends of $C_{12} - C_{14}$ (representing the Cauchy pressure), $G_{(100)[010]}$ (shear modulus $G_{(100)[010]} = C_{44}$ in the $(100)[010]$ direction), $G_{(110)[1\bar{1}0]}$ (shear modulus $G_{(110)[1\bar{1}0]} = (C_{11} - C_{12})/2$ in the $(110)[1\bar{1}0]$ direction), and $E_{\langle 100\rangle}$ (Young's modulus in the <100> direction) as functions of pressure, where $E_{\langle 100\rangle}$ is expressed by $E_{\langle 100\rangle} = (C_{11} - C_{12})[1 + C_{12}/(C_{11} + C_{12})]$ [42,43]. As the pressure increases from 0 GPa to 60 GPa, $E_{\langle 100\rangle}$ exhibits a significant upward trend, steadily rising from its initial value to a high level, which indicates that under high pressure, the ability of Sc_3AlC to resist elastic deformation in the <100> direction is significantly enhanced, with the interatomic bonding forces in this direction markedly increasing with pressure. Both $G_{(100)[010]}$ and $G_{(110)[1\bar{1}0]}$ also increase with increasing pressure, although their growth rates are relatively smaller compared to $E_{\langle 100\rangle}$. This suggests that in a high-pressure environment, the resistance of Sc_3AlC to shear deformation in these two directions is improved. Generally, a negative Cauchy pressure indicates that the material possesses directional bonding characteristics, and a positive Cauchy pressure suggests metallic characteristic [43–45]. The Cauchy pressure $C_{12} - C_{14}$ remains negative throughout the pressure range; however, it gradually increases and asymptotically approaches zero as the pressure rises. This trend indicates a weakening of the directional characteristics of interatomic bonding within the material, accompanied by an enhancement of metallic properties.

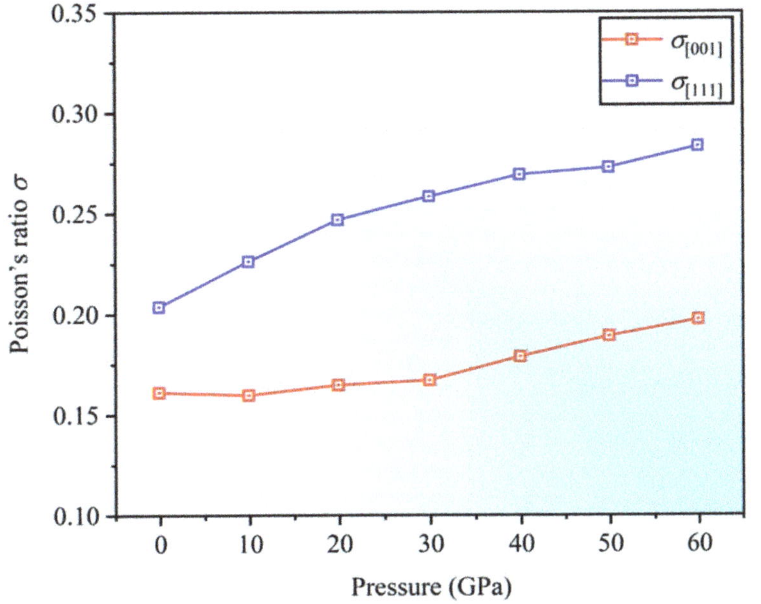

Figure 8. Variation curves depicting Poisson's ratio σ versus applied pressure.

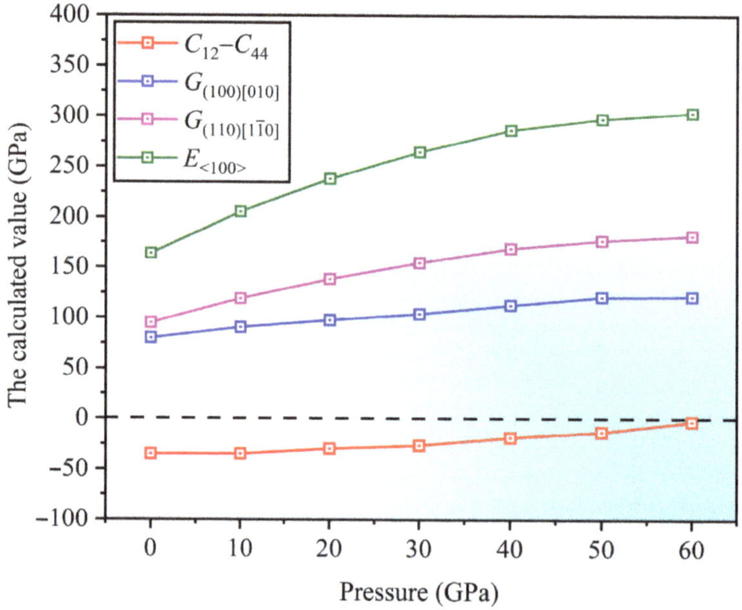

Figure 9. Variation curves depicting Cauchy pressure and material modulus versus applied pressure.

3.4. Hardness and Yield Strength

Hardness and yield strength are essential parameters for characterizing a material's resistance to deformation, and they are closely related to the material's mechanical properties [46]. Specifically, hardness is defined as the ability of a material's surface to resist localized plastic deformation, such as indentation, scratching, or wear. An increased hardness value implies that the material is less susceptible to plastic deformation. On the other hand, yield strength σ_y is defined as the minimum stress where a material begins to undergo permanent deformation during tension or compression. Once the applied stress exceeds the yield strength, the material translates from elastic to plastic deformation. Previous studies [47–50] indicate that the Vickers hardness (H_V) of a material can be determined using Equation (11), where $k = G/B$ [51]. The Vickers hardness (H_V) is approximately three times yield strength (σ_y), as expressed in Equation (12).

To explore the effects of high pressure on the Vickers hardness and yield strength of Sc_3AlC, the relevant data are calculated within the pressure range of 0 to 60 GPa, as illustrated in Figure 10. Figure 10a shows the variation of Vickers hardness versus applied pressure, while Figure 10b presents the changes in yield strength under pressure. From Figure 10a, the value of Vickers hardness is about 17.52 GPa at ambient pressure, aligning well with the calculated result (H_V = 17.81 GPa) of Rayhan et al. [26], and it can be observed that within the pressure range of 0 to 10 GPa, the Vickers hardness increases rapidly from approximately 17.52 GPa to about 18.16 GPa. This indicates that during the initial stage of pressure increase, the resistance to localized plastic deformation significantly improves in the Sc_3AlC MAX phase. Subsequently, from 10 GPa to 50 GPa, H_V fluctuates within the range of 17.52 GPa to 18.16 GPa, remaining relatively stable. This suggests that changes in pressure have a minimal impact on Vickers hardness of Sc_3AlC phase. However, when the pressure exceeds 50 GPa, H_V drops sharply, reaching approximately 16.84 GPa at P = 60 GPa, which demonstrates that the hardness value decreases significantly under high pressure. In Figure 10b, the yield strength shows an increase from 5.84 GPa to 6.05 GPa within the 0 to 10 GPa range, indicating that the stress to resist plastic deformation increases.

From 10 GPa to 50 GPa, σ_y fluctuates between 5.84 GPa and 6.05 GPa, remaining relatively stable, which suggests that the yield characteristics of the material are relatively stable within this pressure range. However, when the pressure reaches 60 GPa, σ_y decreases to about 5.61 GPa, indicating a reduction in the material's yield strength under the conditions of high pressure.

$$H_v = 2\left(k^2 G\right)^{0.585} - 3 \tag{11}$$

$$\sigma_y = H_V/3 \tag{12}$$

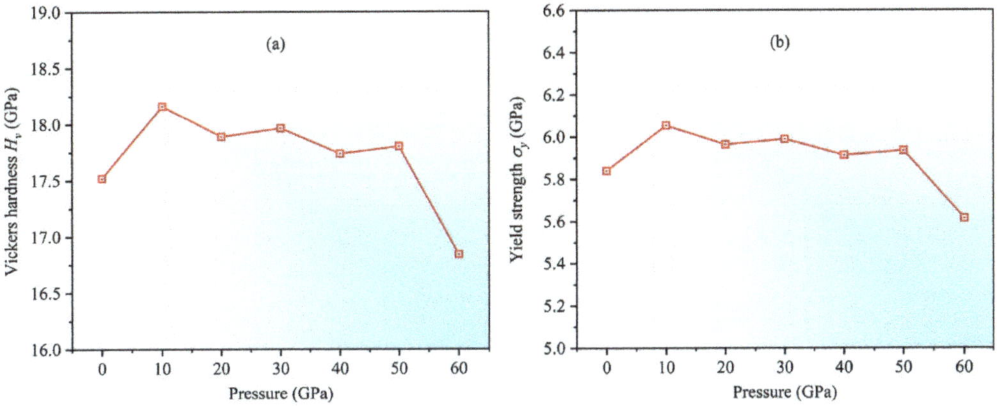

Figure 10. Variation curves depicting H_V (**a**) and σ_y (**b**) versus applied pressure.

3.5. Energy Factor K

In crystalline materials, the ability of dislocation nucleation is closely associated with the material's plasticity. The energy factor K is a crucial parameter that quantifies the dislocation nucleation ability, representing the minimum energy for plastic deformation in a specific direction, and the calculation of energy factor K is related to the elastic constant. For cubic crystals, Equations (13) and (14) give the expressions of K factors of edge and screw dislocations [52,53]. Based on the calculated results, Figure 11 illustrates the variation of the energy factor K for edge and screw dislocations versus applied pressure. As the pressure increases from 0 to 60 GPa, the energy factors K for both edge and screw dislocations increase. For edge dislocations, the energy factor K_{edge} increases steadily from an initial value of 107.45 GPa to 197.68 GPa, which suggests that under high-pressure conditions, the nucleation energy of edge dislocations in Sc_3AlC significantly increases, resulting in a higher energy barrier that must be overcome for dislocation nucleation, thereby inhibiting plastic deformation. For screw dislocations, the energy factor K_{screw} gradually rises from 87.22 GPa to 148.38 GPa, indicating that the nucleation energy also increases with increasing pressure, thus leading to the inhibition of screw dislocations.

$$K_{screw} = \left[\frac{1}{2}C_{44}(C_{11} - C_{12})\right]^{\frac{1}{2}} \tag{13}$$

$$K_{edge} = (C_{11} + C_{12})\left[\frac{(C_{11} - C_{12})C_{44}}{(C_{11} + C_{12} + 2C_{44})C_{11}}\right]^{\frac{1}{2}} \tag{14}$$

The mixed dislocation factor K_{mixed} is a parameter that quantifies the nucleation of mixed dislocations, and this factor can be characterized as a function of the orientation angle θ. For mixed dislocations, the dislocation core width ζ represents the extent of deformation within the dislocation region. The relationships between K_{mixed}, ζ, and other parameters are described by Equations (15) and (16) [53,54], where $C = (C_{11} - C_{12})/2$, with d denoting the distance between adjacent slip planes. Figure 12 shows the calculated curves of K_{mixed} and ζ versus applied pressure. In Figure 12a, as pressure increases, the mixed dislocation factor K_{mixed} also increases, which is consistent with the conclusion in Figure 11. In Figure 12b, it is evident that the dislocation core width gradually decreases as pressure increases, which increases the stacking-fault energy of the Sc_3AlC MAX phase, and then inhibits twinning deformation.

$$K_{mixed} = K_{edge} \cdot \sin^2\theta + K_{screw} \cdot \cos^2\theta \quad (15)$$

$$\zeta = \frac{K_{mixed} \cdot d}{2C} \quad (16)$$

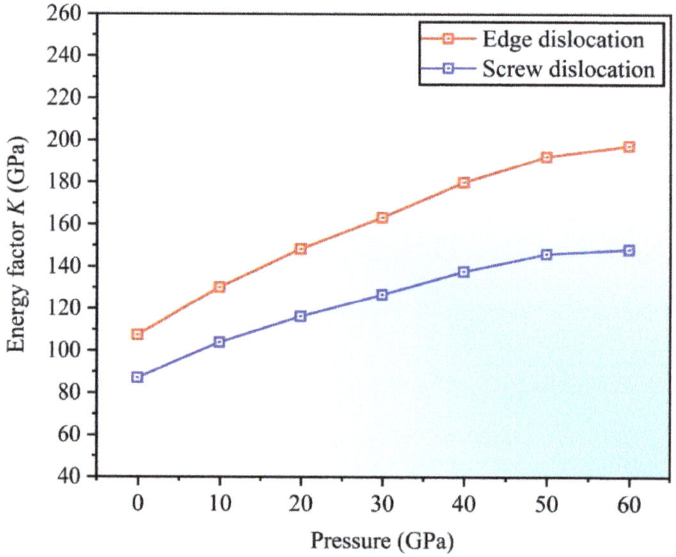

Figure 11. Variation curves depicting energy factors K for screw and edge dislocations versus applied pressure.

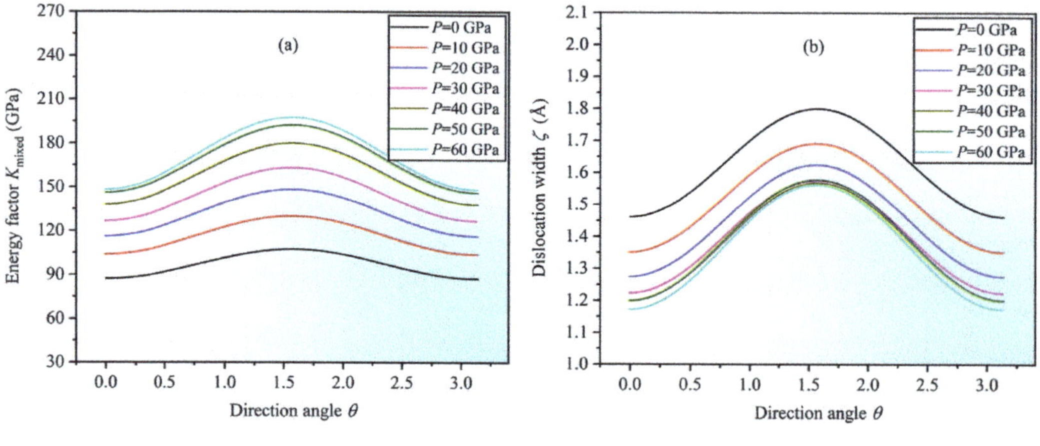

Figure 12. Variation curves depicting K_{mixed} factor of mixed dislocation (**a**) and dislocation width ζ (**b**) versus direction angle $\theta (0 \leq \theta \leq \pi)$ at various pressures.

3.6. Electronic Properties

The electronic structure is intrinsically related to the bonding mechanisms and directly affects the structural stability of a material. To examine the effects of high pressure on the electronic structure of Sc_3AlC, the variation of total density of states (TDOS) is analyzed under pressures ranging from 0 to 60 GPa, as shown in Figure 13. As the pressure increases from 0 to 60 GPa, the TDOS at the Fermi level slightly increases, and the DOS for both the valence and conduction bands near the Fermi level shifts slightly to the right, indicating that high pressure enhances the interatomic bonding in the Sc_3AlC MAX phase, while the enhancement effect is not substantial.

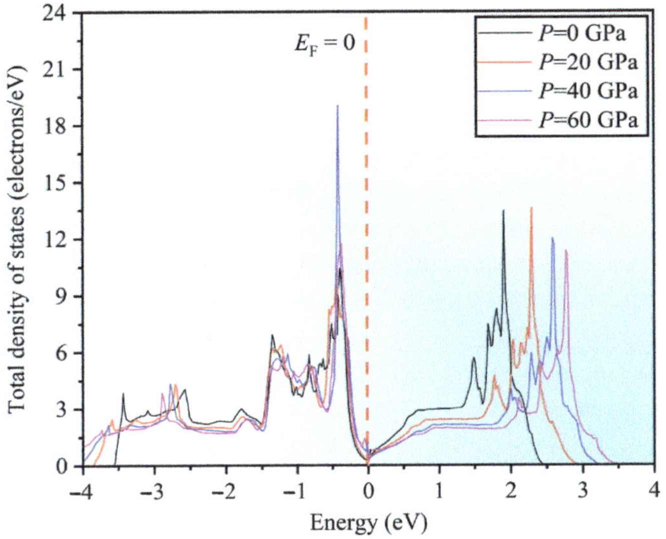

Figure 13. TDOS of Sc_3AlC MAX phase under different pressures.

4. Conclusions

This work systematically investigates the effects of pressure on the structural, mechanical, dislocation, and electronic properties of the Sc_3AlC MAX phase using first-principles calculations. The main conclusions are as follows:

(1) Under pressures ranging from 0 to 60 GPa, the lattice constant ratio and volume ratio decrease. The elastic constants satisfy stability criteria, indicating that the structure remains stable under high pressure.

(2) As the pressure increases, the bulk modulus, Young's modulus, and shear modulus increase, enhancing resistance to material deformation. The B/G ratio rises, approaching 1.75, suggesting a tendency toward ductile behavior, but the material still maintains its brittle characteristics.

(3) As the pressure increases, the anisotropy in the (100)[001] and (110)[001] directions also increases. The Poisson's ratio in the [001] and [111] directions increases, resulting in the enhanced plasticity, with better plasticity along the [111] direction. The negative Cauchy pressure indicates the directional characteristics of interatomic bonding.

(4) Vickers hardness and yield strength increase under pressures of 0 to 10 GPa and remain relatively stable between 10 and 50 GPa, and then decrease sharply above 50 GPa.

(5) High pressure increases the energy factors of edge and screw dislocations, suppressing dislocation nucleation. The mixed dislocation factor increases with increasing pressure, and the dislocation core width reduces, which may inhibit twinning deformation.
(6) High pressure enhances the interatomic bonding in the Sc_3AlC MAX phase, while the enhancement effect is not substantial.

Author Contributions: Methodology, Formal analysis, Investigation, Writing—original draft, J.X., Z.W. and L.Z.; Funding acquisition, Supervision, Visualization, Z.W. and L.M.; Project administration, Data curation, L.Z.; Software, P.T.; Writing—review and editing, J.X., Z.W. and L.M. All authors have read and agreed to the published version of the manuscript.

Funding: This research was funded by the National Natural Science Foundation of China (52162009), State Key Laboratory of Advanced Design and Manufacturing for Vehicle Body, Hunan University, Changsha, China (32175006).

Data Availability Statement: The data presented in this study are available on request from the corresponding authors.

Conflicts of Interest: The authors declare no conflicts of interest.

References

1. Dong, H.; Yu, Y.; Jin, X.; Tian, X.; He, W.; Ma, W. Microstructure and mechanical properties of SiC-SiC joints joined by spark plasma sintering. *Ceram. Int.* **2016**, *42*, 14463–14468. [CrossRef]
2. Shrivastava, V. Microwave processed $SrBi_2Nb_2O_9$ ferroelectric ceramics with controlled dielectric relaxation and metallic conduction. *Ceram. Int.* **2016**, *42*, 10122–10126. [CrossRef]
3. Lu, Y.; Yang, A.; Duan, Y.; Peng, M. Structural stability, electronic and optical properties of MAX-phase ternary nitrides β-TM_4AlN_3 (TM = V, Nb, and Ta) using the first-principles explorations. *Vacuum* **2021**, *193*, 110529. [CrossRef]
4. Zhang, Z.; Duan, X.; Jia, D.; Zhou, Y.; Zwaag, S.V.D. On the formation mechanisms and properties of MAX phases: A review. *J. Eur. Ceram. Soc.* **2021**, *41*, 3851–3878. [CrossRef]
5. Ching, W.Y.; Mo, Y.; Aryal, S.; Rulis, P. Intrinsic mechanical properties of 20 MAX-phase compounds. *J. Am. Ceram. Soc.* **2013**, *96*, 2292–2297. [CrossRef]
6. Sokol, M.; Natu, V.; Kota, S.; Barsoum, M.W. On the chemical diversity of the MAX phases. *Trends Chem.* **2019**, *1*, 210–223. [CrossRef]
7. Sauceda, D.; Singh, P.; Falkowski, A.; Chen, Y.; Doung, T.; Vazquez, G.; Radovic, M.; Arroyave, R. High-throughput reaction engineering to assess the oxidation stability of MAX phases. *NPJ Comput. Mater.* **2021**, *7*, 6. [CrossRef]
8. Badie, S.; Sebold, D.; Vaßen, R.; Guillon, O.; Gonzalez-Julian, J. Mechanism for breakaway oxidation of the Ti_2AlC MAX phase. *Acta Mater.* **2021**, *215*, 117025. [CrossRef]
9. Bao, W.; Wang, X.G.; Ding, H.; Lu, P.; Zhu, C.; Zhang, G.J.; Xu, F. High-entropy M_2AlC-MC (M = Ti, Zr, Hf, Nb, Ta) composite: Synthesis and microstructures. *Scripta Mater.* **2020**, *183*, 33–38. [CrossRef]
10. Nowotny, H.; Boller, H.; Beckmann, O. Alloy phases crystallizing with structures which occur with nonmetallic compounds. *J. Solid State Chem.* **1970**, *2*, 462–471. [CrossRef]
11. Nowotny, V.H. Strukturchemie einiger verbindungen der übergangsmetalle mit den elementen C, Si, Ge, Sn. *Prog. Solid State Chem.* **1971**, *5*, 27–70. [CrossRef]
12. Ghule, A.; Ghule, S.; Garde, C.; Pandey, B.; Ramakrishnan, S. Ab initio study of Sc_3MO (M = Al, Ga, In, Tl) and systematics in Sc_3MZ (Z = B, C, N, O). *Comput. Condens. Matter* **2020**, *25*, e00509. [CrossRef]
13. Chern, M.Y.; Vennos, D.; DiSalvo, F. Synthesis, structure, and properties of anti-perovskite nitrides Ca_3MN, M = P, As, Sb, Bi, Ge, Sn, and Pb. *J. Solid State Chem.* **1992**, *96*, 415–425. [CrossRef]
14. Deng, Z.; Ni, D.; Chen, D.; Bian, Y.; Li, S.; Wang, Z.; Zhao, Y. Anti-perovskite materials for energy storage batteries. *InfoMat* **2022**, *4*, e12252. [CrossRef]
15. Lenggenhager, P.M.; Liu, X.; Neupert, T.; Bzdušek, T. Universal higher-order bulk-boundary correspondence of triple nodal points. *Phys. Rev. B* **2022**, *106*, 085129. [CrossRef]
16. Kresse, G.; Furthmüller, J. Efficient iterative schemes for ab initio total-energy calculations using a plane-wave basis set. *Phys. Rev. B* **1996**, *54*, 11169–11186. [CrossRef]
17. Kresse, G.; Hafner, J. Ab initio molecular dynamics for liquid metals. *Phys. Rev. B* **1993**, *47*, 558–561. [CrossRef]
18. Blöchl, P.E. Projected augmented-wave method. *Phys. Rev. B* **1994**, *50*, 17953–17979. [CrossRef]

19. Perdew, J.P.; Burke, K.; Wang, Y. Generalized gradient approximation for the exchange-correlation hole of a many-electron system. *Phys. Rev. B* **1996**, *54*, 16533. [CrossRef]
20. Pack, J.D.; Monkhorst, H.J. Special Points for Brillonin-Zone Integrations. *Phys. Rev. B* **1976**, *13*, 5188.
21. Birch, F. Finite elastic strain of cubic crystals. *Phys. Rev.* **1947**, *71*, 809. [CrossRef]
22. Gesing, T.M.; Wachtmann, K.H.; Jeitschko, W. The Perovskite Carbides A_3MC (A = Sc, Y, La-Nd, Sm, Gd-Lu; M = AI, Ga, In, Tl, Sn, Pb). *Z. Naturforsch. B* **1997**, *52*, 176–182. [CrossRef]
23. Haddadi, K.; Bouhemadou, A.; Zerarga, F.; Bin-Omran, S. First-principles investigation of the ternary scandium based inverse-perovskite carbides Sc_3AC (A = Al, Ga, In and Tl). *Solid State Sci.* **2012**, *14*, 1175–1185. [CrossRef]
24. Kanchana, V.; Ram, S. Electronic structure and mechanical properties of Sc_3AC (A = Al, Ga, In, Tl) and Sc_3BN (B = Al, In): Ab-initio study. *Intermetallics* **2012**, *23*, 39–48. [CrossRef]
25. Medkour, Y.; Roumili, A.; Maouche, D.; Saoudi, A. A first-principles study on the structural, elastic and electronic properties of $AlCSc_3$ and $AlNSc_3$. *Solid State Commun.* **2009**, *149*, 1840–1842. [CrossRef]
26. Rayhan, M.; Ali, M.; Jahan, N.; Hossain, M.; Uddin, M.; Islam, A.; Naqib, S. Insights into the unrevealed physical properties of $Sc_2Al_2C_3$ compared with other Sc-Al-C systems via ab-initio investigation. *Phys. Open* **2024**, *19*, 100217. [CrossRef]
27. Wang, K.; Zhang, X.; Wang, F. Exploring the electronic, mechanical, anisotropic and optical properties of the Sc-Al-C MAX phases from a first principles calculations. *Chem. Phys. Lett.* **2024**, *836*, 141024. [CrossRef]
28. Patil, S.K.R.; Khare, S.V.; Tuttle, B.R.; Bording, J.K.; Kodambaka, S. Mechanical stability of possible structures of PtN investigated using first-principles calculations. *Phys. Rev. B* **2006**, *73*, 104118. [CrossRef]
29. Wang, J.; Yip, S.; Phillpot, S.R.; Wolf, D. Crystal instabilities at finite strain. *Phys. Rev. Lett.* **1993**, *71*, 4182. [CrossRef]
30. Wallace, D.C.; Callen, H. Thermodynamics of Crystals. *Am. J. Phys.* **1972**, *40*, 1718–1719. [CrossRef]
31. Yang, X.W.; Zhu, J.C.; Lai, Z.H.; Liu, Y.; He, D.; Nong, Z.S. Finite element analysis of quenching temperature field, residual stress and distortion in A357 aluminum alloy large complicated thin-wall workpieces. *T. Nonferr. Metal. Soc.* **2013**, *23*, 1751–1760. [CrossRef]
32. Lu, Y.; Yang, Y.; Zheng, F.; Wang, B.T.; Zhang, P. Electronic, mechanical, and thermodynamic properties of americium dioxide. *J. Nucl. Mater.* **2013**, *441*, 411–420. [CrossRef]
33. Hill, R. The elastic behavior of crystalline aggregate. *Proc. Phys. Soc. A* **1952**, *65*, 349–354. [CrossRef]
34. Iotova, D.; Kioussis, N.; Lim, S.P. Electronic structure and elastic properties of the Ni_3X (X = Mn, Al, Ga, Si, Ge) intermetallics. *Phys. Rev. B* **1996**, *54*, 14413–14422. [CrossRef] [PubMed]
35. Cottrel, A.H. Advanced Structural Materials. In Proceedings of the 2nd European Conference on Advanced Materials and Processes, Cambridge, UK, 22–24 July 1991.
36. Fu, H.; Li, D.; Peng, F.; Gao, T.; Cheng, X. Structural and elastic properties of γTiAl under high pressure from electronic structure calculations. *J. Alloys Comp.* **2009**, *473*, 255–261. [CrossRef]
37. Mattesini, M.; Ahuja, R.; Johansson, B. Cubic Hf_3N_4 and Zr_3N_4: A class of hard materials. *Phys. Rev. B* **2003**, *68*, 184108. [CrossRef]
38. Yoo, M.H. On the theory of anomalous yield behavior of Ni_3Al—Effect of elastic anisotropy. *Scripta Metall.* **1986**, *20*, 915–920. [CrossRef]
39. Fu, H.; Li, X.F.; Liu, W.F.; Ma, Y.; Tao, G.; Hong, X. Electronic and dynamical properties of NiAl studied from first principles. *Intermetallics* **2011**, *19*, 1959–1967. [CrossRef]
40. Lau, K.; McCurdy, A.K. Elastic anisotropy factors for orthorhombic, tetragonal, and hexagonal crystals. *Phys. Rev. B* **1998**, *58*, 8980–8984. [CrossRef]
41. Friák, M.; Šob, M.; Vitek, V. Ab initio calculation of tensile strength in iron. *Philos. Mag.* **2003**, *83*, 3529–3537. [CrossRef]
42. Zhu, Y.M.; Weyland, M.; Morton, A.J.; Oh-ishi, K.; Hono, K.; Nie, J.F. The building block of long-period structures in Mg-RE-Zn alloys. *Scripta Mater.* **2009**, *60*, 980–983. [CrossRef]
43. Fu, H.; Li, D.; Peng, F.; Gao, T.; Cheng, X. Ab initio calculations of elastic constants and thermodynamic properties of NiAl under high pressures. *Comp. Mater. Sci.* **2008**, *44*, 774–778. [CrossRef]
44. Johnson, R.A. Analytic nearest-neighbor model for fcc metals. *Phys. Rev. B* **1988**, *37*, 3924–3931. [CrossRef]
45. Pettifor, D. Theoretical predictions of structure and related properties of intermetallics. *Mater. Sci. Technol.* **1992**, *8*, 345–349. [CrossRef]
46. Tse, J.S. Intrinsic hardness of crystalline solids. *J. Superhard Mater.* **2010**, *32*, 177–191. [CrossRef]
47. Smedskjaer, M.M.; Mauro, J.C.; Yue, Y. Prediction of glass hardness using temperature-dependent constraint theory. *Phys. Rev. Lett.* **2010**, *105*, 115503. [CrossRef]
48. Li, K.; Wang, X.; Zhang, F.; Xue, D. Electronegativity identification of novel superhard materials. *Phys. Rev. Lett.* **2008**, *100*, 235504. [CrossRef] [PubMed]
49. Gao, F.; He, J.; Wu, E.; Liu, S.; Yu, D.; Li, D.; Zhang, S.; Tian, Y. Hardness of covalent crystals. *Phys. Rev. Lett.* **2003**, *91*, 015502. [CrossRef]

50. Šimůnek, A.; Vackář, J. Hardness of covalent and ionic crystals: First-principle calculations. *Phys. Rev. Lett.* **2006**, *96*, 085501. [CrossRef]
51. Chen, X.; Niu, H.; Li, D.; Li, Y. Modeling hardness of polycrystalline materials and bulk metallic glasses. *Intermetallics* **2011**, *19*, 1275–1281. [CrossRef]
52. Savin, M.M.; Chernov, V.M.; Strokova, A.M. Energy factor of dislocations in hexagonal crystals. *Phys. Status Solidi A* **1976**, *35*, 747–754. [CrossRef]
53. Foreman, A.J.E. Dislocation energies in anisotropic crystals. *Acta Metall.* **1955**, *3*, 322–330. [CrossRef]
54. Reid, C.N. Dislocation widths in anisotropic B.C.C. crystals. *Acta Metall.* **1966**, *14*, 13–16. [CrossRef]

MDPI AG
Grosspeteranlage 5
4052 Basel
Switzerland
Tel.: +41 61 683 77 34

Metals Editorial Office
E-mail: metals@mdpi.com
www.mdpi.com/journal/metals

www.ingramcontent.com/pod-product-compliance
Lightning Source LLC
Chambersburg PA
CBHW061557210926
51850CB00009B/2131
* 9 7 8 3 7 2 5 8 5 0 6 1 7 *